Xenobiotic Conjugation
Chemistry

ACS SYMPOSIUM SERIES **299**

Xenobiotic Conjugation Chemistry

Gaylord D. Paulson, EDITOR
U.S. Department of Agriculture

John Caldwell, EDITOR
University of London

David H. Hutson, EDITOR
Shell Research Ltd.

Julius J. Menn, EDITOR
U.S. Department of Agriculture

Developed from a symposium sponsored by
the ACS Division of Pesticide Chemistry and
the International Society for Study of Xenobiotics
at the 189th Meeting
of the American Chemical Society,
Miami Beach, Florida,
April 28–May 3, 1985

American Chemical Society, Washington, DC 1986

Library of Congress Cataloging-in-Publication Data

Xenobiotic conjugation chemistry.
  (ACS symposium series; 299)

  "Developed from a symposium sponsored by the
ACS Division of Pesticide Chemistry and the
International Society for Study of Xenobiotics at the
189th meeting of the American Chemical Society,
Miami Beach, Florida, April 28–May 3, 1985."

  Includes bibliographies and indexes.

  1. Xenobiotics—Metabolism—Congresses.
2. Metabolic conjugation—Congresses.

  I. Paulson, Gaylord D. II. American Chemical
Society. Division of Pesticide Chemistry.
III. International Society for Study of Xenobiotics.
IV. Series.

QP529.X45   1986   574.19'2   85–32553
ISBN 0–8412–0957–X

ACS Symposium Series

M. Joan Comstock, *Series Editor*

Advisory Board

FOREWORD

The ACS SYMPOSIUM SERIES was founded in 1974 to provide a medium for publishing symposia quickly in book form. The format of the Series parallels that of the continuing ADVANCES IN CHEMISTRY SERIES except that, in order to save time, the papers are not typeset but are reproduced as they are submitted by the authors in camera-ready form. Papers are reviewed under the supervision of the Editors with the assistance of the Series Advisory Board and are selected to maintain the integrity of the symposia; however, verbatim reproductions of previously published papers are not accepted. Both reviews and reports of research are acceptable, because symposia may embrace both types of presentation.

CONTENTS

INDEXES

PREFACE

IN THE FIELD OF XENOBIOTIC METABOLISM, the study of xenobiotic conjugation reactions has progressed through three phases. From 1840 (when hippuric acid biosynthesis was discovered) until the First World War, the primary field of study was xenobiotic conjugation reactions. A high point during this era was the formulation of the "chemical defense hypothesis" by C. P. Sherwin. After the 1930s, oxidative reactions became increasingly more important, and in the late 1970s cytochrome P450 and reactive oxidation products seemingly defined the limits of xenobiotic metabolism. Recently, however, interest in the properties and enzymology of formation of xenobiotic conjugates has increased. Numerous studies demonstrate that some xenobiotic conjugates are as active or more active than the parent compound. Thus, a complete assessment of the effects of xenobiotics must include evaluations of the formation and biological properties of xenobiotic conjugates.

Xenobiotic conjugates can be divided into two classes—the principal conjugates that are commonly formed and the novel conjugates that are less commonly formed. Mammals, aquatic organisms, insects, and higher plants (as well as different organisms within each of these broad classes) form xenobiotic conjugates that have similarities as well as differences. Qualitative and quantitative differences in conjugation may profoundly alter the biological activity of certain xenobiotics. These differences in conjugation activity can be beneficial (e.g., herbicide selectivity) or may have adverse consequences (e.g., xenobiotic toxicity due to the absence of a conjugation system). These products must be isolated and unequivocally identified in order to fully evaluate the biological significance of xenobiotic conjugate formation.

Novel xenobiotic conjugates may provide direct or indirect clues concerning the mode of action and the pharmacological and toxicological properties of certain xenobiotics. As the methods of separation and identification of polar conjugates improve and as a larger variety of xenobiotics are studied, spectacular modifications of principal conjugation mechanisms and many examples of novel xenobiotic conjugates are being discovered.

Occasionally, no indication as to the involvement of a particular intermediate conjugation step can be gained by the identification of the structure of the terminal metabolite. These intermediate or "hidden" conjugates may remain undetected because of their transient nature. Questions concerning the biological significance of xenobiotic conjugates and the disposition of these compounds after they are formed must be answered. Therefore, the involvement of xenobiotic conjugate intermediates in carcinogenesis and the binding of certain xenobiotics to macromolecules are being studied.

Finally, the widespread exposure of food plants to a great variety of xenobiotics makes this area one of current significance.

The symposium upon which this book is based was organized to provide a critical review, evaluation, and summary of current information and technology relevant to xenobiotic conjugation chemistry. The highlights of this symposium are presented in this volume.

GAYLORD D. PAULSON
Metabolism and Radiation Research Laboratory
Agricultural Research Service
U.S. Department of Agriculture
Fargo, ND 58105

JOHN CALDWELL
Department of Pharmacology
St. Mary's Hospital Medical School
University of London
London W2 1PG, England

DAVID H. HUTSON
Shell Research Ltd.
Sittingbourne
Kent ME9 8AG, England

JULIUS J. MENN
National Program Staff
Agricultural Research Service
U.S. Department of Agriculture
Beltsville, MD 20705

October 1985

PRINCIPAL XENOBIOTIC CONJUGATES

1

Conjugation Mechanisms of Xenobiotic Metabolism: Mammalian Aspects

John Caldwell

Department of Pharmacology, St. Mary's Hospital Medical School, University of London, London W2 1PG, England

Living organisms of all types exist in a chemical environment, made up of nutrients (macronutrients, micronutrients and trace elements) and anutrients or xenobiotics. These latter comprise those compounds from which organisms are (virtually) unable to utilize for energy-yielding metabolism, and may be broadly divided into two classes, natural and synthetic. Living organisms may be exposed to xenobiotics (a) deliberately, due to their use as drugs, pesticides, herbicides etc., (b) co-incidentally in the diet, most notably in the consumption of plant foods by animals of all types, or (c) accidentally, from industrial, agricultural or other sources. It is the purpose of this review to consider the fates of such xenobiotics in living organisms, as an introduction to the topic of this Symposium volume, concentrating in particular upon the metabolic conjugation reactions as they occur in mammals.

When a xenobiotic enters a living organism, it may undergo one or more of a number of fates, as follows :

1. Enzymic metabolism leading to elimination

2. Enzymic metabolism leading to accumulation

3. Elimination unchanged

4. Spontaneous (non-enzymic) chemical transformation

5. Accumulation unchanged

Although the last two options are very important when they occur, as exemplified by the unfortunate cases of thalidomide (1) and the polyhalogenated biphenyls (2) respectively, the great majority of xenobiotics undergo enzymic metabolism and/or are eliminated unchanged. In animals, in most cases, metabolism favours the elimination of the compound by enhancing its polarity

0097–6156/86/0299–0002$07.50/0

and water solubility. Only rarely does metabolism cause the retention of a xenobiotic within an animal : this may occur in one of two ways (a) the formation of lipophilic metabolites (3) or (b) metabolism to reactive species, which bind covalently to target sites within body constituents (4). Only in very rare cases are such reactions of quantitative importance in the disposition of a xenobiotic, although they are of course often of biological significance, notably in toxicity (5). The situation in plants obviously differs from the above. Lacking the excretion mechanisms of animals, metabolised xenobiotics are retained by the plant, frequently sequestered in lignin and other cell constituents which prevent any interference with cell function (6).

Any survey, however cursory, of the now substantial literature on the metabolic fate of xenobiotics will serve to illustrate the plethora of possible reactions. One of the major contributions of R. T. Williams was to discern that these reactions may be classified into two distinct types which generally occur sequentially (7). The xenobiotic initially undergoes a Phase I, or functionalization, reaction of oxidation, reduction or hydrolysis the product of which is subjected to a Phase II, or conjugation, reaction. These latter are biosyntheses, in which the xenobiotic (or exocon) is linked with an endogenous conjugating agent (or endocon) to give a characteristic product termed a conjugate. In some cases, this biphasic sequence does not operate : some Phase I metabolites may be eliminated without being conjugated, while other xenobiotics may undergo Phase II metabolism directly. The elimination products of xenobiotics may thus comprise the unchanged compound, Phase I metabolites, Phase II metabolites and the products of the biphasic sequence. It is most commonly the case that xenobiotics are eliminated in the form of conjugates.

Compounds undergoing metabolic conjugation must possess within their structures a functional group appropriate for combination with the conjugating agent (8). This is generally introduced by Phase I metabolism, and may either be chemically stable e.g. phenolic, alcoholic or carboxylic hydroxyl, amine, thiol or a cyclic heteroatom, or chemically reactive e.g. arene oxide, or carbonium ion.

It is now generally accepted that there occur in mammals six major conjugation reactions available to xenobiotics (9). In each case, the substrate and co-factor requirements, enzymic basis and phylogenetic distribution are (reasonably) well understood. These six reactions each have a well-defined role in the metabolism of endogenous compounds and in biosyntheses. All involve the participation of a transferase enzyme, which exhibits high specificity for the conjugating agent in question. Each reaction is a biosynthesis, and is energy-requiring : this is provided either by the conjugating agent being present in an activated form (most frequently as a nucleotide) or by the prior activation of the xenobiotic substrate. The major reactions are listed in Table I, which gives details of their energy sources and endogenous role(s).

In addition to the 'classical' conjugation reactions listed in Table 1, there occurs also an increasing number of so-called 'novel' conjugations (10, 11). These are added to regularly, as a consequence of the ever more rigorous investigation of both excreted metabolites and of tissue residues of xenobiotics by increasingly sophisticated analytical methodology. The novel reactions are those presently viewed as being restricted in their occurrence to particular combinations of substrate(s) and species. This status is due simply to the current lack of knowledge (10), and some, at least, of these reactions may be expected to attain the status of major reactions as more information is acquired. Indeed, it already seems that the formation of methylthio conjugates, a reaction first observed in the mid-1970s (12), may be a general reaction of xenobiotic metabolism (13). Recent developments in the discovery of novel reactions, and in the analytical methodology upon which they are based, are covered in detail elsewhere in this volume, and thus will not be dealt with further here.

Historical perspective

Before proceeding to a review of the major conjugation reactions extant in mammals, it is of some interest to consider the historical evolution of our knowledge of this important group of pathways. The conjugation reactions were the first reactions of foreign compound metabolism to be discovered, if only for the simple reason that they produce the final elimination products of xenobiotics. However, the early history of their study derives from a time long before the delineation of our present sub-discipline of 'drug metabolism', when pioneering workers dealt with the application of chemical principles to physiological problems. The study of foreign compound metabolism led to numerous important discoveries in biochemistry (14, 15), beside which their significance, viewed simply in terms of 'drug metabolism', pales in comparison. This is exemplified by the sequence of events leading to the discovery of the conjugation of benzoic acid with glycine, the hippuric acid synthesis, which was not merely the first reaction of drug metabolism, but which was the first biosynthesis of any kind to be discovered and which could not be reproduced in the test tube for some years afterwards.

The great Swedish chemist C. W. Scheele discovered benzoic acid in gum benzoin in 1775, by sublimation of the resin (16). Applying this technique to many natural materials, between 1770 and 1786, he also discovered tartaric, lactic, oxalic, citric, gallic and malic acids, and in 1776 he found benzoic acid in human urine (16). Following this in 1784, Rouelle (17) found benzoic acid in the urine of cows, but in 1799, Fourcroy and Vauquelin (18) showed that the acid obtained by these workers was not benzoic acid, but another, similar acid, which they could not identify.

The founder of modern organic chemistry and pioneer of the application of chemical principles to physiological problems, Justus von Liebig, became interested in these accounts and in 1830 he repeated the experiments of Fourcroy and Vauquelin (19). He found a new nitrogen-containing acid in horse urine, a compound of benzoic acid with an unknown nitrogenous body, which he termed 'hippuric acid' from the Greek for 'horse' and 'urine'. In this paper, Liebig speculated on the origin of hippuric acid, and noted that benzoic acid as such was not present in the diet of horses.

The honour of the discovery of synthesis of hippuric acid from ingested benzoic acid should be accredited to the British surgeon Alexander Ure, at that time on the staff of Guy's Hospital in London. Ure attempted to treat gout with benzoic acid and following the oral administration of one scruple (= 1.3g) he isolated 15 grains (= 0.97g) of hippuric acid from the urine. The publication of Ure's finding in the London Medical Gazette in 1841 (20) stimulated further work in Germany and Liebig's close collaborator Friedrich Wohler prompted his student Wilhelm Keller (21) to repeat the investigations of Ure, by self-experimentation. It was this study which served to prove that hippuric acid is indeed formed from benzoic acid.

The exact nature of hippuric acid was discerned in 1845 by Dessaignes (22) who showed that the nitrogenous body linked with benzoic acid was glycine. The existence of glycine had been known since 1820, when Braconnot found it in gelatine (23). However, the chemical synthesis of hippuric acid from benzoic acid and glycine was not achieved until 1857, when Dessaignes (24) fused the constituents in a sealed tube reaction in a furnace.

Although the work of Ure and Keller had shown that the administration of benzoic acid led to the excretion of hippuric acid, the natural occurrence of the latter in the urine could be considered to render their studies equivocal. The unambiguous proof of the linkage of an exogenous carboxylic acid could be linked with glycine in the body was provided by Cesar Bertagnini, who showed that p-nitrobenzoic and o-hydroxybenzoic (salicylic) acids (25, 26). were both converted in part to the corresponding hippuric acids. In these studies, the nitro and phenolic hydroxyl groups functioned as chemical labels, much as we would use isotopes today.

Following this, from studies of normal and pathological body fluids, and from examination of the metabolic fates of various organic compounds, between 1870 and 1900 the remainder of the major conjugation reactions were discovered. These are listed in Table II. Between 1900 and 1970, new discoveries occurred very infrequently, but in the last 15 years the application of modern analytical techniques (see above) has added new examples at a steady rate.

The more recent history of the conjugation reactions reveals that their study has followed a sequence in which the years following

Table I. Classification of the Major Conjugation Reactions

Reaction	High-energy intermediate	Endogenous roles
Reactions with activated conjugating agents		
Glucuronidation	UDP glucuronic acid	Biosynthesis, detoxication
Sulfation	PAPS	Biosynthesis, detoxication
Methylation	S-Adenosyl methionine	Biosynthesis, detoxication
Acetylation	Acetyl CoA	Biosynthesis, intermediary metabolism
Reactions with activated xenobiotic		
Glutathione conjugation	Epoxides, nitrenium ions etc.	Maintenance of redox potential, leukotriene synthesis
Amino acid conjugation	Xenobiotic acyl CoAs	Biosynthesis, detoxication (esp. in amino acidurias)

Table II. The Discovery of the Major Conjugations

Conjugation	Author and date
Glycine	Keller (1842)
Sulfate	Baumann (1876a)
Glucuronic acid	Jaffe (1874)
Ornithine	Jaffe (1877)
Mercapturic acid	Jaffe (1879); Baumann & Preusse (1879)
Methylation	His (1887)
Acetylation	Cohn (1893)
Glutamine	Thierfelder and Sherwin (1914)
Taurine	James (1971)

Based on ref. 27

their initial discovery have been devoted to the consolidation of knowledge concerning their occurrence (27), involving the search for new xenobiotic substrates and the examination of various animal species for the occurrence of the reaction in question. Following this, the post-World War 2 years have seen the development of knowledge concerning the biochemistry of these reactions (28, 29), both from the viewpoints of the enzymology of the transferases and of the supply and regulation of the conjugating agent. Some of these discoveries have had fundamental importance outside drug metabolism alone, notably the discovery of Coenzyme A by Lipmann (30) arising from work on the acetylation of sulfonamides.

Two characteristic features mark current activity in the study of the conjugations (a) the application of the techniques of molecular biology and (b) an enhanced appreciation of their pharmacological and toxic consequences. It is to be expected that the next few years will see many developments arising from the former innovation (31), but it is the latter area that most progress has been made in recent years.

From the 1870s onwards, various early workers concerned themselves with the consequences of xenobiotic metabolism, in particular noting that conjugates were markedly less toxic than their parent foreign compounds (27). This led to the idea that this process resulted in 'detoxication' (a translation of the German 'entgiftung'), and this became very widely accepted. The concept of detoxication reached its most sophisticated development in the chemical defence hypothesis of C.P. Sherwin (27). Since the 1920s, it has been increasingly realized that the Phase I reactions are often involved in increasing the activity of xenobiotics, and thus the idea of these reactions being detoxications was abandoned. However, this concept was much longer lasting in the case of the conjugations : these give rise to stable, readily excreted, inactive metabolites and it was not until the late 1970s that this view came to be modified. It was realized that due to inter-species differences in the occurrence of these reactions or the limited capacity of certain conjugations, the detoxication function of the conjugations failed more frequently than had been appreciated hitherto. In addition, a number of instances of conjugates being active metabolites had accumulated in the literature, and the collation of information on these two aspects served to focus the attention of pharmacologists and toxicologists on conjugative metabolism The last five years have seen a great increase in work in this area, which continues to grow.

The major conjugation reactions

It is the purpose of the remainder of this review to survey the major conjugation reactions, in terms of their substrate versatility, enzymic mechanism and distribution amongst mammalian

species, and to comment briefly upon their biological
consequences.

Glucuronic acid conjugation

The glucuronic acid conjugation is the most versatile of the
conjugation reactions in terms of the range of xenobiotic
substrates it may accept and its widespread distribution through
species and tissues. The glucuronic acid residue incorporated
into the conjugate derives from the nucleotide uridine
diphosphate glucuronic acid (UDPGA) and is transferred to the
xenobiotic under the influence of the UDP glucuronyl transferases
(UDPGTs) (32). The glucuronic acid residue may be attached to
one of a wide range of functional groups (33), which are listed
in Table III. Although the main classes of glucuronide have been
known for more than a century, it is of interest to note that new
classes of glucuronide are still being discovered e.g. the
quaternary amino glucuronides of drugs such as cyproheptidine and
the C-glucuronides of the pyrazolones.

In UDPGA, the pyranose ring of glucuronic acid is in the alpha
form i.e. the protons on C-1 and C-2 are cis to each other.
During the enzymic transfer to the acceptor substrate, the
anomeric centre undergoes inversion such that, in the conjugate,
glucuronic acid is of the beta-configuration. The xenobiotic
residue is always linked to the hydroxyl group on C-1, although
in the case of the ester glucuronides there may occur subsequent
acyl group migration to the hydroxyl groups on C-2, C-3 and C-4.
Any such migration occurs after formation of the 1-O-acyl
glucopyranosiduronates, and other isomers are not formed
enzymatically (34).

UDPGT is a membrane-bound enzyme, found particularly in the
endoplasmic reticulum of the liver and numerous other tissues
(32). The activity of the enzyme is at least partly latent (35),
being enhanced by a variety of agents able to disrupt membrane
structure (lipolytic enzymes, detergents, organic solvents etc.).
UDP-N-acetylglucosamine plays a role in determining the latency
of the enzyme in vivo. Evidence showing that UDPGT exhibits at
least functional heterogeneity has been accumulated over many
years, and includes findings of species and strain differences,
substrate specificity, differential ontogenesis, induction and
inhibition, tissue distribution and in vivo activation (32). The
molecular basis of this heterogeneity has been hard to establish
owing to the difficulties of working with membrane-bound enzymes.
At the present time, three clearly-defined, separate forms have
been purified to apparent heterogeneity, these being
differentially inducible and having different, albeit
overlapping, substrate specificities (31). The existence of
further forms has been suggested by studies of the induction of
UDPGT and of its substrate specificity.

The occurrence of multiple UDPGTs with overlapping substrate
specificities makes assays of glucuronidation activity of
particular substrates by subcellular fractions unreliable as

Table III.   Types of Compounds Giving Rise
to Glucuronic Acid Conjugates

Functional group	Example
Hydroxyl	
Primary alcohol	trichloroethanol
Secondary alcohol	propranolol
Tertiary alcohol	<u>tert</u>-butanol
Alicyclic alcohol	cyclohexanol
Terpenoid alcohol	menthol
Phenol	phenol
Terpenoid phenol	eugenol
Enol	4-hydroxycoumarin
Alphatic hydroxylamine	<u>N</u>-hydroxychlorphentermine
Aromatic hydroxylamine	2-naphthylhydroxylamine
Hydroxamic acid	<u>N</u>-hydroxy-2-acetamidofluorene
Carboxylic acid	
Alkyl	2-ethylhexanoic acid
Aromatic	benzoic acid
Heterocyclic	nicotinic acid
Arylacetic	Indole-3-acetic acid
Arylpropionic	hydratropic acid
Aryloxybutyric	clofibric acid
Carbamic acid	tocainide carbamate
Amino functions	
Aromatic	aniline
Azaheterocycle	sulphisoxazole
Carbamate	meprobamate
Sulphonamide	sulphadimethoxine
Hydroxylamine <u>N</u>-	<u>N</u>-hydroxy-2-acetamidofluorene
Tertiary aliphatic	cyproheptadine
Urea	dulcin
Sulphur functions	
Thiol	2-mercaptobenzothiazole
Dithioic acid	<u>N</u>,<u>N</u>-diethyldithiocarbamic acid
Carbon centres	
Pyrazolone ring	phenylbutazone

Adapted from ref. <u>33</u>

indicators of the presence of particular isozymes. This situation will remain until immunochemical or molecular biology techniques permit the assay of individual isozymes, or until isozymic-specific substrates are discovered. To reduce confusion in the interim, standardized assay conditions have recently been recommended (36).

Glucuronic acid conjugation is widely distributed through the animal kingdom. It is involved in the metabolism of foreign compounds in mammals, birds, fish, amphibians and reptiles, but not in insects or molluses (33). In these latter instances, it is replaced by the analogous reaction of glucose conjugation (33). Glucuronidation has also been reported to occur in plants (33).

Virtually all mammals are able to form glucuronic acid conjugates, but there do exist certain species, and genetically stable mutants within species, which are marked by an (at least partial) inability to effect this reaction (33). These include three examples of substrates whose glucuronidation only occurs in primate species (a) cyproheptadine quaternary N-glucuronidation (b) C-glucuronidation of phenylbutazone and other pyrazolones (both only found in human and chimpanzee) and (c) the N^1-glucuronidation of sulfadimethoxine and some other sulfonamides (found in all primates).

Probably the best example of a 'species defect' in drug metabolism is the inability of the domestic cat to form glucuronides of many compounds which are extensively metabolised along this pathway in most other species. This was originally noted by Robinson and Williams (37) in 1956, and many subsequent studies have confirmed and extended their observations. However, it is now evident that the defect is only partial, and must be qualified with respect to the substrate in question (33). Consideration of the substantial amount of comparative data now available shows that the cat is unable to glucuronidate small, water-soluble substrates but that the conjugation of larger, more lipid-soluble aglycones proceeds in the same way as in other species.

In addition to the domestic cat, a number of related carnivores exhibit an inability to glucuronidate a range of aglycones, and this defect is apparently a biochemical characteristic of the cat-like carnivores, the Feloidea. Studies from this laboratory have documented the occurrence of the defect in the lion, lynx, civet and genet, but not the hyena (33).

There also occur intra-species, or strain, differences in glucuronidation capacity (33). In man, these include the virtually complete defect of the Crigler-Najjar syndrome and the much less serious Gilbert's disease, while in the rat the Gunn mutant of the Wistar strain provides an animal model of certain aspects of the Crigler-Najjar syndrome. Outbred Wistar rat populations also contain mutants defective in androsterone glucuronidation. These intra-species differences are in general indicative of the occurrence of independently regulated forms of UDPGT.

Sulfate conjugation

In the sulfate conjugation reaction, a hydroxyl (or occasionally an amino) group present in a xenobiotic is linked with the sulfate ion giving a highly polar, highly ionized sulfate ester. The inorganic sulfate used in this conjugation is first activated by conversion to the high energy sulfate donor 3'-phosphoadenosine-5'-phosphosulfate (PAPS) (38), and the transfer of sulfate to the xenobiotic is catalysed by a sulfotransferase (39). The enzymes of sulfate activation and transfer are all located in the cytosol of liver and other tissues.

The sulfotransferases are a family of enzymes (39), separable by ammonium sulfate fractionation into a group conjugating phenolic substrates (39) and another group responsible for steroid sulfation (40). The phenol sulfotransferase fraction contains at least four distinct enzyme forms, specifically catalysing the sulfation of various substrates including phenols, N-hydroxyacetamides and oestrone.

A wide range of xenobiotics may undergo sulfation, principally those containing hydroxyl groups but also certain aromatic amines. Among the hydroxylic substrates are alcohols, phenols, catechols and hydroxylamines (see Table IV).

In general terms, sulfate conjugation represents an alternative to glucuronidation for the metabolism of a wide range of substrates (41). Two principal factors determine the relative extents to which these two reactions may contribute to the metabolism of a particular substrate (1) its structural features and (2) the dose given. Sulfation is a feature of the metabolism of relatively small, hydrophilic substrates in which the centre undergoing conjugation is sterically unhindered, and whose intracellular distribution favours the cytosol (41, 42).

It has long been known that the capacity of the sulfate conjugation mechanism is limited, and that the proportion of the dose of a compound undergoing sulfation falls with increasing dose size (43). The sulfate mechanism is thus more prominent at lower doses. It is generally accepted that the low capacity of the reaction is due to limitations upon the size of the pool of PAPS, and perhaps also inorganic sulfate, available for conjugation (44). However, while this may be the case with respect to interactions between compounds both undergoing sulfation, which may be reversed by the administration of PAPS precursors such as cysteine, cystine or inorganic sulfate (44), it may not underly the restricted capacity of the sulfation mechanism for compounds given singly. The pool of inorganic sulfate feeding PAPS synthesis is readily repleted from other cellular sulfate precursors, and the capacity limitation of

Table IV. Types of Compounds Forming Sulfates

Functional group	Example
Primary alcohol	Ethanol
Secondary alcohol	Butan-2-ol
Phenol	Phenol
Catechol	alpha-Methyl-DOPA
Alicyclic alcohol	Dehydroepiandrosterone
Heterocyclic alcohol	3-Hydroxycoumarin
Hydroxyamide	N-hydroxy-2-acetamidofluorene
Aromatic hydroxylamine	2-Naphthylhydroxylamine
N-oxide	Minoxidil

Drawn from ref. 57

sulfation instead may well originate from the kinetic features of
the sulfotransferases themselves (45).

Although sulfate conjugation is widely distributed amongst
mammals, the pig has a deficiency in the sulfation of certain
phenols (42).

Methylation

The transfer of methyl groups to various hydroxyl, thiol and
amino functions, although common in the metabolism of endogenous
compounds (46), is only rarely important quantitatively in the
metabolism of xenobiotics (42, 47). Functional groups of
xenobiotics undergoing methylation (42) include primary amines
(e.g. amphetamine), secondary amines (e.g. desmethylimipramine),
tertiary amines, (e.g. dimethylaminoethanol), azaheterocycles
(e.g. pyridine), phenols (e.g. 4-hydroxy-3,5-diiodobenzoic acid),
catechols (e.g. alpha-methylDOPA) and thiols (e.g. thiouracil).
In the majority of these cases, methylation serves to increase
lipophilicity and reduce polarity of the functional group in
question (42), and thus the function of methylation is apparently
not simple facilitation of excretion. However, the methylation
of tertiary amines and azaheterocycles gives quaternary ammonium
compounds with markedly increased water solubility (48).

In the greatest majority of cases, the methyl group transferred
to a xenobiotic derives from the nucleotide S-adenosyl methionine
(46), but 5-methyltetrahydrofolic acid may be the methyl group
donor to primary and secondary amines in the brain (49). A
variety of N-methyltransferases are known (50) and
thiolmethyltransferase has been the subject of much recent
attention (51). Xenobiotic phenols undergoing methylation are
generally either catechols or phenols with bulky ortho
substituents (42).

Comparatively little is known about the species distribution of
the various methylation reactions. Catechol and N-methylations
apparently occur throughout the Mammalia (46) : in the case of
azaheterocycle pyridine there occurred at 10 fold variation in
the extent of methylation (48), from the rabbit and cat which are
extensive methylators (up to 50% of the dose), to rat, mouse and
man, in which methylation only accounts for 2-10% of an
administered dose. It is well known that the O-methylation of 4-
hydroxy-3,5-diiodobenzoic acid is far more extensive in primate
species than non-primate mammals (52).

Acetylation

A wide variety of compounds containing the primary amino group (-
NH_2), including amines, amino acids, sulfonamides, hydrazines and
hydrazides, may undergo N-acetylation in the body (42). Although
endobiotics may be acetylated at -OH (e.g. choline) or -SH
(e.g. Coenzyme A) centres, there are no reports of such centres
in xenobiotics being so conjugated (42, 47).

The acetyl group used to form the amide bond with the amino group derives from the ubiquitous high energy intermediate acetyl CoA, and the formation of the bond is catalysed by the enzyme N-acetyltransferase (53). The reaction appears to involve a ping-pong Bi-Bi mechanism, in which acetyl CoA first acetylates the enzyme, followed by acetylation of the amine and regeneration of the enzyme (53).

Substrates for acetylation may be conveniently divided into two groups (53), monomorphic (unimodal), as typified by para-aminobenzoic acid and sulfanilamide, and polymorphic (bimodal), such as sulfamethazine, isoniazid and dapsone. This nomenclature refers to the ability of the second group of substrates to reveal the existence of two distinct forms of the N-acetyltransferase, under separate genetic (regulatory) control, which catalyse the acetylation at markedly different rates. The rate of acetylation is a trait exhibiting autosomal Menedelian inheritance, controlled by a pair of alleles, one for fast acetylation (Hf) and one for slow (Hs) acting at a single locus (54). Thus the population contains three genotypes, homozygous fast (HfHf), homozygous slow (HsHs) and heterozygotes (HfHs). The phenotypic expression of these genotypes is such that the heterozygotes may only with difficulty be distinguished from homozygous fast acetylators (55). Many attempts have been made to demonstrate the existence of variant forms of N-acetyltransferase in the liver of man and other species exhibiting the polymorphism, but these have been largely unsuccessful (53). Indeed, it seems likely that a single enzyme acetylates both the 'monomorphic' and 'polymorphic' substrates, and that the difference between the phenotypes occurring in populations lies in a differential flexibility of the active site. There appears to be an induced fit of substrate to the active site, which is more flexible in the case of fast acetylator phenotype, permitting the access of bulkier substrates.

The genetic polymorphism controlling N-acetylation may be demonstrated readily in human populations, where the distribution of the phenotypes shows marked ethnic variability (54). In the U.S.A., the incidence is close to 50% fast / 50% slow, while in other parts of the world the fast acetylator phenotype predominates up to 90% fast/ 10% slow in Orientals. The polymorphism has also been demonstrated in outbred animal populations of rabbits and squirrel monkeys (56), while animal models of the fast and slow acetylator phenotypes occur in various inbred mouse and hamster strains (56).

As well as the genetic polymorphism of N-acetylation, there are two marked instances of species defects in such reactions. The dog (57) and related canine carnivores, such as the fox and hyena (57, 58), are unable to acetylate most of the nitrogen centres which are substrates for this reaction, the exception being the alpha-amino group of S-substituted cysteines and the sulfomanido-N of sulfonamides. Secondly, it is well-known that the guinea pig does not excrete mercapturic acids (57), although it is able to form glutathione conjugates (59), which are the precursors of mercapturic acids (see below). The defect apparently arises from an inability to N-acetylate the S-substituted cysteines formed by catabolism of glutathione conjugates (60). However, the acetylation of other nitrogen centres in the guinea pig occurs in the same way as most other mammals (57).

Glutathione Conjugation

A large number of xenobiotics undergo conjugation with the abundant nucleophilic tripeptide glutathione (gamma-glutamylcysteinylglycine), at its free -SH group (59, 61). The S-substituted glutathiones undergo metabolism by a number of pathways giving rise to various excretory products, the most important of which are the mercapturic acids (S-substituted N-acetylcysteines). These matters will be dealt with in more detail later in the chapter by Dr. Bakke.

Two distinct types of xenobiotic substrates for the glutathione conjugation may be discerned (59), (a) those which are sufficiently electrophilic to undergo conjugation directly and (b) those which first undergo metabolic activation to produce an electrophile which is then conjugated. The first group of substrates includes halo- and nitro-alkanes and benzenes and sulfonic acid esters, from which glutathione displaces an electron-withdrawing group (59), and alpha, beta-unsaturated ketones and other compounds with activated carbon-carbon double bonds e.g. maleic acid diesters. In the latter cases, glutathione adds across the double bond (62).

A variety of reactive intermediates produced by metabolic oxidation may undergo glutathione conjugation, including oxiranes (arene oxides and aliphatic and alicyclic epoxides) and reactive N-oxidation products. Glutathione opens the highly strained oxirane ring in an addition reaction (59), or attacks an electron-deprived centre in the molecule. In addition, glutathione can also interact with metabolically-formed free radical species (63).

These various glutathione conjugations are catalysed by a family of glutathione S-transferase isozymes (64), located in the cytosol of many tissues. The multiplicity of this enzyme was recognised by Boyland and Chasseaud some 20 years ago (65), and now some 12 forms have been separated and characterized. The number of forms varies from tissue to tissue. The glutathione transferases are dimeric proteins : at present seven different subunits have been recognised and the various isozymes are homo- or hetero-dimers of these (66). In addition to catalysing the formation of glutathione conjugates, the glutathione S-transferases are also involved in the metabolism of nitrite esters, peroxides and alkyl thiocyanates (59). In the case of the nitrite esters, a reaction which is seemingly a hydrolysis yielding nitrite and an alcohol is actually a two step reaction, consuming two molecules of glutathione and giving rise to the homoconjugate oxidized glutathione (67).

Relatively little is known of the zoological distribution of
glutathione conjugation. However, glutathione S-transferase
activity is very widely distributed throughout mammalian species
and tissues, no instance of a deficiency being recorded (59). As
noted above, the isozymic make-up of particular tissues may be
highly variable (68). This extensive distribution of these
enzymes is probably a reflection of their increasingly recognized
role in the metabolism of endogenous substrates such as steroids
and leukotrienes (68), and as a 'defence mechanism' against lipid
peroxides (63).

There is no properly documented instance of the excretion of a
glutathione conjugate in the urine, although these metabolites
are found intact in the bile (69). This latter route of
elimination is favoured by their molecular weight and polarity
(70). The urinary excretion products which derive from
glutathione conjugates are numerous, and all stem from the S-
substituted cysteine produced by the action of gamma-glutamyl
transferase and a peptidase, successively removing the gamma-
glutamyl and glycine moieties (71). The principal routes of
metabolism of the S-substituted cysteines are (a) N-acetylation,
giving mercapturic acids (71), and (b) the so-called beta-lyase
pathway, in which the cysteine conjugate is cleaved so as to give
rise to a thiol group in the xenobiotic together with, pyruvate
and ammonia (72). These thiols are then methylated and the
thiomethyl conjugates, together with the corresponding sulfoxides
and sulfones produced by subsequent oxidation, are found in the
excreta and tissues (13, 73). This pathway is unique in
introducing a simple sulfur functional group into a xenobiotic,
albeit by a rather roundabout route. Finally, S-substituted
cysteines may undergo transamination yielding the corresponding
thiopyruvic acids, which are further metabolised to thiolactic
and thioacetic (thioglycolic) acids (74). Of these pathways, the
formation of mercapturic acids is quantitatively the most
important, but the beta-lyase pathway is now realised to be
toxicologically highly significant in a number of cases. The
defect of cysteine conjugate N-acetylation in the guinea pig has
been noted above.

The glutathione conjugation mechanism is of great significance in
protecting the body against the harmful effects of electrophiles
(53) and radical species (63). In addition, it facilitates the
excretion of many xenobiotics, since most of the products of the
catabolism of conjugates are highly polar and water soluble (73).
However, in some cases the conjugates themselves or certain of
their breakdown products may be involved in the expression of
toxicity (56, 73).

Epoxide hydration

One important group of substrates for glutathione conjugate are
oxiranes, which are produced by the oxidation of aromatic rings
or alkenes. Arene oxides generally undergo spontaneous
rearrangement via the 'NIH shift' mechanism to give phenols (75),
but both they and the alkene oxides may also be hydrated to

dihydrodiols, by the action of the enzyme epoxide hydrolase (76).
A great number of oxiranes are substrates for this reaction,
ranging from simple compounds like styrene 7,8-oxide and benzene
1,2-oxide to very complex structures such as the various
benzo(a)pyrene oxides.   Two distinct epoxide hydrolases exist in
the cell (77), one in the microsomes (smooth endoplasmic
reticulum) and one in the cytosol.   They exhibit differential
induction and inhibition and in general show different, clear-cut
substrate specifities.   The cytosolic enzyme shows a marked
preference for aliphatic epoxides, especially those on fatty
acids, but is poor at hydrating the arene oxides typically
produced by the oxidation of xenobiotics (77).   Phylogenetic
differences in the activities of these two enzymes are documented
: the microsomal enzyme is most active in the rhesus monkey and
least active in the mouse, whereas the cytosolic enzyme is most
active in the mouse and rabbit and least active in the rat (77).
However, the in vivo significance of these differences remains to
be discerned.

The amino acid conjugations

Many xenobiotic carboxylic acids undergo conjugation with one of
a variety of amino acids, in which the carboxyl group of the
xenobiotic is linked in an amide (peptide) bound with the alpha-
amino group of the amino acid (78).   The chemical classes of acid
involved in amino acid conjugation are relatively few in number,
and the reactions are restricted to certain aliphatic, aromatic,
heteroaromatic, cinnamic and arylacetic acids (78).   The
occurrence of the reaction is markedly dependent on the steric
hindrance around the carboxyl group by substituents on the aryl
moiety or the side chain bearing the acid function (79).

The major amino acid conjugations involve glycine, glutamine,
taurine and ornithine (56), while isolated instances of
conjugation with a number of other amino acids have been reported
(56), including glutanic acid, serine, histidine, alamine and
aspartic acid.   In addition to these instances of the conjugation
of a single amino acid, there are examples of the formation of
dipeptide conjugates involving glycylglycine, glycyltaurine and
glycylvaline (80).

The particular amino acid used in the conjugation of an acid is a
function of its structure and the animal species under
consideration (80).   For benzoic, heterocyclic and cinnamic
acids, most species use glycine, which is replaced by ornithine
in birds such as the chicken (galliform) and duck (anseriform).
For aryl- and aryloxy-acetic acids, subprimate mammals use
glycine, but this is replaced by glutatamine in primates.   In
addition, these acids are conjugated with taurine, a reaction
generally found at a low level but which is especially well
developed in carnivores.

Nature of interspecies differences in xenobiotic conjugation

During the course of evolution, living organisms have developed an immense diversity. The Mammalia alone comprise over 4000 species. It is remarkable to note that the biphasic pattern of drug metabolism is found to occur in all mammals (and also most other organisms): however, there occur widespread qualitative and quantitative variations within this fundamental pattern (9, 57), and these are documented at length in the drug metabolism litera ture Qualitative differences between species may arise in one of two ways (a) a species may be (relatively) defective in a reaction of otherwise widespread occurrence, or (b) reactions being restricted in their occurrence to particular species or groups of species. Quantitative species differences arise from variations in the relative activities of two or more alternative metabolic options which a given compound may undergo. It is this last case which is encountered most frequently.

A number of so-called 'species defects' of metabolic conjugation have been documented, and a list is presented in Table V.

The best known of these examples are the defects of glucuronidation in the cat and related feline species and of N-acetylation in the dog. It is important to appreciate that these defects are not absolute, but must be qualified with reference to the substrate(s) in question. Thus, the cat is unable to glucuronidate simple, relatively water-soluble phenols and carboxylic acids, but conjugation of more complex, lipid-soluble substrates proceeds in the cat to the same extent as in other species. Similarly, dogs are unable to N-acetylate aromatic amino groups and hydrazides, but do acetylate the S-substituted cysteines which are the penultimate intermediates in the conversion of glutathione conjugates to mercapturic acids. Guinea pigs, on the other hand, apparently cannot N-acetylate these substituted cysteines, but do acetylate a variety of other amines.

For many years, there has been much interest in the possibility that certain metabolic reactions have been restricted by evolutionary pressures to specific groups of species. In particular, the close similarities between man and primate species has led to numerous comparative metabolic studies which have thus far revealed the existence of five conjugation reactions which only occur in these species. These are (i) the glutamine conjugation of arylacetic and aryloxyalkyl acids (found in man, apes and Old and New World monkeys), (ii) the O-methylation of 4-hydroxy-3,5-diiodobenzoic acid (man, Old and New World monkeys) (52) (iii) the N^1-glucuronidation of certain methoxysulfonamides (all primates) (33), (iv) the quaternary N-glucuronidation of tertiary aliphatic amines (man and apes only) (33) and (v) the C-glucuronidation of pyrazolone rings (man and apes only) (33). Another example of a metabolic reaction being largely restricted in its occurrence to a group of species is that of taurine conjugation of arylacetic acids, which occurs at low levels in many species but is particularly well-developed only in carnivores (80).

Quantitative species differences in the relative extents of

competing metabolic options can be seen in the cases of phenols, which may undergo either sulfation or glucuronidation, and carboxylic acids, which are conjugated with amino acids or glucuronic acid. Table VI illustrates this in the case of phenol itself (57).

A major reason for the study of species differences in xenobiotic metabolism is the interpretation of species differences in pharmacological and toxic responses to chemicals. In addition, such data may be of value in the zoological classification of animals, for which the term 'pharmacotaxonomy' was coined (58). The occurrence of two characteristic defects in conjugation reactions in carnivorous species has been mentioned earlier, those of glucuronidation in the cats and of N-acetylation in dogs. Comparative studies of suitable probe substrates in a range of carnivores have provided important information about the true classification of the hyena, an animal generally considered by zoologists as feline. Table VII presents data on the glucuronidation of l-naphthylacetic acid and N-acetylation of sulfadimethoxine in six carnivores, and this shows that the feline species (cat, lion, lynx, civet) are completely distinct from the canines (dog and hyena). The status of the hyena with respect to these two biochemical 'marker reactions' of carnivores strongly suggest that this species should be regarded as a canine, rather than a feline.

Biological significance of the metabolic conjugation reactions

The various metabolic conjugation reactions of drugs and other chemicals are of great importance for their biological properties, and reasons for this are presented in Table 8. The majority of conjugates are more polar and have markedly greater water solubility than their parent compounds, so that they result in both the loss of specific receptor interactions and facile elimination from the body. However, there occur instances where either the detoxication function of conjugation may fail, either as a result of a species defect or by saturation, or a conjugate contributes to biological activity. In addition, the conjugation reactions have a number of pharmacokinetic implications.

The capacity of the principal metabolic conjugations depends upon the affinities of the transferase enzymes involved both for the xenobiotic substrate and the endogenous conjugating agent, and upon the availability of the conjugating agent, which may be limited. This, together with the widespread distribution of conjugation activity amongst the tissues of the body (notably in the absorptive and excretory organs) leads to a number of pharmacokinetic consequences of conjugation which are listed in Table VIII. Space does not permit more than a brief mention of these, but full descriptions are to be found elsewhere. Although these matters are largely outside the scope of this review, they are covered in detail elsewhere (42, 82, 83).

Species variations in metabolic conjugation, notably originating from species defects, are of great significance as determinants

Table V.   Some "Species Defects" in Metabolic Conjugation Reactions

Reaction	Affected species (reference)
Glucuronidation	Cat and related species (33)
N-Acetylation of aromatic amines	Dog and related species (57)
N-Acetylation of S-substituted cysteines	Guinea pig (57)
Sulfation	Pig (42)
Glutamine conjugation of arylacetic acids	Non-primates (80)
Glycine conjugation of salicylate (but not benzoate)	Horse (81)

Table VI.   Species Variations in Competing Conjugation Options:
The Fate of Phenol

| Species | % dose conjugated with | | Ratio S/G |
	Sulfate	Glucuronide	
Man and Old World monkeys	80	12	7
New World monkeys	25	50	0.5
Rat and mouse	45	40	1
Cat	93	1	93
Pig	2	95	< 0.1

adapted from ref. 58

Table VII.   Comparative Glucuronidation and Acetylation
of Various Substrates Amongst Carnivores

	Glucuronidation of 1-Naphthyl-acetic acid	N-Acetylation of sulfadimethoxine
Cat	0	18
Lion	0	48
Lynx	0	--
Civet	0	66
Hyena	40	0
Dog	20	0

Table VIII. Biological Significance of the Conjugation Reactions

They result in :

1.   Readily excreted end-products of xenobiotic metabolism.

2.   Metabolic activation, in certain cases.

3.   Detoxication, which may however be defective, due to :

     (a) species defects

     (b) saturation (capacity limitations)

4.   Pharmacokinetic implications, including :

     (a) capacity limitations

     (b) enterohepatic recirculation

     (c) presystemic elimination

     (d) determination of route-of-elimination

     (e) drug-drug interactions

of biological response. Cats are far more susceptible to the toxic effects of many glucuronidogenic substrates than are rats, rabbits and other species (56). Aromatic amines gives rise to methaemoglobinaemia in the dog far more easily than in other cspecies which are able to N-acetylate these compounds. In place of this conjugation, dogs are able to activate these amines by N-oxidation (9).

N-Acetylation of carcinogenic aromatic amines has a major site-directing influence upon their tumorigenicity (9). In the dog, these amines are potent bladder carcinogens, with little or no effect on other organs, but in the mouse, rat and other species in which they are acetylated, they produce tumours at multiple sites, with little or no effect on the urinary bladder. There exist complex inter-relationships between the oxidation, acetylation, sulfation and glucuronidation of aromatic amines and their carcinogenicity, but the N-acetylation defect in the dog has a major role in determining the target for carcinogenicity.

Although conjugates are generally devoid of biological activity, for the reasons stated above, there are a number of well-documented examples of conjugative metabolites contributing towards the actions of the parent compounds. Three classes of active conjugates may be discerned (a) stable conjugates with activity at defined receptor sites, (b) conjugates acting as reactive intermediates, and (c) conjugates active after further metabolism, and important examples of each may be cited (9). Thus, products of glucuronidation, acetylation and methylation reactions may have receptor activity : in some cases, the metabolites may become drugs in their own right e.g. A-acetylprocainamide (9). The contribution of sulfation to the metabolic activation of N-hydroxyarylamines and 1'-hydroxallylbenzenes is well documented (5, 9, 84), while the formation of episulfonium ions from the glutathione conjugates of dihaloalkanes underlies their chemical reactivity (85). Various examples are to be found in the literature of conjugates giving rise to active or reactive products upon further metabolism. One of the best known of these is the antitubercular drug isoniazid, whose hepatoxicity depends upon the metabolism of its N-acetyl conjugate (9). The further metabolism of glutathione conjugates, normally thought of as detoxication products, may similarly result in the formation of reactive intermediates, as in the cases of hexachlorobutadiene and dichlorovinylcysteine (86), or of metabolites whose accumulation leads to toxicity : this is well illustrated by the various methylsulfoxides and sulfones of polyhalogenated aromatic compounds (73).

Concluding remarks

It will be clear to the reader from the above sections that the conjugation reactions are of great importance in the metabolic disposition of xenobiotics of all types in the animal body. These reactions are of historical significance in this field, but are nowadays increasingly realized as having pharmacological and toxicological implications. Any attempt to relate together the chemical structure, metabolic fate and biological actions of a chemical must take conjugative metabolism into account.

The conjugation reactions constitute the second part of the biphasic sequence of xenobiotic metabolism in the body. Although the general features of this sequence are the same throughout mammals (and most other living organisms), the details are highly variable between species. Many examples of quantitative and qualitative differences in metabolic conjugation between species have been illustrated, and these are frequently of functional significance for the toxicity of substrates. By using endogenous conjugation agents, which have well-defined roled in biosynthesis and intermediary metabolism, the conjugation reactions represent an interface between the metabolism of xenobiotics and the biochemistry of endogenous compounds. The limited supply of certain conjugating agents underlies the ready saturability of some conjugating reactions, but the broader implications of the relationships at this interface remain to be discerned.

Since the discovery of the hippuric acid synthesis some 145 years ago, the conjugation reactions have proved a rewarding field of study for successive generations, and continue to generate much interest. In the future, our awareness of these reactions may be expected to develop in various ways (a) enhanced knowledge of both the major and the novel reactions, (b) discovery of more novel reactions, (c) further recognition of the biological consequences of conjugation, and (d) development of the concept of the conjugation reactions as interfaces between xenobiotic and endobiotic biochemistry. Prospects in many of these areas are to be found throughout the present volume.

Acknowledgments

I am grateful to Professor R. L. Smith for helpful discussion, and to Ms. Irene Ross for the preparation of the typescript.

Literature Cited

1. Williams, R.T. Lancet 1963, 1, 723.

2. Opperhuizen, A., Gobas, F.A.P.C., Hutzinger, O. In : 'Foreign Compound Metabolism'; Caldwell, J., Paulson, G.D., Eds., Taylor & Francis; London, 1984; p. 109.

3. Caldwell, J, Marsh, M.V. Biochem. Pharmacol. 1983, 32, 1667.

4. Gillette, J.R. In 'Biological Reactive Intermediates'; Jollow, D.J.; Kocsis, J.J.; Snyder, R.; Vainio, H.; Eds.; Plenum : New York, 1977; p. 25.

5. Miller, J.A.; Miller, E.C. In 'Biological Reactive Intermediates'; Jollow, D.J.'; Kocsis, J.J.; Snyder, R.; Vainio, H.; Eds.; Plenum : New York, 1977, p. 6.

6. Baldwin, B.C.   In 'Drug Metabolism – from Microbe to Man';
    Taylor and Francis : London, 1977; p. 191.

7. Williams, R.T.   'Detoxication Mechanisms, 2nd Edition';
    Chapman and Hall, London, 1959.

8. Dutton, G.J.   In 'Drug Metabolism in Man'; Taylor and
    Francis : London, 1978; p. 81.

9. Caldwell, J. Drug Metab. Rev., 1982, 13, 745.

10. Israili, Z.H.; Dayton, P.G.; Kiechel, J.R. Drug Metab.
    Disp., 1977, 5, 411.

11. Eadsforth, C.V.; Hutson, D.H. In 'Foreign Compound
    Metabolism'; Caldwell, J.; Paulson, G.D., Eds.; Taylor &
    Francis : London, 1984; p. 171.

12. Tateishi, M.; Shimizu, H. Xenobiotica, 1976, 6, 431.

13. Jakoby, W.B.; Stevens, J. Biochem. Soc. Trans. 1984, 12,
    33.

14. Hopkins, F.G. Rep. Brit. Ass., 1913, p. 652.

15. Dakin, H.D. 'Oxidations and Reductions in the Animal Body,
    2nd Edition'; Longmans, Green : London, 1922.

16. Scheele, C.W. 'Chemical Essays', Scott, Greenwood : London,
    1901.

17. Berzelius, J.   'Lehrbuch der Chemie, 3rd Edition'; Dresden
    and Leipzig, 1840; vol. 9, p. 425.

18. Fourcroy, A.F.; Vauquelin, L.N. Ann. Chim. 1799, 31, 63.

19. Liebig, J. Ann. Chim. 1830, 43, 188.

20. Ure, A.   London Medical Gazette, 1841, 27(I) (New Series),
    73.

21. Keller, W. Ann., 1842, 43, 108.

22. Dessaignes, V. C.R. Acad. Sci. (Paris). 1845, 21, 1224.

23. Braconnot, M. Ann. Chim. Phys. 1820, 13, 113.

24. Dessaignes, V. J. Pharm. (Paris) 1857, 32, 44.

25. Bertagnini, C. Ann., 1856, 97, 248.

26. Bertagnini, C. Ann., 1851, 78, 100.

27. Smith, R.L.; Williams, R.T. In 'Metabolic conjugation and Metabolic Hydrolysis'; Fishman, W.H., Ed.; Academic : New York, 1970; Chap. 1.

28. Jakoby, W.B., Ed. 'Enzymatic Basis of Detoxication, Volume 2' Academic : New York, 1980.

29. Jakoby, W.B.; Ed. 'Methods in Enzymology Volume 77'; Academic : New York, 1981.

30. Lipman, F. J. Biol. Chem., 1945, 160, 173.

31. Burchell, B.; Jackson, M.R.; McCarthy, L.; Barr, G.C. In 'Advances in Glucuronide Conjugation'; Matern, S.; Bock, K.W.; Gerok, W.; Eds.; MTP : Lancaster, 1985; p. 119.

32. Dutton, G.J. 'Glucuronidation of Drugs and Related Compounds'; CRC : Boca Raton, Fl, 1980.

33. Caldwell, J. In 'Advances in Glucuronide Conjugation'; Matern, S.; Bock, K.W.; Gerok, W.; Eds; MTP : Lancaster, 1985; p. 7.

34. Faed, E.M. Drug Metab. Rev. 1984, 15, 1213.

35. Berry, C.S. In 'Conjugation Reactions in Drug Biotransformation'; Aitio, A., Ed.; Elsevier : Amsterdam, 1978; p. 233.

36. Bock, K.W.; Burchell, B.; Dutton, G.J.; Hanninen, O.; Mulder, G.J.; Owens, I.S.; Siest, G.; Tephly, T.R. Biochem. Pharmacol. 1983, 32, 953.

37. Robinson, D.; Williams, R.T. Biochem. J. 1956, 68, 23P.

38. Siegel, L.M. In 'Metabolic Pathways, Volume 7, 3rd Edition', Academic : New York, 1975; Ch. 7.

39. Jakoby, W.B.; Sekura, R.D.; Lyon, E.S.; Marcus, C.J.; Wang, C.J. In 'Enzymatic Basis of Detoxication, Volume 2'; Jakoby, W.B., Ed.; Academic : New York, 1980, p. 199.

40. Singer, S.S. Biochem. Soc. Trans. 1984, 12, 35.

41. Mulder, G.J. In 'Metabolic Basis of Detoxication'; Jakoby, W.B.; Bend, J.R.; Caldwell, J., Eds.; Academic : New York, 1982, p. 248.

42. Caldwell, J. In 'Concepts in Drug Metabolism, Part A'; Jenner, P.; Testa, B., Eds.; Marcel Dekker : New York; 1980, p. 211.

43. Bray, H.G.; Humphries, B.G.; Thorpe, W.V.; White, K.; Wood, P.B.S. Biochem. J. 1952, 52, 416.

44. Levy, G. In 'Conjugation Reactions in Drug Biotransformation'; Aitio, A., Ed.; Elsevier : Amsterdam, 1978; p. 469.

45. Krijgsheld, K.R.; Mulder, G.J. In 'Sulfate Metabolism and Sulfate Conjugation'; Mulder, G.J.; Caldwell, J.; Van Kempen, G.M.J.; Vonk, R.J., Eds. Taylor & Francis : London, 1982; p.59.

46. Usdin, E.; Borchardt, R.T.; Creveling, C.R.; Eds. 'Transmethylations'; American Elsevier : New York, 1979.

47. Williams, R.T. In 'Biogenesis of Natural Compounds, 2nd. Edition', Bernifeld, P.; Ed. Pergamon : Oxford, 1967, p. 590.

48. D'Souza,; Caldwell, J.; Smith, R.L. Xenobiotica, 1980, 10, 51.

49. Laduron, P. Biochem. Pharmacol. 1974, 23 (Suppl.), 75.

50. Borchardt, R.T. In 'Enzymatic Basis of Detoxication, Volume 2'; Jakoby, W.B.; Ed.; Academic : New York, 1980, p. 43.

51. Weisiger, R.A.; Jakoby, W.B. In 'Enzymatic Basis of Detoxication, Volume 2'; Jakoby, W.B.; Ed.; Academic : New York, 1980; p. 131.

52. Wold, J.S.; Smith, R.L.; Williams, R.T. Biochem. Pharmacol. 1973, 22, 1865.

53. Weber, W.W.; Hein, D.W.; Hirata, M; Patterson, E. In 'Conjugal Reactions in Drug Biotransformation'; Aitio, A., Ed.; Elsevier : Amsterdam, 1978; p. 145.

54. Ellard, G.A. Clin Pharmacol. Ther. 1976, 19, 610.

55. Chapron, D.J.; Kramer, P.A.; Mercik, S.A. Clin. Pharmacol. Ther. 1980, 27, 104.

56. Caldwell, J. In 'The Liver : Biology and Pathobiology'; Arias, I.M.; Popper, H.; Schachter, D.; Shafritz, D.A.; Eds.; Raven : New York, 1982; p. 281.

57. Caldwell, J. In 'Enzymatic Basis of Detoxication, Volume 1'; Jakoby, W.B.; Ed.; Academic : New York, 1980; p. 85.

58. Caldwell, J.; Williams, R.T.; French, M.R.; Bassir, O. Eur. J. Drug Metab. Pharmacokin. 1978, 3, 61.

59. Jerina, D.M.; Bend, J.R. In 'Biological Reactive Intermediates' Jollow, D.J.; Kocsis, J.J.; Snyder, R.; Vainio, H.; Eds.; Plenum : New York, 1977; p. 207.

60. Bray, H.G.; Franklin, T.J.; James, S.P. Biochem. J. 1959, 73, 465.

61. Chasseaud, L.F. Adv. Cancer Res. 1979, 29, 175.

62. Boyland, E.; Chasseaud, L.F. Rep. Brit. Emp. Cancer Camp. 1968, 46, 27.

63. Sies, H.; Cadenas, E. In 'Biological Basis of Detoxication'; Caldwell, J.; Jakoby, W.B.; Eds.; Academic : New York, 1983, p.182.

64. Jakoby, W.B.; Habig, W.H. In 'Enzymatic Basis of Detoxication, Volume 2'; Jakoby, W.B.; Ed.; Academic : New York, 1980; p. 63.

65. Boyland, E.; Chasseaud, L.F. Biochem. J. 1967, 104, 95.

66. Jakoby, W.B.; Ketterer, B.; Mannervik, B. Biochem. Pharmacol 1984, 33, 2539.

67. Habig, W.H.; Keengood, J.H.; Jakoby, W.B. Biochem. Biophys. Res. Commun. 1975, 64, 501.

68. Mannervik, B; Alin, P.; Guthenberg, C.; Jensson, H.; Warholm, M. In 'Microsomes and Drug Oxidations'; Boobis, A.R.; Caldwell, J.; De Matteis, F.; Elombe, C.R.; Eds.; Taylor & Francis : London, 1985; p. 221.

69. Malnoe, A.; Strolin Benedetti, M.; Smith, R.L.; Frigerio, A. Biological Reactive Intermediates'; Jollow, D.J.; Kocsis, J.J.; Snyder, R.; Vainio, H.; Eds.; Plenum : New York, 1977; p. 387.

70. Smith, R.L. 'The Excretory Function of Bile'; Chapman and Hall : London, 1973.

71. Tate, S.S. In 'Enzymatic Basis of Detoxication, Volume 2'; Jakoby, W.B., Ed.; Academic : New York, 1980; p. 95.

72. Tateischi, M.; Shimizu, H. In 'Enzymatic Basis of Detoxication, Volume 2'; Jakoby, W.B.; Ed.; Academic : New York, 1980; p. 121.

73. Bakke, J.E. Chemosphere 1983, 12, 793.

74. Lertratanangkoon, K.; Horning, M.; Middleditch, B.; Tsang, W.; Griffin, G. Drug Metab. Disp. 1982, 10, 614.

75. Boyd, D.R.; Jerina, D.M. In 'Small Ring Heterocycles - Part 3'; Hassner, A.; Ed.; John Wiley : New York, 1985, p. 197.

76. Oesch, F. In : 'Enzymatic Basis of Detoxication, Volume 2'; Jakoby, W.B., Ed.; Academic : New York, 1980; p. 277.

77. Wixtrom, R.; Hammock, B.D. In 'Biochemical Pharmacology and Toxicology, Volume 1'; Zakim, D.; Vessey, D.A.; Eds.; Wiley-Interscience : New York, 1985; p. 1.

78. Caldwell, J.; Idle, J.R.; Smith, R.L. In 'Extrahepatic Metabolism of Drugs and Other Foreign Compounds'; Gram, T.E., Eds.; SP Medical and Scientific : New York, 1980; p. 435.

79. Caldwell, J. In 'Conjugation Reactions in Drug Biotransformation'; Aitio, A., Ed.; Elsevier : Amsterdam, 1978; p. 111.

80. Caldwell, J. In "Metabolic Basis of Detoxication'; Jakoby, W.B.; Bend, J.R.; Caldwell, J., Eds.; Academic Press : New York, 1982; p. 271.

81. Marsh, M.V.; Caldwell, J.; Smith, R.L.; Horner, M.W.; Houghton, E.; Moss, M.S. Xenobiotica, 1981, 11, 655.

82. Caldwell, J. In 'Conjugation Reactions in Drug Biotransformation'; Aitio, A., Ed.; Elsevier : Amsterdam, 1978, p. 477.

83. Caldwell, J. Life Sci. 1979, 24, 571.

84. Flammang, T.J.; Kadlubar, F.F. In 'Microsomes and Drug Oxidations'; Boobis, A.R.; Caldwell, J.; De Matteis, F.; Elcombe, C.R.; Eds.; Taylor & Francis : London, 1985; p. 190.

85. Van Bladeren, P.J.; Breimer, D.D.; Rotteveel-Smijs, G.M.T.; Mohn, G.R. Mut. Res. 1980, 74, 341.

RECEIVED November 20, 1985

Xenobiotic Conjugation in Fish and Other Aquatic Species

Margaret O. James

Department of Medicinal Chemistry, University of Florida, Gainesville, FL 32610

A number of in vivo studies have established the importance of conjugation reactions in determining the fate of foreign chemicals in aquatic animals. While most of the major pathways of xenobiotic conjugation have been reported in fish and crustacean species, detailed in vivo and in vitro studies have focussed on a few pathways, namely glycosylation, glutathione conjugation and conjugation with amino acids. Several fish species, notably trout, have been shown to excrete glucuronide conjugates of a variety of aglycones. The properties of UDP-glucuronosyltransferase have been studied in hepatic microsomes from marine and fresh-water fish, although the enzyme has not yet been purified. Glutathione S-transferase activity with several substrates, e.g. aryl halides, epoxides, arene oxides, has been found in a wide range of aquatic species. Several forms of glutathione S-transferase have been purified from hepatic cytosol of the skate, shark and trout. One of the skate transferases has extremely high activity with arene oxides of benzo(a)pyrene. Amino acid conjugation of carboxylic acids has been studied in fish and crustacea, and the only amino acid conjugates identified to date have been taurine conjugates. In some instances taurine con-jugates do not appear to be excreted more readily than the parent xenobiotic carboxylic acid.

Interest in the phylogenetic aspects of xenobiotic metabolizing en-zymes has motivated several investigators to study xenobiotic con-jugation in aquatic animals. In addition, evidence of extensive chemical pollution of some bodies of water which results in the exposure of aquatic animals to xenobiotics has provided a practical reason to study the ability of aquatic animals to biotransform lipo-philic foreign chemicals to more polar metabolites which should be more readily excreted.

Although Dutton (1) and Arias (2) demonstrated more than two

decades ago that livers of trout (1) and dogfish shark (2) contained
UDP-glucuronosyltransferase activity, there has been considerable
confusion in the literature concerning the ability of aquatic animals to metabolize xenobiotics. As recently as 1974, a widely read
Pharmacology text stated that drug metabolizing enzymes were absent
in fish and further that fish did not need drug metabolizing enzymes,
because lipid soluble materials could be eliminated across their
gills by simple diffusion (3). Examination of the primary literature showed that neither of these statements was true, and several
reviews (e.g. 4-6) provided documentation that, under the proper
experimental conditions, both phase 1 and phase 2 pathways (as
defined by Williams, 7) were present in a variety of aquatic species.
      In the past decade, several aspects of xenobiotic conjugation
have been studied in aquatic animals. This paper will summarize and
review recent findings, with particular emphasis on glycosylation,
glutathione conjugation and mercapturic acid biosynthesis, and amino
acid conjugation.

Glycoside Conjugation

Xenobiotic molecules containing OH, COOH or other nucleophilic
groups are potential substrates for glycosylation (8). Among fish
species studied to date, the major glycoside conjugates formed in
vivo are glucuronides, whereas among invertebrates, glucosides are
the major sugar conjugates.

In Vivo Studies. In vivo metabolism of a number of aglycones has
been studied in marine and freshwater fish as summarized in Table I.
With few exceptions, glucuronides were identified as metabolites in
bile, not urine or tankwater. In the goldfish there is evidence
that phenol glucuronide is excreted in bile, then hydrolyzed in the
gastrointestinal tract, and the phenol aglycone is reabsorbed and
ultimately excreted into the surrounding water as a sulfate conjugate (11). Thus, identification of a glucuronide as a major biliary
metabolite does not always mean that the major excreted metabolite
is the glucuronide. Studies of the rates of renal excretion of
benzo(a)pyrene metabolites in fish have been carried out by
Pritchard and Bend (27). They showed that 7,8-dihydro-7,8-
dihydroxybenzo(a)pyrene was metabolized by southern flounder to
sulfate and glucuronide conjugates which were secreted into urine
by the organic anion transport system, the sulfate conjugate being
better transported than the glucuronide. Other studies showed that
4-methylumbelliferone sulfate was better transported by southern
flounder renal tubules than 4-methylumbelliferone glucuronide
(Pritchard, J. B., unpublished data).
      Studies by Lech have shown clearly that for some toxic chemi-
cals, glucuronidation is an effective detoxication mechanism; he
showed that differences in the rate of glucuronidation of
3-trifluoromethyl-4-nitrophenol between trout (good rate of
glucuronidation) and lamprey (low rate of glucuronidation) correlated with the lethal concentration (LC_{50}) of this compound (13).
Further, Lech showed that if trout were pretreated with salicyl-
amide, which inhibits glucuronidation, the LC_{50} of 3-trifluoromethyl-
4-nitrophenol to trout decreased.

Table I. In Vivo Studies Of Glucuronide Formation In Fish

Aglycone	Fish Species	Where Glucuronide Conjugate Was Found	Reference
Phenol	Bream, Abramis brama	Water	9
	Perch, Perca fluviatilis	Water	9
	Roach, Rutilus rutilus	Water	9
	Rudd, Scardinius erythropthalmus	Water	9
	Goldfish, Carassius auratus	Bile	10,11
2,4,6-Trinitrophenol	Trout, Salmo gairdnerii	Intestinal contents	12
2-Amino-4,6-dinitrophenol	Trout	Bile, Intestinal contents	12
3-Trifluoromethyl-4-nitrophenol	Trout	Bile	13
Pentachlorophenol	Trout	Bile	14
Permethrin	Trout	Bile	15
Phenolphthalein	Trout	Bile	16
Diethylhexylphthalate	Trout	Bile	17
1-Naphthol	Coho salmon, Oncorhynchus kisutch	Liver, Bile, Brain Muscle	18
	Trout	Bile	19
1,2-Dihydrodihydroxynaphthalene	English sole, Parophrys vetulus	Liver, Bile	20
	Rock sole, Lepidopsetta bilineata	Liver, Bile, Skin	21

Continued on next page

Table I. Continued

Aglycone	Fish Species	Where Glucuronide Conjugate Was Found	Reference
2-Hydroxymethylnaphthalene	Trout	Bile	22
Dihydrodiols (3,4-,5,6-,7,8-) of 2-methylnaphthalene	Trout	Bile	22
Aflatoxicol M_1	Trout	Bile	23
7,8-Dihydrodihydroxybenzo(a)-pyrene	English sole	Bile, liver, Muscle, Gonads	20 24-26
7-Hydroxybenzo(a)pyrene	Southern flounder, _Paralichthyes lethostigma_	Urine	27
Tetrachlorobiphenyl	Trout	Bile	28
2',5-Dichloro-4'-nitro salicylamide	Trout	Bile	29
Testosterone, 11-ketotestosterone	Trout	Testes	30
Estradiol-17β-	Trout	Bile	31
Bilirubin	Dogfish shark, _Squalus acanthias_	Bile	32

Few detailed studies of xenobiotic metabolism in vivo have been carried out in invertebrates, and these few have concentrated on identification of phase I metabolites. It is presumed that invertebrates, like insects, use glucose rather than glucuronic acid (36).

Studies In Isolated Organ Preparations Or Cells. The conjugation of a few xenobiotics has been studied in the isolated perfused trout liver and in trout hepatocytes (Table II). In these examples, the glucuronide conjugates were the major metabolites found, with little or no evidence for the presence of sulfate conjugates.

Table II. Glucuronide Formation In Isolated Organs
Or Cells Of Fish

Aglycone	Species	Site	Reference
4-Nitrophenol (from 4-nitroanisole)	Trout	Isolated Perfused Liver	33
7-Hydroxycoumarin (from 7-ethoxycoumarin)	Trout	Isolated Perfused Liver	34
Acetaminophen	Trout	Hepatocytes	35

In Vitro Studies. Because glucuronidation is quantitatively a very important pathway for metabolism of xenobiotics, as well as endogenous molecules, with nucleophilic centers, considerable attention has been paid to understanding this pathway at a molecular level. Studies in mammals have shown that UDP-glucuronosyltransferase (UDPGT) activity is located on the lumenal side of hepatic microsomal vesicles and that activity can be enhanced by assaying for UDPGT activity in the presence of a detergent or other agent which disrupts the vesicular structure (8, 36). In the rat and other mammals there is evidence for at least 2 forms of UDPGT which have activity with xenobiotic aglycones (37). One form, sometimes termed form A or GT_1, is inducible by pretreatment of rats with 3-methylcholanthrene and reaches adult levels prenatally. Substrates for this form of UDPGT include 4-nitrophenol, 4-methylumbelliferone, clofibrate, 1-naphthol, and 3-hydroxybenzo(a)pyrene. A second form, sometimes termed D or GT_2 is inducible by phenobarbital, reaches adult levels postnatally and preferentially catalyzes glucuronidation of substrates such as morphine, 4-hydroxybiphenyl, chloramphenicol, and valproic acid. In early studies it seemed that bilirubin was also a substrate for form D (GT_2), but more recent studies suggest that glucuronidation of bilirubin is catalyzed by yet another UDPGT (form B) in the rat (37, 38). In fish, hepatic microsomal UDPGT activity has been demonstrated in several species with a variety of substrates. In trout, hepatic microsomal activity has been found with 4-nitrophenol (39-44), 1-naphthol (42, 44) and testosterone (44) and higher activities were measured in the presence of

surfactants (42,44) or after freezing and thawing microsomes (43) in
accordance with the expected latency of UDPGT. UDPGT activity with
4-methylumbelliferone has been demonstrated in hepatic microsomes in
the bluegill (Lepomis machrochirus) and channel catfish (Ictalurus
punctatus) and in each species a 4-5-fold activation was effected by
addition of detergent to incubation vials (Ankley, G.T., personal
communication). UDPGT activity with 4-nitrophenol and bilirubin has
been studied in hepatic microsomes from the dogfish shark (32).
    The possibility that there are multiple forms of UDPGT has been
studied in the dogfish shark and trout. In the dogfish shark, it was
shown that there was no difference in bilirubin UDPGT or 4-nitrophenol
UDPGT of hepatic microsomes from fetal (20-22 months gestation) or
adult dogfish shark (32) as is the case in mammals (45). Thus it was
not possible to determine from this study (32) whether or not multiple forms of UDPGT exist in shark liver. Recent studies in trout
(44) do provide evidence in favor of the existence of multiple forms
of UDPGT. Andersson et al (44) showed that the time course of
β-naphthoflavone induction of UDPGT activity with testosterone as
substrate differed from the time course of induction with 4-nitro-
phenol or 1-naphthol as substrates. UDPGT activity with testosterone,
in digitoxin activated hepatic microsomes from β-naphthoflavone
treated trout was increased 1.5-fold by 1 week after treatment, and
thereafter fell to control values. By contrast, activity with
4-nitrophenol as substrate was increased 2-fold in β-NF treated trout
by 1 week after treatment, 3-fold by 3 week posttreatment and
activity remained elevated up to 6 weeks after the dose (44).
    There have been few recent studies of glycosylation in aquatic
invertebrates. Conjugation of 1-naphthol with UDP-glucose, but not
UDP-glucuronic acid was catalyzed by hepatopancreas, green gland, gut
and gills of the crayfish, Astacus astacus, (42). Dutton (36) reported that molluscs contained UDP-glucosyltransferase activity.

Sulfate Conjugation

Xenobiotic substrates which contain hydroxy groups may be excreted as
sulfate or glycoside conjugates. Thus in vivo studies of the metabolism of xenobiotics which contain hydroxyl groups should also provide information on the relative importance of the sulfate conjugation
pathway in aquatic species. Phenol was shown to be excreted solely
as a sulfate conjugate by goldfish, guppy, Poecilia reticulata,
minnow, Phoxinus phoxinus and tench, Tinca tinca and as a mixture of
glucuronide and sulfate conjugates by bream, perch, roach and rudd
(9, see also Table I). In southern flounder, benzo(a)pyrene 7,8-diol
was conjugated with sulfate and the conjugate rapidly excreted in
urine (27). The trout appears to be relatively defective in sulfate
conjugation, as there was little or no evidence in vivo or in isolated
perfused organ systems or hepatocytes for the presence of sulfate
conjugates, even where this has been directly sought (34,35). In
many experiments, glucuronide and sulfate conjugates have not been
clearly differentiated, but have been hydrolyzed to the parent
compound by a mixture of β-glucuronidase and sulfatase.

Glutathione Conjugation And Mercapturic Acid Biosynthesis

Xenobiotic molecules with electrophilic centers are potential substrates for conjugation with the thiol group of the tripeptide

glutathione (GSH). Formation of a glutathione conjugate is the first step in the synthesis of mercapturic acids (N-acetylcysteine conjugates of xenobiotics) (46,47). Because electrophilic chemicals can also react with cellular nucleophiles such as DNA, RNA and protein, thereby possibly initiating cytotoxicity, conjugation of electrophiles with GSH is usually an important detoxication mechanism (48). Considerable attention has been paid to studying GSH conjugation in aquatic animals, in part because several common aquatic pollutants, e.g. polycyclic aromatic hydrocarbons, are metabolized to electrophiles which are detoxified by conjugation with GSH. Most studies have been carried out in vitro, but some excellent in vivo studies have also been conducted.

In Vivo Studies of Glutathione Conjugation and Mercapturic Acid Biosynthesis in Fish. Numerous studies have shown that benzo(a)-pyrene is metabolized by the cytochrome P-450 system to arene oxides and other electrophilic metabolites which can bind to DNA or be detoxified by several pathways including conjugation with GSH. Varanasi et al (26) have shown that after administration of ^{14}C-benzo(a)pyrene to English sole, bile contains high concentrations of ^{14}C of which more than 50 percent is in the form of ninhydrin positive conjugates which comigrate with known glutathione and cysteinylglycine conjugates.

Evidence that glutathione and cysteinylglycine conjugates can be further metabolized to the cysteine and N-acetylcysteine (mercapturic acid) conjugates has been obtained in two species of fish, the Japanese carp (49) and the winter flounder (50). In Japanese carp, Cyprinus carpis, the herbicide molinate (S-ethylhexahydro-1H-azepine-1-carbothioate) was found in bile as a mercapturic acid conjugate (49). The winter flounder were injected at three dose levels, with the GSH conjugates of 8-^{14}C-styrene-7,8-oxide [S-(1-phenyl-2-hydroxyethyl)glutathione and S-(2-phenyl-2-hydroxyethyl)glutathione] and urine and bile samples were collected for 24 hours (50). The 24 hour urine contained 57-86% of the administered ^{14}C and bile about 2%. Analysis of urine and bile revealed that the major urinary metabolite was a mixture of S-(1-phenyl-2-hydroxyethyl)cysteine and S-(2-phenyl-2-hydroxyethyl)-cysteine, while the major biliary metabolite was the unchanged GSH conjugate mixture (see Table III). Urine also contained some of the mercapturic acid, especially at the lowest dose, but this study suggests that winter flounder are relatively deficient in N-acetyl transferase activity for the cysteinyl conjugates.

In Vitro Studies of the Enzymes Involved in Mercapturic Acid Bio-Synthesis. Studies in mammals have shown that mercapturic acid biosynthesis is a four step process, as follows:

$$R-X + GSH \longrightarrow R-S-CH_2-\underset{\underset{HNCOCH_2CH_2CH(NH)_2COOH}{|}}{C}HCONHCH_2COOH \qquad (1)$$

Step (1) is catalyzed by glutathione S-transferases, or can occur non-enzymatically.

$$R-S-CH_2-\underset{\underset{HNCOCH_2CH_2CH(NH_2)COOH}{|}}{C}HCONHCH_2COOH \longrightarrow R-S-CH_2-\underset{\underset{NH_2}{|}}{C}HCONHCH_2COOH \qquad (2)$$

Table III. Urinary And Biliary Metabolites Of Styrene Oxide-
            Glutathione In Winter Flounder

| | % analysed ^{14}C identified as | | | |
	SO-GSH	SO-CYS-GLY	SO-CYS	SO-NAc-CYS
1 mg/fish				
Urine, 7-8hr	ND	ND	33.3	21.0
Bile, 0-24hr	–	–	–	–
3.9 mg/fish				
Urine, 7-8hr	ND	ND	48.5	11.6
Bile, 0-24hr	39.8	19	7.1	13.3
24.4 mg/fish				
Urine, 7-8hr	2.0	ND	64	6.0
Bile, 0-24hr	30.4	7.4	8.2	8.2

ND - not detectable

SO-GSH;   a mixture of S-(1-phenyl-2-hydroxyethyl)- and (2-phenyl-
            2-hydroxyethyl)glutathione.

SO-CYS-GLY; a mixture of S-(1-phenyl-2-hydroxyethyl)- and (2-phenyl-
            2-hydroxyethyl)-cysteinylglycine.

SO-CYS;   a mixture of S-(1-phenyl-2-hydroxyethyl)- and (2-phenyl-
            2-hydroxyethyl)-cysteine.

SO-NAc-CYS; the N-acetyl conjugates of SO-CYS.

Data from Ref. 50.

Step (2) is catalyzed by γ-glutamyltranspeptidases.

$$R\text{-}SCH_2CH(NH_2)CONHCH_2COOH \longrightarrow R\text{-}SCH_2CH(NH_2)COOH \qquad (3)$$

Step (3) is catalyzed by aminopeptidases.

$$R\text{-}SCH_2\underset{\underset{NH_2}{|}}{C}HCOOH \longrightarrow R\text{-}SCH_2\underset{\underset{NHCOCH_3}{|}}{C}HCOOH \qquad (4)$$

Step (4) is catalyzed by N-acetyltransferases. A review of mercapturic acid biosynthesis is given in Ref. 51.

Glutathione S-transferases. Enzyme activity has been found in cytosol from liver, hepatopancreas, gill, kidney, green gland, gonads, digestive tract and brain from a variety of different aquatic organisms (5, 52-58). The range of activities found with four different substrates in hepatic or hepatopancreatic cytosol from several aquatic species, compared with rat, is shown in Table IV.

Table IV. GSH-S Transferase Activities In Cytosol From
Several Species

	CDNB	DCNB	BP-4,5-O	SO
		nmole/min/mg protein		
Rat Liver	1256	53	24	190
Teleost Liver (19 species)	250-4000	ND-60	0.3-60	1.3-30
Elasmobranch Liver (7 species)	320-650	0.4-12	0.6-110	1.9-18
Crustacean Hepatopancreas (4 species)	50-700	0.2-8	0.1-5	0.1-2
Mollusc (Whole Body) (3 species)	-	-	0.3-0.4	0.6-0.7

CDNB - 1-chloro-2,4-dinitrobenzene; DCNB - 1,2-dichloro-4-nitrobenzene; BP-4,5-O - (±)-benzo(a)pyrene 4,5-oxide; SO - (±)-styrene 7,8-oxide.

Data from Refs. 5, 54, 58.

It is clear from Table IV that the metabolic capability for this pathway varies very widely between species.
    Recently a microsomal GSH S-transferase has been characterized in rat liver, which has the distinctive property of being activated by N-ethylmaleimide (59). The existence of this GSH S-transferase in fish species is not documented, although one study with pike liver microsomes showed that there was no N-ethylmaleimide-

activatable activity, and that pike liver microsomes did not contain
any proteins which cross-reacted with antibodies to the rat microsomal
GSH S-transferase (60).

One unusual finding from the study of cytosolic GSH S-transferase
activity was the ability of hepatic cytosol from a few fish species,
notably the teleost fish, sheepshead, Archosargus probatocephalus
and the elasmobranch fish, little skate, Raja erinacea, to conjugate
benzo(a)pyrene 4,5-oxide with GSH at higher rates than the rat
(Table IV and Ref. 54). In order to further investigate the conjugation of arene oxides with GSH, Foureman and Bend purified the GSH
transferases from little skate hepatic cytosol (61). Five transferases (designated E1 to E5) with different retentions on DEAE-cellulose
were further purified by chromatography on a GSH affinity column. By
this procedure, the major transferases, E1 and E4, were purified to
apparent homogeneity (61). As was the case with hepatic cytosolic
GSH S-transferases from rat (48), the skate transferases were all
homo- or heterodimers with subunit molecular weights in the region of
25,000 to 28,000. Results from sodium dodecyl sulfate polyacrylamide
gel electrophoresis and immunological studies provided evidence that
transferase E4 was a homodimer which had one subunit in common with
each of the heterodimers E2 and E3, but not with E1 or E5. All
5-transferases had good activity with 1-chloro-2,4-dinitrobenzene as
substrate, but only transferases E2, E3 and E4 had good activity with
benzo(a)pyrene 4,5-oxide as substrate. Transferase E4 had the highest
activity of all skate transferases with 1-chloro-2,4-dinitrobenzene
and benzo(a)pyrene 4,5-oxide as substrates. Transferase E4 also
functioned as a binding protein, with dissociation constants for
hematin and bilirubin in the micromolar range (61). In further
elegant studies of the transferase E4-catalyzed conjugation of
(±)benzo(a)pyrene 4,5-oxide with GSH, it was shown that the transferase E4-catalyzed reaction was enantioselective and stereospecific
(62-64). Other arene oxides, including (±)-7β,8α-dihydroxy-9α,10α-
epoxy-7,8,9,10-tetrahydrobenzo(a)-pyrene, an ultimate carcinogen of
benzo(a)pyrene, were also good substrates for the little skate GSH
S-transferase, E4 (64). These studies suggest that certain marine
fish possess efficient detoxification pathways for the potentially
toxic arene oxide metabolites of polycyclic aromatic hydrocarbons.
It is worth noting here that although many fish, e.g. scup (65),
English sole (24) and trout (66) do not appear to form benzo(a)pyrene
4,5-oxide from benzo(a)pyrene, as judged from the lack of BaP 4,5-
dihydrodiol identified from incubations with hepatic microsomes,
there is evidence that the little skate and the sheepshead can metabolize benzo(a)pyrene to benzo(a)pyrene 4,5-oxide (67,68). All marine
fish studied produce benzo(a)pyrene 7,8-oxide and 9,10-oxide as major
metabolites from benzo(a)pyrene (69).

GSH S-transferases have been purified from hepatic cytosol of
two other fish species, the thorny-back shark, Platyrhinoides tri-
seriata (70) and the rainbow trout (71,72). Two transferases were
isolated from the thorny-back shark, termed PLATY1 and PLATY2 and
the binding properties of these transferases were compared to those
of rat ligandin (70). Two of the ligands studied, sulfobromophthalein
and bilirubin had poorer affinity for the shark transferase than for
rat ligandin, but other ligands, including Rose Bengal and 1-anilino-
8-naphthalene sulfonate had high binding affinities for both rat and

shark transferase. Evidence for at least seven GSH S-transferases was found in the rainbow trout (72). Five of these trout transferases were purified and shown to be homo- or heterodimers with subunit molecular weights between 22,000 and 25,000. The catalytic activity of these transferases with a variety of ligands was reported (72).

The kinetic parameters for GSH S-transferase-catalyzed conjugation of 1-chloro-2,4-dinitrobenzene with GSH were reported by the groups who purified fish GSH S-transferases (61,70,72) and are summarized in Table V, with the parameters for rat ligandin included for comparison, as reported by Sugiyama et al (70).

Table V. Kinetic Parameters Of Purified GSH-S-Transferases

Enzyme	Apparent K_m (mM)		Reference
	CDNB	GSH	
Rat Ligandin	0.04	0.07	70
Shark Y1	0.38	0.22	70
Shark Y2	0.28	4.3	70
Skate E4	0.42	1.1	61
Trout C1	0.2	0.5	72
Trout C2	0.2	0.4	72
Trout C4	0.3	0.4	72
Trout C5	0.2	0.4	72
Trout A	0.7	1.9	72

Data from Refs. 61, 70, 72.

Table V shows that higher concentrations of both substrate and GSH were needed for the fish transferases to exhibit maximal activity, compared with rat. Indeed the fish transferase K_m values for GSH are in the same region as the reported concentrations of GSH in liver of several fish species as shown in Table VI. Thus, small changes in the GSH content of liver could markedly alter the ability of the fish to detoxify chemicals by conjugation with GSH.

Other Enzymes Involved in Mercapturic Acid Biosynthesis. These enzymes have not been as extensively studied in fish as the GSH-S-transferases. Gamma glutamyl transpeptidase activity has been demonstrated in kidney and intestinal caeca but not liver of the rainbow trout (55), and in kidney, but not liver, red blood cells, brain or

Table VI. Glutathione Concentration In Fish Liver

Species	GSH Concentration mM	Reference
Little skate	2.33 + 1.11	73
Large skate, Raja ocellata	1.38	73
Thorny skate, Raja radiata	2.52	73
Thorny-back shark	1.31 + 0.21	70
Sea bass, Centropristis striata	1.32 + 0.25	57
Rainbow trout	1.8	55

Values are mean + S.D., n = 3-5 individuals, except where single values are reported.

Data from Refs. 55, 57, 70, 73.

muscle of the black sea bass (57). This organ distribution is also found in mammals. There are no reports of cysteinylglycinase activity in fish species, although there is evidence from in vivo studies (49,50) that the enzyme must be present. N-acetyl transferase activity has been found in several fish species (74), but cysteine conjugates have not been used as substrates.

Amino Acid Conjugation.

Xenobiotic carboxylic acids can be excreted as amino acid conjugates, where the COOH group forms an amide bond with the $-NH_2$ group of an amino acid. In mammals, the amino acid most commonly used for conjugation is glycine. This pathway has been studied in vitro and in vivo in several fish species and a marine invertebrate.

In Vivo Studies. The metabolism and urinary excretion of 2-, 3- and 4-aminobenzoic acids were studied in goosefish, winter flounder and dogfish shark by Huang and Collins (75). The administered acids were not radiolabelled. It was reported that unchanged acid and metabolites consisting of glycine and glucuronic acid conjugates and, for the 3- and 4-aminobenzoic acids, acetylated products, were excreted (75). Very low recoveries were reported, and the assumed glycine conjugates were not isolated or rigorously identified. Thus, this paper cannot be considered to provide definitive evidence that glycine conjugation occurs in these fish species. To my knowledge, the only other report that glycine conjugates were found in fish was

in a study of the effects of outboard motor exhaust on goldfish (76)
in which it was reported that goldfish metabolized toluene to hippuric
acid (the glycine conjugate of benzoic acid). Again, the supposed
glycine conjugate was not isolated, and the criteria for metabolite
identification were not rigorous.

   In several fish species and one marine crustacean, taurine conjugates of xenobiotic carboxylic acids have been isolated from urine
(fish) or hepatopancreas (crustacea) and subjected to unequivocal
chemical characterization (77-82). Table VII shows which acids are
conjugated with taurine in some marine species.

Table VII. Taurine Conjugation in Marine Species

Species	Acids Which Are Metabolized To Taurine Conjugates	Reference
Winter flounder	Phenylacetic	77
	2,4-Dichlorophenoxyacetic	79,81
Southern flounder	Phenylacetic	79
	2,4-Dichlorophenoxyacetic	79
	2,4,5-Trichlorophenoxyacetic	James, un-published
	Bis(4-chlorophenyl)acetic	79
	Benzoic	80
Dogfish shark	Phenylacetic	77
	2,4-Dichlorophenoxyacetic	77,78
	2,4,5-Trichlorophenoxyacetic	78
Spiny lobster, Panulirus argus	Phenylacetic	82
	2,4-Dichlorophenoxyacetic	82
	2,4,5-Trichlorophenoxyacetic	82
	Bis(4-chlorophenyl)acetic	82

The finding that benzoic acid was metabolized to benzoyltaurine by
southern flounder was very interesting. In most of the vertebrate
species studied to date, benzoic acid is metabolized to hippuric acid
(benzoylglycine) (7), a metabolite which is readily excreted in urine,
in part because it is an excellent substrate for the renal organic
anion transport system (83). After administration of ^{14}C-benzoic
acid (10 mg/kg) by i.m. injection to southern flounder, radioactivity
was recovered in urine at a slow, constant rate (zero order excretion
kinetics) so that $55 \pm 11\%$ (mean $\pm$ S.D., n = 5) was excreted in 120
hours. Analysis of urine from each 24 hour collection period showed
that overall more than 90% of the ^{14}C was a single polar metabolite.
The metabolite was isolated from urine by chromatography on XAD-2
resin, and a portion hydrolyzed in 10M hydrochloric acid. After
hydrolysis the ^{14}C was benzoic acid and the hydrolysate contained an
equimolar amount of a ninhydrin and fluram positive substance which
co-chromatographed with taurine in two TLC systems and on amino acid
analysis (80). Studies *in vitro* showed that in the presence of

sonicated kidney mitochondria from the southern flounder and other marine fish, benzoyl CoA and phenylacetyl CoA would combine with ^{14}C-taurine to give the taurine conjugate, but in the presence of ^{14}C-glycine, no conjugate was formed (James, unpublished). Thus, in the marine fish studied, no glycine conjugates were excreted in vivo or formed in vitro. Several of the carboxylic acids studied, however, were not extensively metabolized by fish, but were excreted unchanged. As may be seen from Table VIII, conjugation with taurine did not always lead to enhanced excretion of xenobiotic acids by southern flounder, as might be expected by analogy with the glycine conjugates, which are usually excreted better by the kidney than the unchanged acids because they are secreted into urine by the renal organic anion transport system (83).

Table VIII. Metabolism And Excretion Of Carboxylic Acids
By Southern Flounder

Acid	% Dose Excreted In 24 Hr. Urine	% Urinary ^{14}C Present As	
		Taurine Conjugate	Free Acid
Phenylacetic	74±3(4)[a]	95±5	2±1
4-Aminobenzoic	72,74	8,7	85,88
2,4-Dichlorophenoxyacetic	41,36	40,50	60,50
2,4,5-Trichlorophenoxyacetic	42±5(3)	70±20	25±10
Bis(4-chlorophenyl)acetic	27±14(4)	97±2	2±1
Benzoic	11±4(6)	85±7	8±4

[a]Mean ± S.D. (n), or individual results if n < 3.

Data from Ref. 79.

For example, 4-aminobenzoic acid, which is excreted unchanged is eliminated much more rapidly by the southern flounder than benzoic acid, which is extensively metabolized to the taurine conjugate (Table VIII). ^{14}C-labelled 2,4-dichlorophenoxyacetic acid was rapidly excreted by the winter flounder (85% in 24 hr.), but less than 10% of the urinary ^{14}C was present as the taurine conjugate (81) whereas in the southern flounder 40-50% of the urinary ^{14}C was taurine conjugate and only 36-41% of the dose was excreted in urine in 24 hr. (Table VIII). In comparing the rate of excretion of 2,4-dichlorophenoxy-acetic acid and phenylacetic acid by winter flounder, Pritchard et al (84) noted that phenylacetic acid, which was found in urine extensive-ly (> 90%) conjugated with taurine, was excreted more slowly than 2,4-dichlorophenoxyacetic acid, which was excreted largely (> 90%) unchanged in urine. It was shown, however, that in vitro renal tubular transport of phenylacetic acid or phenylacetyltaurine in winter flounder was 3-4 times slower than transport of 2,4-dichloro-phenoxyacetic acid (84).

A number of factors influence the rate of renal excretion of xenobiotics in fish as well as other species. The apparent inverse relationship between the extent of taurine conjugation and the rate of urinary excretion of organic acids may be due to factors other than the relative ease with which unchanged acid or its taurine conjugates are transported into urine. Differences in extent of binding to plasma proteins affect the amount of free acid available for glomerular filtration into urine. Some of the differences in percent dose excreted in 24 hr. urine shown in Table VIII can be explained by differences in plasma protein binding. Thus, phenyl-acetic acid and 4-aminobenzoic acid, which are rapidly excreted in urine, are not bound to southern flounder plasma proteins, whereas 2,4,5-trichlorophenoxyacetic acid and bis(4-chlorophenyl)acetic acid, which are slowly excreted, are extensively (> 75%) bound to southern flounder plasma proteins (Pritchard, J. B., unpublished). Benzoic acid and 2,4-dichlorophenoxyacetic acid did bind to southern flounder plasma proteins, but > 65% of each was unbound so it is unlikely that the slow excretion of benzoic acid by southern flounder is due solely to plasma protein binding. Studies are ongoing by James and Pritchard to try to determine the role of taurine conjugation in influencing the rate of renal excretion of carboxylic acids.

In Vivo Studies In Invertebrates. The metabolism and excretion of phenylacetic acid, 2,4-dichlorophenoxyacetic acid, 2,4,5-trichloro-phenoxyacetic acid and bis(4-chlorophenyl)acetic acid have been studied in a marine crustacean, the spiny lobster (82). Each ^{14}C-labelled acid was administered into the hemolymph at a dose of 10mg/kg, and groups of lobsters were sacrificed and dissected at various times after the dose. Terminal urine was collected in each case. In the first 24 hr. period following the dose, there was substantial urinary elimination of each acid, as the parent compound. The ^{14}C remaining in the animal after 24 hours was present largely in the hepatopancreas as the taurine conjugate. It appeared that all four acids were rapidly excreted into urine by the green gland (the equivalent of mammalian kidney), but that phenylacetic acid was also rapidly taken up by hepatopancreas where it was metabolized to phenylacetyltaurine (82). The competition between green gland and hepatopancreas for uptake of the four acids from hemolymph can be seen from the data presented in Table IX.

For each acid, the portion taken up into hepatopancreas was metabolized to the taurine conjugate. There was indirect evidence that the taurine conjugates were excreted into the intestinal tract, then hydrolyzed and the parent acid recirculated in a process similar to enterohepatic recirculation of glucuronide conjugates discussed earlier. Thus, in the spiny lobster, taurine conjugation does not facilitate urinary excretion.

In Vitro Studies. An assay for measuring taurine N-acyltransferase activity with ^{14}C-phenylacetyl CoA as substrate has been described (79). Enzyme activity was located in the matrix of the mitochondrion and was usually higher in kidney than liver for the nine fish species studied. Studies with isolated renal tubules from the winter flounder and the southern flounder (84 and James and Pritchard, unpublished) have shown that added phenylacetic acid and benzoic acid were exten-sively metabolized to their taurine conjugates during a 60 minute incubation.

Table IX. Tissue: Hemolymph Concentrations of 2,4-D,
2,4,5-T, DDA And PAA In Spiny Lobster After
Injection Into The Percardial Sinus

	2,4-D	2,4,5-T	DDA	PAA
6 Hr. After Dose				
Concentration in hemolymph, pmol/mg	73.9	100.4	35	17.2
HP/HL Total [14]C	1.8	1.2	10	79.8
HP/HL Free acid	0.7	0.8	3.9	10.0
Green Gland/HL	12.1	7.5	9.0	19.5
Urine/HL	12.5	3.9	2.7	25.4
Gill/HL	0.5	0.6	1.2	5.8
24 Hr. After Dose				
Concentration in hemolymph, pmol/mg	7.0	18.4	22.6	5.7
HP/HL Total [14]C	18.4	7.1	10.3	301
HP/HL Free acid	3.8	1.2	3.4	38
Green Gland/HL	41.8	15.8	4.7	26
Urine/HL	40.7	18.7	3.6	25
Gill/HL	1.6	0.7	1.1	4.0

Abbreviations: 2,4-D - 2,4-dichlorophenoxyacetic acid;
2,4,5-T - 2,4,5-trichlorophenoxyacetic acid; DDA - bis(4-chlorophenyl)-
acetic acid; PAA - phenylacetic acid; HP - hepatopancreas;
HL - hemolymph.
The ^{14}C in hemolymph and urine was unchanged acid at 6 hr. and 24 hr.

Data from Ref. 82.

Overview

To summarize, several studies have demonstrated the importance of
conjugation reactions in determining the fate of xenobiotics in
aquatic animals. Among fish, most information had been gained in
the trout, the southern flounder, the winter flounder and the dogfish shark. It is evident that there are considerable interspecies
differences in the ability of fish to conjugate xenobiotics, as is
true for other species, but the major pathways of conjugation are
similar in all animal species studied.

Acknowledgments

I wish to thank T. Andersson and L. Forlin for supplying preprints
of pertinent papers, and G. Ankley for communicating some results.
I am grateful to J. B. Pritchard for supplying unpublished results,
and for continued stimulating discussion on the role of metabolism
in excretion of xenobiotics.

Literature Cited

1. Dutton, G. J.; Montgomery, J. P. Biochem. J. 1958, 70, 17P.
2. Arias, I. M. Bull. Mt. Desert Island Biol. Lab. 1962, 10, 614.
3. Goldstein, A.; Aronow, L.; Kaplan, S. "Principles of Drug Action, The Basis of Pharmacology," Wiley: New York, 1974; pp. 283-284.
4. Chambers, J. E., Yarbrough, J. D. Comp. Biochem. Physiol. 1976 55C, 77-84.
5. Bend, J. R.; James, M. O. In "Biochemical and Biophysical Perspectives in Marine Biology"; Malins, D. C.; Sargent, J. Academic: New York, 1978; Vol. 4, pp. 125-187.
6. Lech, J. J.; Bend, J. R. Environ. Health Perspect. 1980; 34, 115-131.
7. Williams, R. T. "Detoxication Mechanisms"; Chapman and Hall: London, 1959.
8. Kasper, C. B.; Henton, D. In "Enzymation Basis of Detoxication"; Jakoby, W. B., Ed.; Academic: New York, 1980, Vol. II, pp. 4-35.
9. Layiwola, P. J.; Linnecar, D. F. C. Xenobiotica 1981, 11, 167-171.
10. Kobayashi, K.; Kimura, S.; Shimizu, E. Bull. Jap. Soc. Sc. 1976, 42, 1365.
11. Layiwola, P. J.; Linnecar, D. F. C.; Knights, B. Xenobiotica, 1983, 13, 27-29.
12. Cooper, K. R.; Burton, D. T.; Goodfellow, W. L.; Rosenblatt, D. H. J. Toxicol. Environ. Health 1984, 14, 731-747.
13. Lech, J. J. Toxicol. Appl. Pharmacol. 1973, 24, 114-124.
14. Glickman, A. H.; Statham, C. N.; Wu, A.; Lech, J. J. Toxicol. Appl. Pharmacol. 1977, 41, 649-658.
15. Glickman, A. H.; Hamid, A. R.; Rickert, D. C.; Lech, J. J. Toxicol. Appl. Pharmacol. 1981, 57, 88-98.
16. Curtis, L. R. Comp. Biochem. Physiol. 1983, 76C, 107-111.
17. Melancon, M.; Lech, J. J. Drug Metab. Disp. 1976, 4, 112-118.
18. Roubal, W. T.; Collier, T. K.; Malins, D. C. Arch. Environ. Contam. Toxicol. 1977, 5, 513-529.
19. Krahn, M. M.; Brown, D. W.; Collier, T. K.; Friedman, A. J.; Jenkins, R. G.; Malins, D. C. J. Biochem. Biophys. Methods 1980, 2, 233-246.
20. Varanasi, U.; Gmur, D. J. Aquatic Toxicol. 1981, 1, 49-67.
21. Varanasi, U.; Gmur, D. J.; Tresler, P. A. Arch. Environ. Contam. Toxicol. 1979, 8, 673-692.
22. Melancon, M. J., Lech, J. J. Comp. Biochem. Physiol. 1984 79C, 331-336.
23. Loveland, P. M.; Nixon, J. E.; Bailey G. S. Comp. Biochem. Physiol. 1984, 78C, 13-19.
24. Gmur, D. J.; Varanasi, U. Carcinogenesis 1982, 3, 1397-1403.
25. Varanasi, U.; Nishimoto, M.; Reichert, W. L.; Stein, J. E. Xenobiotica 1982, 12, 417-425.
26. Varanasi, U.; Nishimoto, M.; Stover, J. In "Polynuclear Aromatic Hydrocarbons: Eighth International Symposium on Mechanisms, Methods and Metabolism"; Cooke, M. W.; Dennis, A. J., Eds.; Battelle: Columbus, Ohio, 1983, pp. 1315-1328.
27. Pritchard, J. B.; Bend, J. R. Drug Metab. Rev. 1984, 15, 655-671.
28. Melancon, M. J., Jr.; Lech, J. J. Bull Environ. Contam. Toxicol. 1976, 15, 181-188.

29. Statham, C. N.; Lech, J. J. J. Fish Res. Bd. Can. 1975, 32,
    515-522.
30. Kime, D. E. General Comp. Endocrinol. 1980, 41, 164-172.
31. Förlin, L.; Haux, C. Aquatic Toxicol. 1985, In Press.
32. Chowdhury, N. R.; Chowdhury, J. R.; Arias, I. M. Comp. Biochem.
    Physiol. 1982, 73B, 651-653.
33. Förlin, L.; Andersson, T. Comp. Biochem. Physiol. 1981, 68C,
    239-242.
34. Andersson, T.; Förlin, L.; Hansson, T. Drug Metab. Disp. 1983,
    11, 494-498.
35. Parker, R. S.; Morrissey, M. T.; Moldeus, P.; Selivonchick,
    D. P. Comp. Biochem. Physiol. 1981, 70B, 631-633.
36. Dutton, G. J.; Glucuronidation of Drugs and Other Compounds,
    CRC Press, Boca Raton, 1980.
37. Burchell, B. Rev. Biochem. Toxicol. 1981, 3, 1-32.
38. Burchell, B.; Blanckaert, N. B. Biochem. J. 1984, 223, 461-465.
39. Koivusaari, U.; Harri, M.; Hanninen, O. Comp. Biochem. Physiol.
    1981, 70C, 149-157.
40. Koivusaari, U. J. Exp. Zool. 1983, 227, 35-42.
41. Castren, M.; Oikari, A. Comp. Biochem. Physiol. 1983, 76C,
    365-369.
42. Hänninen, O.; Lindstrom-Seppä, P.; Koivusaari, U.; Väisänen, M.;
    Julkunen, A.; Juvonen, R. Biochem. Soc. Trans. 1984, 12, 13-17.
43. Förlin, L.; Andersson, T. Comp. Biochem. Physiol. 1985, 80B,
    569-572.
44. Andersson, T.; Pesonen, M.; Johansson, C. Biochem. Pharmacol.
    1985, In Press.
45. Wishart, G. J. Biochem. J. 1978, 174, 485-489.
46. Boyland, E.; Chasseaud, L. F. Adv. Enzymol. 1969, 32, 173-219.
47. Arias, I. M.; Jakoby, W. B., Eds., "Glutathione: Metabolism
    and Function", Raven: New York, 1976.
48. Jakoby, W. B.; Habig, W. H. In "Enzymatic Basis of Detoxication",
    Jakoby, W. B., Ed.; Academic: New York, 1980; Vol. II, pp. 95-120.
49. Lay, M. M.; Menn, J. J. Xenobiotica 1979, 11, 669-673.
50. Yagen, B.; Foureman, G. L.; Ben-Zvi, Z.; Ryan, A. J.; Hernandez,
    O.; Cox, R. H.; Bend, J. R. Drug Metab. Disp. 1984, 12, 389-395.
51. Tate, S. In "Enzymatic Basis of Detoxication"; Jakoby, W. B.,
    Ed.; Academic: New York, 1980, Vol. II, pp. 95-120.
52. Bend, J. R.; Fouts, J. R. Bull. Mt. Desert Isl. Biol. Lab.
    1973, 13, 4-8.
53. Bend, J. R., James, M. O.; Dansette, P. M. Ann. N.Y. Acad. Sci.
    1977, 298, 505-521.
54. James, M. O.; Bowen, E. R.; Dansette, P. M.; Bend, J. R.
    Chem. Biol. Interact. 1979, 25, 321-344.
55. Bauermeister, A., Lewendon, A.; Ramage, P.I.N., Nimmo, I. A.
    Comp. Biochem. Physiol. 1983, 74C, 89-93.
56. Nimmo, I. A. Comp. Biochem. Physiol. 1985, 80B, 365-369.
57. Braddon, S. A., McIlvaine, C. M., Balthrop, J. E. Comp. Biochem.
    Physiol. 1985, 80B, 213-216.
58. Balk, L.; Meijer, J.; Seidegärd, J.; Morgenstern, R.; DePierre,
    J. W. Drug Metab. Disp. 1980, 8, 98-103.
59. Morgenstern, R.; DePierre, J. W. Eur. J. Biochem. 1983, 134,
    591-598.
60. Morgenstern, R.; Lundqvist, G.; Andersson, G.; Balk, L.;
    DePierre, J. W. Biochem. Pharmacol. 1984, 33, 3609-3614.
61. Foureman, G. L.; Bend, J. R. Chem. Biol. Interact. 1984, 49,
    89-103.

62. Hernandez, O.; Walker, M.; Cox, R. H.; Foureman, G. L.; Smith, B. R.; Bend, J. R. Biochem. Biophys. Res. Commun. 1980, 96, 1494-1502.

63. Hernandez, O.; Foureman, G. L.; Cox, R. H.; Walker, M.; Smith, B. R.; Bend, J. R. In "Polynuclear Aromatic Hydrocarbons, Fifth International Symposium on Chemical Analysis and Biological Fate"; Cooke, M.; Dennis, A. J., Eds. Battelle: Columbus, Ohio, 1981, p. 667.

64. Foureman, G. L.; Bend, J. R. Bull. Mt. Desert Isl. Biol. Lab. 1982, 22, 35-37.

65. Stegeman, J. J.; Klotz, A. V.; Woodin, B. R.; Pajor, A. M. Aquatic Toxicol. 1981, 1, 197-212.

66. Egaas, E.; Varanasi, U. Biochem. Pharmacol. 1982, 31, 561-566.

67. Bend, J. R.; Ball, L. M.; Elmamlouk, T. H.; James, M. O.; Philpot, R. M. In "Pesticide and Xenobiotic Metabolism in Aquatic Organisms", Khan, M. A. Q.; Lech, J. J.; Menn, J. J., Eds.; ACS Symposium Series No. 99, American Chemical Society: Washington, D.C., 1979; pp. 297-318.

68. James, M. O.; Little, P. J. Chem. Biol. Interact. 1981, 36, 229-248.

69. Stegeman, J. J. In "Polycyclic Hydrocarbons and Cancer", Gelbain, H. V.; T'so, P.O.P., Eds.; Academic: New York, 1981, Vol. 3, pp. 1-60.

70. Sugiyama, Y.; Yamada, T.; Kaplowitz, N. Biochem. J. 1981, 199, 749-756.

71. Ramage, P.I.N.; Nimmo, I. A. Biochem. J. 1983, 211, 523-526.

72. Ramage, P.I.N.; Nimmo, I. A. Comp. Biochem. Physiol. 1984, 78B, 189-194.

73. James, M. O.; Fouts, J. R.; Bend, J. R. Bull. Mt. Desert Isl. Lab. 1974, 14, 41-46.

74. Allen, J. L.; Dawson, V. K.; Hunn, J. B. In "Pesticide and Xenobiotic Metabolism in Aquatic Animals", Khan, M. A. Q., Lech, J. J. and Menn, J. J., Eds.; ACS Symposium Series No. 99, American Chemical Society: Washington, D.C., 1979; pp. 121-129.

75. Huang, K. C.; Collins, S. F. J. Cell. Comp. Physiol. 1962, 60, 49-52.

76. Brenniman, G. R.; Anver, M. R.; Hartung, R.; Rosenburg, S. H. J. Environ. Pathol. Toxicol. 1979, 2, 1267-1281.

77. James, M. O.; Bend, J. R. Xenobiotica 1976, 6, 393-398.

78. Guarino, A. M.; James, M. O.; Bend, J. R. Xenobiotica 1977, 7, 623-631.

79. James, M. O. In "Conjugation Reactions in Drug Biotransformation", Aitio, A., Ed.; Elsevier: North Holland, 1978; pp. 121-129.

80. James, M. O. Fed. Proc. 1980, 39, 758.

81. Pritchard, J. B.; James, M. O. J. Pharmacol. Exp. Therap. 1979, 208, 280-286.

82. James, M. O. Drug Metab. Disp. 1982, 10, 516-522.

83. Pritchard, J. B.; James, M. O. In "Metabolic Basis of Detoxication"; Jakoby, W. B.; Bend, J. R.; Caldwell, J., Eds.; Academic: New York, 1982; pp. 339-357.

84. Pritchard, J. B.; Cotton, C. U.; James, M. O.; Giguere, D.; Koschier, F. J. Bull. Mt. Desert Isl. Biol. Lab. 1978, 18, 58-60.

RECEIVED October 25, 1985

3

Xenobiotic Conjugation in Insects

C. F. Wilkinson

Institute for Comparative and Environmental Toxicology, Cornell University, Ithaca, NY 14853

Insect species possess a diverse spectrum of enzymatic conjugation capabilities that allow them to effect the secondary metabolism of a wide variety of pesticides and other xenobiotics containing appropriate hydroxyl, carboxyl or amino groups. The range of reactions catalyzed by insects includes glycoside, sulfate, and phosphate formation and a variety of conjugations involving glutathione and amino acid conjugation. A major difference between insects and mammals is that the former utilize glucose rather than glucuronic acid in the formation of glycoside conjugates. There is also evidence that enzymatic conjugation in insects may constitute an important mechanism for the regulation of insect steroid hormones such as the ecdysteroids.

Insects, like most other living organisms, have evolved a remarkable battery of enzyme-catalyzed reactions that provides an effective biochemical defense against the potentially toxic effects of a large number of naturally occurring and synthetic chemicals (1-5). In addition to having a versatile cytochrome P-450-mediated mixed-function oxidase system that is responsible for the primary (phase I) metabolism of xenobiotics, insects also possess a variety of conjugation (phase II) mechanisms that catalyze the all-important final step in the conversion of lipophilic xenobiotics to polar, water-soluble, readily-excretable products (3-6).

Major phase II reactions that have been demonstrated to occur in insects include xenobiotic conjugations with glycoside, sulfate, phosphate, amino acid and glutathione (3-7). With the exception of the utilization of glucose rather than glucuronic acid in glycoside formation, the conjugation reactions occurring in insects are qualitatively similar to those present in mammals and higher organisms. Unfortunately, few attempts have been made to isolate and characterize the enzymes catalyzing these reactions in insects and in many cases the sum total of the evidence indicating the

0097-6156/86/0299-0048$06.00/0
© 1986 American Chemical Society

existence of a particular mechanism rests entirely on the isolation
of the conjugate in the excreta or body tissues of treated insects.

The following constitutes a brief review of our rather woeful
state of knowledge of insect conjugation reactions and their role
in xenobiotic and intermediary metabolism.

Glycoside formation

In insects, as in mammals, glycoside formation constitutes the
major mechanism for the conjugation of xenobiotic phenols.   In
contrast to mammals, however, where the major sugar conjugates are
β-glucuronides, insects predominantly form the corresponding
β-glucosides (3,6,7). Since this also appears to be the case for
all plants and other invertebrates, β-glucoside formation undoubtedly constitutes a more ubiquitous conjugation mechanism than the
better known mammalian glucuronide system (7). Early reports that
houseflies (Musca domestica) and blowflies (Lucilia sericata) were
capable of forming glucuronides as well as glucosides following
treatment with naphthalene and α-naphthol (8), have not been
confirmed and it is now generally accepted that glucosides are the
only sugar conjugates produced by insects (3).

O-and S-glucosides from phenols (or alcohols) and thiols,
respectively, are those most commonly isolated from insect species
although the identification of benzoylglucoside as a normal product
in cockroaches (9) suggests that ester glucosides of xenobiotic
acids may also be produced. To date, no N-glucosides have been
isolated from insects.

By analogy with the glucuronide conjugation mechanism in
mammals (6,10), it is probable that uridine diphosphoglucose (UDPG)
constitutes the glucosyl donor in insects and that transfer of
glucose to an appropriate hydroxy or mercapto acceptor is catalyzed
by a UDP-glucosyltransferase (Figure 1) (3,6). UDPG is a normal
component of carbohydrate metabolism in locust species (11) and has
been detected in silkworms (Bombyx mori) (12).

Few attempts have been made to characterize insect glucosyltransferases and the reports that do exist are quite equivocal.
Thus, in contrast to mammalian glucosyltransferases that are
localized in the microsomal fraction of tissue homogenates (6),
those from the housefly, cockroach (Periplaneta americana) and
locust (Schistocerca cancellata) are reported to be associated with
the 15,000-20,000g pellet (13,14) and that from the tobacco hornworm (Manduca Sexta) with the high speed supernatant (14). The
enzyme has been identified in gut and fat body tissues of the few
insect species studied and may have a fairly broad tissue distribution.

Judging from its broad distribution among plants and invertebrates and its relatively rare occurrence in mammals, glucosylation
appears to have evolved as a conjugation mechanism prior to glucuronidation. Smith (7), speculating on why evolution might have

favored the development of glucuronide conjugation over glucoside formation, suggested that active secretion of acidic glucuronides in the vertebrate kidney tubule or in the bile might have proved advantageous in the evolution of higher species.

Sulfate conjugation

The ability to form sulfate esters of xenobiotic phenols is widely distributed throughout the animal kingdom (6,17,15) and along with glucoside formation appears to constitute a major conjugation mechanism in most insect species (3,4,6,7). After studying the conjugation of m-aminophenol, 8-hydroxyquinoline and 7-hydroxycoumarin in 15 species of insects representing five different orders, Smith (16) concluded that although sulfate conjugation occurred in all species examined, glucoside formation remained the predominant metabolic pathway.

Sulfate formation has also been demonstrated in a variety of other invertebrates including spiders, scorpions and ticks (7), although the arachnids appear to possess no alternative glycoside mechanism for phenol conjugation.

As a result of its ubiquitous distribution and its presence in the primitive arthropod peripatus (Peripatoides novazealandiae), it was suggested that sulfate conjugation might constitute a primitive detoxication pathway that was of considerable importance prior to the evolution of glucoside formation (17). This contrasts sharply with the views of Dodgson and Rose (18) who suggested that the metabolism of xenobiotics via sulfate conjugation may be a relatively recent evolutionary development that has become increasingly sophisticated in higher mammals.

Most studies demonstrating sulfate conjugation in insects have been limited to the identification of conjugates excreted by insects treated in vivo with appropriate phenols (16,17,19). Based on the few in vitro studies that have been conducted, however, the insect enzyme system appears to be similar to that found in mammalian tissues (20,21), and probably follows the same three-step reaction sequence established in mammals (Figure 2) (3,6,15). To date, however, formation of adenosine 5'-phosphosulfate (APS) and the high energy sulfate donor, 3'-phosphoadenosine 5'-phosphosulfate (PAPS) (Figure 3) have not been unequivocally established in insect species.

An active sulfotransferase system requiring ATP, inorganic sulfate and Mg^{2+}, has been demonstrated in the high speed (100,000g) supernatant fractions of tissue homogenates from eight species of insects representing four major orders (Diptera, Hymenoptera, Orthoptera and Lepidoptera) (21). Sulfotransferase activity was particularly high in preparations from the gut tissues of the southern armyworm (Spodoptera eridania) and other lepidopterous larvae and, in addition to catalyzing the sulfation of p-nitrophenol, the enzyme was active towards a variety of plant, insect

$$\text{D-glucose 1-phosphate} + \text{UTP} \xrightarrow[\text{pyrophosphorylase}]{\text{UDPG}} \text{UDP-}\alpha\text{-D-glucose} + P_2$$

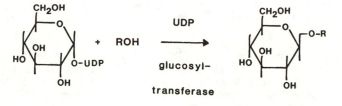

UDP-α-D-glucose   RO-β-D-glucoside

Figure 1.   Formation of β-D-glucoside from xenobiotic alcohol
            or phenol (ROH).

$$\text{ATP} + SO_4^{2-} \underset{\text{SULFURYLASE}}{\overset{\text{ATP}}{\rightleftharpoons}} \text{APS} + P_2O_7^{2-}$$

$$\text{APS} + \text{ATP} \underset{\text{5'-PHOSPHOSULFATE KINASE}}{\overset{\text{ADENOSINE}}{\rightleftharpoons}} \text{PAPS} + \text{ADP}$$

$$\text{PAPS} + \text{ROH} \xrightarrow[\text{SULFOTRANSFERASE}]{} \text{ROSO}_3\text{H} + \text{ADP}$$

Figure 2.   Three-step reaction sequence in formation of sulfate
            esters.

and mammalian steroids including cholesterol, α-ecdysone and β-sitosterol (20,21).

The ability of insect sulfotransferases to catalyze the sulfation of plant and insect steroids may simply reflect the broad substrate specificity of the enzymes. On the other hand, it may be indicative of a more important physiological function of the enzymes. Sulfate esters of cholesterol, campesterol and β-sitosterol have been identified in the meconium of tobacco hornworm (M. sexta) pupae (22), and these steroids are known precursors of α- and β-ecdysone and other molting hormones in this species (23). Further, there is evidence that houseflies (M. domestica) and diapausing pupae of M. sexta convert 22,25-bisdeoxyecdysone, α-ecdysone and 20-hydroxyecdysone into sulfate and glucoside conjugates (24).

Since insects are incapable of de novo synthesis of the steroid ring, the insect steroid hormones must be synthesized from a variety of phytosteroids ingested in the diet (25,26). It is entirely possible, therefore, that insects might utilize the sulfoconjugation pathway to store phytosteroids or other precursors of α- and β-ecdysone. Furthermore, by analogy with the suggested role of sulfoconjugation in mammalian steroid hormone metabolism (18) insects may use this reaction in concert with an appropriate arylsulfatase to regulate titers of their steroid hormones (19,20). This would constitute a readily reversible mechanism whereby the required balance between active (free) and inactive (conjugated) forms of α- and β-ecdysone and other ecdysteroids could be achieved. The possible existence of such a mechanism is supported by the age-dependent changes in sulfotransferase and arylsulfatase activities observed in the midgut tissues of late larval southern armyworm larvae (27). Thus, arylsulfatase activity (probably of lysosomal origin) was found to be highest during the larval molt (Figure 4) and could regulate the titer of free ecdysteroid present at this time.

Although several sulfotransferases with different substrate specificities have been isolated from mammlian tissues, the presence of isozymic forms has not yet been established in insects.

Phosphate conjugation

While the biosynthesis of phosphate esters is common in intermediary metabolism, the conjugation of foreign compounds with phosphate is encountered only rarely in vertebrate species (3,7,28). Indeed, to date, insects appear to be the only major group of organisms that utilize this pathway to any significant extent in the metabolism of foreign compounds (3,6,7).

Phosphate conjugates have been identified in the bodies and excreta of larvae of New Zealand grass grubs (Costelytra zealandica), and of adult houseflies (M. domestica) and blowflies (L. sericata) treated in vivo with 1-naphthol, 2-naphthol and p-nitrophenol (29).

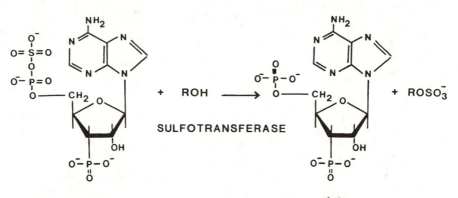

3′-Phosphoadenosine
5′-phosphosulfate (PAPS)

Adenosine 3′,5′-diphosphate

Figure 3.   Transfer of sulfate from PAPS to xenobiotic alcohol
or phenol (ROH).

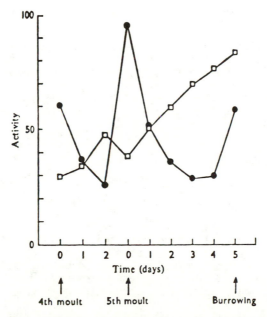

Figure 4.   Arylsulfatase activity (●-●) and sulfotransferase
activity (□-□) in developing larvae of the southern
armyworm (S. eridania).

An active phosphotransferase, requiring ATP and Mg^{2+}, and catalyzing the phosphorylation of p-nitrophenol has been demonstrated in the 100,000g supernatants of gut tissues of the Madagascar cockroach (Gromphadorhina portentosa), tobacco hornworm (M. sexta) larvae and in whole houseflies (M. domestica) (21,30); in G. portentosa it appears to constitute a major pathway for phenol metabolism (21,30). Although the mechanism of phosphorylation has not been elaborated, it is probable that ATP serves as the direct donor of the phosphate moiety transferred to the phenol (Figure 5).

As has been discussed in relation to sulfate conjugation, there is increasing evidence that phosphate formation may play a critical role in regulating insect steroids; this is particularly true for steroids present in eggs and during early stages of embryonic development.

The 22-phosphates of ecdysone and 2-deoxyecdysone have been identified as major steroid conjugates in newly laid eggs of Schistocera gregaria (31) and are also present in early embryos of Locusta migratoria (32). In the latter species these conjugates appear to be derived from mononucleotide precursors [the 22-adenosine monophosphoric ester of 2-deoxyecdysone and the $22-N^6-$(isopentenyl)-adenosine monophosphoric ester of ecdysone] identified in newly laid eggs (33,34). More recently, the major ecdysteroid in 48- to 64-hr-old eggs of M. sexta has been identified as the 26-phosphate of 26-hydroxyecdysone (35).

The formation of phosphate esters of ecdysteroids constitutes a potentially valuable mechanism for storing high titers of insect steroids in a form that is inactive but that provides a readily available source of the active ecdysteroid. Embryos of S. gregaria contain a phosphatase that can hydrolyze ecdysone 22-phosphate (36) and this enzyme could effect the controlled release of ecdysone and possibly other ecdysteroids at appropriate stages of insect development.

Amino acid conjugation

Although, as already discussed, aromatic carboxylic acids can be excreted as ester glucosides by many organisms, the major alternative mechanism for metabolizing such xenobiotics is through conjugation with an amino acid.

In insects, glycine is the amino acid of choice for xenobiotic conjugation and glycine conjugates of a variety of substituted benzoic acids (p-aminobenzoic acid, p-nitrobenzoic acid, salicylic acid and anthranilic acid) have been reported in the housefly (M-domestica), mosquito (Aedes sp.), locust (L. gregaria) and silkworm (B. mori) (3,4,6,7,37).

Although not studied in insects the mechanism of amino acid conjugation usually occurs in two stages (3,6). The first of these, requires ATP and coenzyme A and results in the activation of the xenobiotic acid, and the second involves condensation of the activated xenobiotic acid with the appropriate amino acid.

It is of interest to note that the formation of amino acid conjugates in several other invertebrates is not restricted to glycine as appears to be the case in insects. Thus, in arachnids (spiders, ticks, scorpions) the major primary conjugate of substituted benzoic acids appears to be with arginine. However, as a result of a variety subsequent reactions a number of other conjugates are ultimately formed including those with glutamine, glutamic acid and ornithine (7,38,39).

Glutathione conjugation

Conjugation with reduced glutathione (GSH) constitutes an extremely important mechanism of xenobiotic metabolism and it is catalyzed by a group of enzymes referred to collectively as GSH-S-transferases (6,40,42). The reaction is different from the other reactions discussed above in that it does not involve the formation of a "high energy" or "active" intermediate requiring ATP. In fact, GSH conjugation is frequently considered as a phase I rather than a phase II reaction since the GSH-S-transferases catalyze the primary reaction of GSH (probably in the form of the GSH-thiolate ion, GS$^-$) with a wide variety of lipophilic xenobiotics containing appropriate electrophilic centers. GSH-S-transferases probably contain a hydrophobic, relatively nonspecific substrate binding site(s) in close juxtaposition to a site for binding and activating GSH. It has been argued that the enzyme plays a relatively passive role in its interactions with xenobiotics and that its major catalytic role may be the activation of GSH by providing an environment that promotes its dissociation to GS$^-$ (40,42). If this occurs close to where the xenobiotic binds, and if the xenobiotic has an appropriate electrophilic center, enzymatic reaction with GSH will occur.

In most cases, the initial product of GSH conjugation is a GSH thioether adduct but this often undergoes a series of subsequent reactions involving a γ-glutamyl transferase, a dipeptidase and an acetyl-coenzyme A linked acetylase to yield the corresponding mercapturic acids (43) (Figure 6).

In addition to their catalytic role GSH-S-transferases also serve a binding function in the cell, having a relatively high affinity for several endogenous materials as well as potentially harmful foreign compounds. This is typically referred to as ligandin activity and undoubtedly results in part from the nonspecificity of the hydrophobic binding site of the enzyme.

The structural diversity of the compounds that can serve as substrates for the GSH-S-transferases includes a wide variety of aliphatic, aromatic and alicyclic materials containing numerous unrelated functional groups (40,42). The reaction mechanism may involve a simple displacement of some group by sulfur (for example, the displacement of halide in halomethane) or an addition-elimination reaction as appears to be the case with 1,2-dichloro-4-nitrobenzene (Figure 7). In a few cases, such as with organic thiocya-

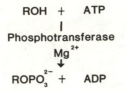

$$ROH \ + \ ATP$$

Phosphotransferase

$$Mg^{2+}$$

$$ROPO_3^{2-} \ + \ ADP$$

Figure 5. Conjugation of xenobiotic phenols with phosphate.

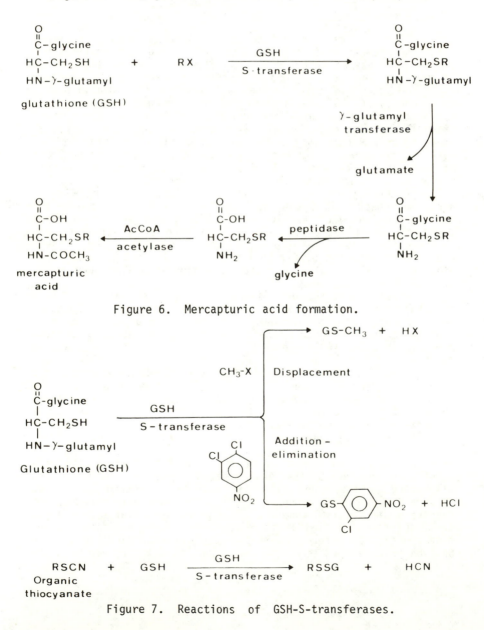

Figure 6. Mercapturic acid formation.

Figure 7. Reactions of GSH-S-transferases.

nates the reaction with GSH-S-transferase does not result in a GSH-thioether but instead yields cyanide and the corresponding mixed disulfide, probably indicating attack on the sulfur of the thiocyano group.

Insect GSH-S-transferases have received a good deal of attention in recent years because of their ability to catalyze the dealkylation and dearylation of a wide variety of organophosphorus insecticides (3,6,41) (Figure 8) as well as γ-hexachlorocyclohexane (44). The detoxication reactions that result have been implicated in insect resistance to these insecticides (45,46).

GSH-S-transferase activity has been found in all insect species examined although the level of activity varies with the species and strain of insect and with the substrate employed (3,6,41). It is evident that, as in mammals, the insect enzymes are present as a group of structurally similar isozymes with varying degrees of overlapping substrate specificity; the enzymes occur in the 100,000g supernatants of a variety of insect tissues. Enzyme purification studies indicate that all GSH-S-transferases have a similar molecular weight (∼50,000) and consist of two approximately equal subunits (40).

During the last two decades there has been continuing controversy concerning the classification of the enzyme DDT-dehydrochlorinase as a GSH-S-transferase (6). DDT-dehydrochlorinase is the enzyme that converts DDT to the relatively non-toxic DDE and has been intensively studied in houseflies and other insects because of its established importance in insect resistance to DDT.

DDT dehydrochlorinase has been highly purified from DDT-resistant houseflies (47) and has been characterized as a soluble lipoprotein with a molecular weight of about 120,000 and consisting of four equal subunits. Formation of the tetramer reportedly occurs only in the presence of DDT, and GSH is required for dehydrochlorination (48). In contrast to most GSH-S-transferases there is no evidence for the formation of a DDT-GSH adduct and no evidence that GSH is depleted during the course of the reaction.

More recently DDT-dehydrochlorinase has been isolated and purified (∼660-fold) to apparent homogeneity from houseflies (49). In contrast to that described in earlier studies, this enzyme was found to be a dimer with subunits of molecular weights of 23,000 and 25,000. It was also found to possess substantial GSH-S-transferase activity towards 2,4-dinitrochlorobenzene and 3,4-dichloronitrobenzene. Based on its structure, catalytic activity and chromatographic behavior it was concluded that the purified heterodimeric DDT-dehydrochlorinase was indeed a GSH-S-transferase isozyme (49). It was proposed that instead of the nucleophilic substitution usually observed in GSH-S-transferase activity, DDT-dehydrochlorination by this enzyme involves an E2 elimination reaction in which the GS⁻ thiolate anion abstracts the hydrogen on the C-2 of DDT and this initiates the departure of the chlorine atom from C-1 (49) (Figure 9).

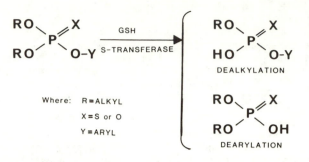

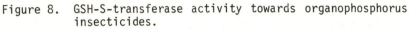

Figure 8.   GSH-S-transferase activity towards organophosphorus
            insecticides.

GSH S-TRANSFERASE   **DDT DEHYDROCHLORINASE**

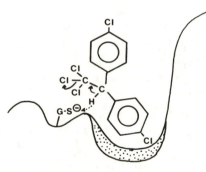

Nucleophilic substitution   E2 elimination

                                    1. Abstraction of benzylic hydrogen (C-2)

                                    2. Elimination chlorine (C-1)

Figure 9.   Proposed mechanisms of GSH-S-transferase towards
            2,4-dinitrochlorobenzene and DDT.

It is, of course, entirely possible that the DDT-dehydro-
chlorinases purified by Dinamarca's group (47) and by Clark (49)
are distinct enzyme proteins since they apparently differ in
several respects. Of particular interest in the case of the former
is that, in the absence of DDT, the enzyme catalyzes the oxidation
of GSH to the corresponding glutathione disulfide GSSH. Although
the mechanism of GSH oxidation has not been established, it is of
interest that GSH-S-transferases having peroxidase activity have
been reported where it is apparent that GSH is attacking an electrophilic oxygen (50).

Summary

It should be clear from the foregoing discussion that our current
state of knowledge concerning the mechanisms of xenobiotic conjugation in insects leaves much to be desired. To date, studies on
xenobiotic metabolism in insect species have been focused mainly on
the pathways of primary metabolism and have been stimulated by the
importance of such pathways in insect resistance to insecticide
chemicals. Conjugation reactions are not considered to be rate
limiting in insecticide metabolism and consequently have been
largely neglected.

However, it is now becoming increasingly apparent that conjugation reactions may play an important physiological role in the
regulation of insect ecdysteroids. Perhaps this possibility will
provide the necessary stimulus for more comprehensive studies in
the future.

Literature cited

1. Wilkinson, C. F. In "Foreign Compound Metabolism"; Caldwell,
   J.; Paulson, G. D., Eds.; Taylor and Francis: London, 1984;
   pp. 133-147.
2. Brattsten, L. B. In "Herbivores: Their Interactions with
   Secondary Plant Metabolites"; Rosenthal, A.; Janzen, D. H.,
   Eds.; Academic: New York, 1979; pp. 199-270.
3. Dauterman, W. C.; Hodgson, E. In "Biochemistry of Insects";
   Rockstein, M., Ed.; Academic: New York, 1978; Chap. 13.
4. Smith, J. N. In "Comparative Biochemistry"; Florkin, M.;
   Mason, H. S., Eds.; Academic: New York, 1964; Vol. VI, pp.
   403-448.
5. Wilkinson, C. F. and Brattsten, L. B. Drug Metab. Revs. 1972,
   1, 153-228.
6. Yang, R. S. H. In "Insecticide Biochemistry and Physiology";
   Wilkinson, C. F., Ed.; Plenum: New York. 1976; Chap. 5.
7. Smith, J. N. Advan. Comp. Physiol. Biochem. 1968, 3,
   173-232.
8. Terriere, L. C.; Boose, R. B.; Roubal, W. T. Biochem. J.
   1961, 79, 620-623.
9. Quilico, A.; Piozzi, F.; Pavan, M.; Mantica, E. Tetrahedron
   Lett. 1959, 5, 10-14.
10. Kasper, C. B.; Henton, D. In "Enzymatic Basis of Detoxica-
    tion"; Jacoby, W. B., Ed.,; Academic: New York, 1980; Chap. 1.

11. Candy, D. J.; Kilby, B. Nature. 1959, 183, 1594-95.
12. Carey, F. G.; Wyatt, G. R. Biochem. Biophys. Acta. 1960, 41, 178-79.
13. Dutton, D. J. Comp. Biochem. Physiol. 1962, 7, 39-46.
14. Mehendele, H. M.; Dorough, H. W. J. Insect Physiol. 1972, 18, 981-987.
15. Jakoby, W. B.; Sekura, R. D.; Lyon, E. S.; Marcus, C. J.; Wang, J. L. In "Enzymatic Basis of Detoxication"; Jakoby, W. B., Ed.; Academic: New York, 1980; Vol. 11, Chap. 11.
16. Smith, J. N. Biochem. J. 1955, 60, 436-442.
17. Jordon, T. W.; McNaught, R. W.; Smith, J. N. Biochem. J. 1970, 118, 1-10.
18. Dodgson, J. S.; Rose, F. A. In "Metabolic Conjugation and Metabolic Hydrolysis"; Fishman, W. H., Ed.; Academic: New York, 1970; Vol. I, pp. 239-325.
19. Yang, R. S. H.; Wilkinson, C. F. Pestic. Biochem. Physiol. 1972, 1, 327-339.
20. Yang, R. S. H.; Wilkinson, C. F. Biochem. J. 1972, 130, 487-493.
21. Yang, R. S. H.; Wilkinson, C. F. Comp. Biochem. Physiol. 1973, 46B, 717-726.
22. Hutchins, R. F. N.; Kaplanis, J. N. Steroids, 1969, 13, 605-614.
23. Robbins, W. E.; Kaplanis, J. N.; Svoboda, J. A.; Thompson, M. J. Ann. Rev. Entomol. 1971, 16, 53-72.
24. Thompson, M. J.; Svoboda, J. A.; Kaplanis, J. N.; Robbins, W. E. Proc. Roy. Soc. Ser. B. 1972, 180, 203-221.
25. Riddiford, L. M.; Truman, J. W. In "Biochemistry of Insects"; Rockstein, M., Ed.; Academic: New York, 1978; Chap. 7.
26. Kaplanis, J. N.; Weirich, G. F.; Svoboda, J. A.; Thompson, M. J.; Robbins, W. E. In "Progress in Ecdysone Research"; Hoffman, J. A., Ed.; Elsevier/North Holland: Amsterdam, 1980; pp. 163-186.
27. Yang, R. S. H.; Pelliccia, J. G.; Wilkinson, C. F. Biochem. J. 1973, 817-820.
28. Mulder, G. J. In "Metabolic Basis of Detoxication"; Jakoby, W. B.; Bend, J. R.; Caldwell, J., Eds.; Academic: New York, 1982, Chap. 13.
29. Binning, A.; Darby, F. J.; Hennan, M. P.; Smith, J. N. Biochem. J. 1967, 103, 42-48.
30. Gil, D. L.; Rose, H. A.; Yang, R. S. H.; Young, R. G.; Wilkinson, C. F. 1974, 47B, 657-662.
31. Isaac, R. E.; Rose, M. E.; Rees, H. H.; Goodwin, T. W. Biochem. J. 1983, 533-541.
32. Sall, C.; Tsoupras, G.; Kappler, C.; Lageux, M.; Zachary, D.; Luu, B.; Hoffmann, J. A. J. Insect Physiol. 1983, 29, 491-498.
33. Tsoupras, G.; Hetru, C.; Luu, B.; Lageux, M.; Constantin, E.; Hoffmann, J. A. Tetrahedron Lett. 1982, 23, 2045.
34. Tsoupras, G.; Luu, B.; Hoffmann, J. A. Science. 1983, 220, 507-508.
35. Thompson, M. J.; Weirich, G. F.; Rees, H. H.; Svoboda, J. A.; Feldlaufer, M. F.; Wilzer, K. R. Arch. Insect Biochem. Physiol. 1985, in press.

36. Isaac, R. E.; Sweeney, F. P.; Rees, H. H. Biochem. Soc. Trans. 1983, 11, 379.
37. Shyamala, M. B. J. Insect Physiol. 1964, 10, 385-391.
38. Hitchcock, M; Smith, J. N. Biochem. J. 1964, 93, 392-400.
39. Hitchcock, M.; Smith, J. N. Biochem. J. 1966, 98, 736-741
40. Jakoby, W. B.; Habig, W. H. In "Enzymatic Basis of Detoxication"; Jacoby, W. B., Ed.; Academic: New York, 1980; Vol. II, pp. 63-94.
41. Motoyama, N.; Dauterman, W. C. Rev. Biochem. Toxicol. 1980, 2, 49-69.
42. Jakoby, W. B. Adv. Enzymol. 46, 383-414.
43. Tate, S. S. In "Enzymatic Basis of Detoxication"; Jacoby, W. B., Ed.; Academic: New York, 1980; Vol. II, pp. 95-120.
44. Tanaka, K.; Kurihara, N.; Nakajima, M. Pestic. Biochem. Physiol. 1976, 6, 392-401.
45. Dauterman, W. C. In "Pest Resistance to Pesticides"; Georghiou, G. P.; Saito, T., Eds.; Plenum: New York, 1983; 229-257.
46. Tanaka, K.; Nakajima, M.; Kurihara, N. Pestic. Biochem. Physiol. 1981, 16, 149-155.
47. Dinamarca, M. L.; Saavedra, I.; Valdes, E. Comp. Biochem. Physiol. 1969, 31, 269-82.
48. Dinamarca, M. L.; Levenbook, L. L.; Valdes, E. Arch. Biochem. Biophys. 1971, 147, 374-83.
49. Clark, A. G.; Shamaan, N. A. Pestic, Biochem, Physiol. 1984, 22, 249-261.
50. Prohaska, J. R.; Ganther, H. E. Biochem. Biophys. Res. Commun. 1977, 76, 437-45.
51. Wilkinson, C. F. In "Agricultural Chemicals of the Future"; Hilton, J. L., Ed.; Rowman and Allanheld: Totowa, New Jersey, 1985; Chap. 24.

RECEIVED October 21, 1985

4

Xenobiotic Conjugation in Higher Plants

G. L. Lamoureux and D. G. Rusness

Metabolism and Radiation Research Laboratory, Agricultural Research Service, U.S. Department of Agriculture, Fargo, ND 58105

Xenobiotic conjugation in plants is extensively reviewed. Where appropriate, the following data are tabulated: xenobiotic, class of conjugate, species, and methods of isolation and identification. The following conjugates are discussed: simple and complex glucose conjugates formed from HOOC-, HO-, H_2N-, HON-, and HS-functional groups; glutathione or homoglutathione conjugates formed from various electrophiles; amino acid conjugates of carboxylic acids; malonic acid conjugates of amines; several lipophilic conjugates; and bound residues. The relationship between metabolism and herbicide selectivity in various plant species is also discussed.

The metabolism of xenobiotics in plants has been an area of intense research for approximately 20 years. This interest has been motivated by concern over the widespread use of pesticides in our environment and by the desire to produce pesticides that are more bio-degradable and more selective. As a result, most of the available information concerning xenobiotic metabolism in plants pertains to pesticides or pesticide analogs. Numerous reviews have been written on plant metabolism of pesticides (1-7) and herbicides (8-11). In addition, more specific reviews have dealt with glycoside conjugation (12), amino acid conjugation (13), glutathione conjugation (14), catabolism of glutathione conjugates in plants (15, 16), bound residues (17), in vitro methods for studying xenobiotic metabolism in plants (18), metabolism in cell culture (19,20), plant enzymes involved in xenobiotic metabolism (21,22), and oxidative enzymes in plants (23).

The mechanisms utilized by plants and mammals in the metabolism of xenobiotics are remarkably similar. Similar classes of compounds or functional groups are frequently metabolized by comparable mechanisms. Oxidation, reduction, hydrolysis, and conjugation reactions occur with similar frequency in both. In most instances, however,

considerably more is known about the enzymes and mechanisms involved
in xenobiotic transformations in mammals than in plants. This is
particularly true in regards to the oxidative reactions and amino
acid conjugation. Although it is commonly stated that xenobiotic
metabolism occurs more slowly in plants than in mammals, many xenobiotic transformations in plants are completed within 6- to 24-hr
following exposure to the chemical. Hydrolysis, glucose conjugation, and glutathione conjugation can occur very rapidly in
plants. The titer of glutathione S-transferase enzyme in corn
appears to be comparable to that in rat liver (24).

Several fundamental differences between plants and mammals may
be responsible for some of the differences observed in xenobiotic
metabolism. Plants lack a well-developed excretory system, and as
a result, xenobiotics may be subjected to a greater array of metabolic transformations over a longer period of time than would occur in
mammals. The autotrophic nature of higher plants may also result in
some differences in metabolism, e.g., reincorporation of CO_2 produced from the metabolism of xenobiotics. Plant cells are characterized by a cell wall that can be highly lignified. Xenobiotics
may be incorporated into the cell wall in nonselective free radical
reactions utilized in the synthesis of lignin. Xenobiotics may also
be incorporated into hemicellulose or other carbohydrate components
of the cell wall in more selective reactions. As a result, bound
residues tend to be much more common in xenobiotic metabolism in
plants than in mammals. Plant cells are also characterized by the
presence of large cell vacuoles. Xenobiotics may be metabolized in
such a manner that they become sequestered from further metabolic
processes by storage in this organelle.

Glycoside conjugation appears to be the most common xenobiotic
conjugation reaction in both plants and mammals; however, plants
form glucose rather than glucuronic acid conjugates and glucoside
conjugation tends to be far more complex in plants. Plants rarely
form sulfate ester conjugates. Glutathione conjugation occurs in
both plants and mammals, probably with similar frequency and with a
similar range of compounds. Glutathione conjugates undergo further
catabolism in both plants and mammals. Although some differences
have been observed, the overall process of glutathione conjugate
catabolism is very similar in both. Amino acid conjugation is not
common in plants, but occurs as a fairly general reaction with a
restricted class of compounds. Compounds that form amino acid conjugates in both plants and mammals have structurally similar characteristics. Xenobiotics are rarely acetylated in common plant
species, but the formation of malonyl conjugates is relatively common. This appears to be a striking difference between plants and
mammals. Several recent reports suggest that plants occasionally
form lipophilic conjugates, as has also been reported in mammals.
In some plant species, these conjugates may be highly unusual in
structure.

In this review, conjugation reactions utilized in xenobiotic
metabolism in plants will be discussed in reference to functional
groups, phase I reactions necessary to produce a functional group
suitable for conjugation, relative rates of reactions, competing
metabolic pathways, frequency of occurence, plant species, stability
of conjugates, and the relationship between metabolism and herbicide
selectivity. Pesticides discussed herein are listed in Table I.

TABLE I.   Nomenclature of Xenobiotics Mentioned in Tables/Text

Abscisic Acid	5-(1-Hydroxy-2,6,6-trimethyl-4-oxo-2-cyclo-hexen-1-yl)-3-methyl-2,4-pentadienoic acid
Acifluorfen	5-(2-Chloro-4-trifluoromethylphenoxy)-2-nitrobenzoate
Alachlor	2-Chloro-2',6'-diethyl-N-(methoxymethyl)-acetanilide
Atrazine	2-Chloro-4-(ethylamino)-6-(isopropylamino)-s-triazine
Barban	4-Chloro-2-butynyl-3'-chlorocarbanilate
BAY NTN 9306	O-Ethyl O-[4-(methylthio)phenyl] S-propyl phosphorodithioate
Bentazon	3-Isopropyl-(1H)-benzo-2,1,3-thiadiazin-4-one 2,2-dioxide
Bidisin	Methyl 2-chloro-3-(4-chlorophenyl)propionate
Botran	2,6-Dichloro-4-nitroaniline
BPMC	2-sec-Butylphenyl N-methylcarbamate
Butachlor	2-Chloro-2',6'-diethyl-N-butoxymethyl-acetanilide
Buthidazole	3-[5-(1,1-Dimethylethyl)-1,3,4-thiadiazol-2-yl]-4-hydroxy-1-methyl-2-imidazolidinone
Buturon	3-(4'-Chlorophenyl)-1-methyl-1-(1-methylprop-2-ynyl)urea
Captan	1,2,3,6-Tetrahydro-N-(trichloromethylthio)-phthalimide
Carbaryl	1-Naphthyl N-methylcarbamate
Carbofuran	2,3-Dihydro-2,2-dimethylbenzofuran-7-yl methylcarbamate
Carboxin	5,6-Dihydro-2-methyl-1,4-oxathiin-3-carboxanilide
CDAA	N,N-Diallyl-2-chloroacetamide
Chloral Hydrate	2,2,2-Trichloro-1,1-ethanediol
Chloramben	3-Amino-2,5-dichlorobenzoic acid
Chlorpropham	Isopropyl-3'-chlorocarbanilate
Chlorsulfuron	2-Chloro-N-[(4-methoxy-6-methyl-1,3,5-triazin-2-yl)aminocarbonyl]benzenesulfonamide
Chlortoluron	3-(3-Chloro-4-methylphenyl)-1,1-dimethylurea
Cisanilide	cis-2,5-Dimethyl-1-pyrrolidinecarboxanilide
Credazine	3-(2'-Methylphenoxy)pyridazine

Table I.   Continued

Cypermethrin	Cyano(3-phenoxyphenyl)methyl 3-(2,2-dichloro-ethenyl)-2,2-dimethylcyclopropanecarboxylate
Cyprazine	2-Chloro-4-(cyclopropylamino)-6-(isopropyl-amino)-s-triazine
Cytrolane	Diethyl (4-methyl-1,3-dithiolan-2-ylidene)-phosphoramidate
2,4-D	(2,4-Dichlorophenoxy)acetic acid
Diamidafos	N,N'-Dimethylphenylphosphorodiamidate
Diazinon	O,O-Diethyl O-(2-isopropyl-6-methylpyrimidin-4-yl)phosphorodithioate
DIB	2-(2,4-Dichlorophenoxy)isobutyric acid
Dichlobenil	2,6-Dichlorobenzonitrile
Dichlofluanid	N-Dichlorofluoromethylthio-N',N'-dimethyl-N-phenylsulphamide
Diclofop-methyl	Methyl 2-[4-(2',4'-dichlorophenoxy)phenoxy]-propanoate
Dieldrin	1,2,3,4,10,10-Hexachloro-6,7-epoxy-1,4,4a,5,-6,7,8,8a-octahydro-1,4-endo-5,8-exo-dimethanonaphthalene
Dimethametryn	2-(1,2-Dimethylpropylamino)-4-ethylamino-6-methylthio-s-triazine
Dinoben	2,5-Dichloro-3-nitrobenzoic acid
Diphenamid	N,N-Dimethyl-2,2-diphenylacetamide
Diuron	3-(3,4-Dichlorophenyl)-1,1-dimethylurea
EPN	O-Ethyl O-(4'-nitrophenyl) phenylphosphonothioate
EPTC	S-Ethyl N,N-dipropylthiocarbamate
Flamprop-isopropyl	Isopropyl N-benzoyl-N-(3-chloro-4-fluorophenyl)-2-aminopropionate
Fluorodifen	2,4'-Dinitro-4-trifluoromethyl diphenylether
Fluvalinate	α-Cyano-3-phenoxybenzyl 2-[2-chloro-4-(trifluoromethyl)anilino]-3-methylbutanoate
GS-13529	2-Chloro-4-(ethylamino)-6-(tert-butylamino)-s-triazine
Hymexazol	3-Hydroxy-5-methylisoxazole
IAA	2-(Indol-3-yl)acetic acid
Isouron	3-(5-tert-Butyl-3-isoxazolyl)-1,1-dimethylurea
Isoxathion	O,O-Diethyl O-(5-phenyl-3-isoxazolyl)-phosphorothionate
Isoxazolinone	Isoxazolin-5-one
Maleic Hydrazide	1,2-Dihydro-3,6-pyridazinedione

Continued on next page

Table I. Continued

MCPA	(4-Chloro-2-methylphenoxy)acetic acid
Mephosfolan	Diethyl(4-methyl-1,3-dithiolan-2-ylidene)-phosphoramidate
Methazole	2-(3,4-Dichlorophenyl)-4-methyl-1,2,4-oxadiazolidine-3,5-dione
Methidathion	S-2,3-Dihydro-5-methoxy-2-oxo-1,3,4-thiadiazol-3-ylmethyl O,O-dimethyl phosphorodithioate
Metolachlor	α-Chloro-2'-ethyl-6'-methyl-N-(1-methyl-2-methoxyethyl)acetanilide
Metribuzin	4-Amino-6-tert-butyl-4,5-dihydro-3-methylthio-1,2,4-triazin-5-one
Mobam	4-Benzothiophene N-methylcarbamate
Molinate	S-Ethyl N,N-hexamethylenethiocarbamate
Monolinuron	3-(4-Chlorophenyl)-1-methoxy-1-methylurea
NAA	1-Naphthaleneacetic acid
Nitrofen	2,4-Dichlorophenyl 4-nitrophenyl ether
Oxamyl	N,N-Dimethyl-2-methylcarbamoyloxyimino-2-(methylthio)acetamide
PCNB	Pentachloronitrobenzene
Perfluidone	1,1,1-Trifluoro-N-[2-methyl-4-(phenylsulfonyl)-phenyl]methanesulfonamide
Permethrin	3-Phenoxybenzyl (1RS)-cis, trans-3-(2,2-dichlorovinyl-2,2-dimethylcyclopropanecarboxylate
Picloram	4-Amino-3,5,6-trichloropicolinic acid
Prometryn	2,4-bis(Isopropylamino)-6-methylthio-s-triazine
Pronamide	3,5-Dichloro-N-(1,1-dimethyl-2-propynyl)-benzamide
Propachlor	2-Chloro-N-isopropylacetanilide
Propanil	3',4'-Dichloropropionanilide
Propazine	2-Chloro-4,6-bis(isopropylamino)-s-triazine
Propham	Isopropyl carbanilate
Pyrazon	5-Amino-4-chloro-2-phenylpyridazin-3-one
R-25788	N,N-Diallyl-2,2-dichloroacetamide
Ro 12-0470	2-Naphthylmethyl cyclopropanecarboxylate
Simazine	2-Chloro-4,6-bis(ethylamino)-s-triazine
Solan	N-(3-Chloro-4-methylphenyl)-2-methylpentanamide
Sweep	Methyl N-(3,4-dichlorophenyl)carbamate
2,4,5-T	(2,4,5-Trichlorophenoxy)acetic acid

Table I. Continued	
Tridiphane	2-(3,5-Dichlorophenyl)-2-(2,2,2-trichloroethyl)oxirane
Triforine	1,4-bis(2,2,2-Trichloro-1-formamidoethyl)-piperazine
Triton	Polyethylene glycol 4-isooctylphenyl ether
Zectran	4-Dimethylamino-N-methyl-3,5-xylylcarbamate

Glycoside conjugates

Higher plants have an extremely well-developed capacity to convert various endogenous and xenobiotic substrates to glucose conjugates (4-6,11,12,21,22). The functional groups most frequently involved in glucose conjugation in plants are HO-X, HOOC-X, H_2N-X, or HN=X; in addition, there are several reports of xenobiotic-glucosides formed from HS-X and HON-X intermediates. Many xenobiotics that do not contain the functional groups described can be metabolized to glucose conjugates after a functional group is introduced by a phase I reaction. Phase I reactions that lead to glucose conjugation in plants are described in Table II.

TABLE II. Phase I Reactions that Produce Metabolites Susceptible to Glucose Conjugation in Higher Plants

Phase I Reaction	Class of Xenobiotic	Metabolite Susceptible to Glucoside Conjugation
Hydrolysis	Carbamate	Phenols
	Anilide	Anilines, Carboxylic Acids
	Phosphorothioate	Phenols
	Ester	Carboxylic Acids, Alcohols
	N-Hydroxyl Deriv.	N-Hydroxyls
GSH Conjugation	Diphenyl ether	Phenols
Reduction	Nitroaromatic	Anilines
Oxidation	Alkyl	Alcohols
	Aryl	Phenols
Isomerization	Cyclic amides	Alcohols, Amines

O-Glucosides

Phenols and alcohols, or xenobiotics that are metabolized through phenols and alcohols as intermediates, are most commonly metabolized to β-O-D-glucosides (3,4,11,12). The ability of different plant families to form glucosides from phenols was first extensively investigated by Pridham (25), who showed that 20 out of 23 species

of angiosperms assayed with p- and m-dihydroxybenzene formed high
levels of glucosides--the only angiosperms that did not form glucosides were three aquatic species. It was also shown that five
species of gymnosperms had a high capacity to form glucosides, but
generally lower or variable rates of glucosylation were observed in
ferns, mosses and a liverwort. No glucosylation was observed in ten
species of algae and two species of fungi. More recent studies with
a wide variety of pesticides in various agronomic and horticul-
turally important plant species has clearly established O-glucosides
as the most common class of xenobiotic metabolites in plants (Table
III). Generally, these O-glucosides have been characterized as β-

Table III. Xenobiotics Metabolized to O-Glucosides in Plants

Xenobiotic	Phase I Reaction[a]	Species	Methods of Isolation[b] and Characterization	REF
Acifluorfen	GSH	Soybean	^{14}C,HPLC,TLC,β-Glcase, Glc-Anal,Acet-Conj-MS	94
BAY NTN 9306	HYD	Cotton	^{14}C,TLC,Synth,H$^+$-Hyd, β-Glcase,Aglycone-Co-TLC	164
Bentazon	OXD	Rice	^{14}C,TLC,β-Glcase,H$^+$-Hyd, Glc-Anal,Aglycone-MS,NMR	165
BPMC (Bassa)	OXD	Rice	^{14}C,TLC,H$^+$-Hyd,β-Glcase, Aglycone-Co-TLC,MS	166
Carbofuran	OXD	Bean	^{14}C,TLC,H$^+$-Hyd,β-Glcase, Aglycone-Co-TLC	167
Carboxin	HYD OXD	Peanut	^{14}C,CC,HPLC,Synth, Acet-Conj-MS	121
Chloral Hydrate	---	Gourd	Synth,Acetylate,MP,Spec. Rotation,Elemental-Anal	168
Chlorpropham	OXD	Soybean	^{14}C,CC,H$^+$-Hyd,β-Glcase, Acet-Aglycone-GC,NMR,MS	161
Chlorsulfuron	OXD	Wheat,Wild Oat,Grasses	^{14}C,HPLC,H$^+$-Hyd,β-Glcase, Aglycone-MS,NMR,IR	31
Chlortoluron	OXD	Wheat,Wild Oat,Cotton	^{14}C,TLC,β-Glcase, Aglycone-Co-TLC	41
Cisanilide	OXD	Carrot, Cotton	^{14}C,CC,TLC,β-Glcase,Glc- Anal, Aglycone-IR,MS	169
Credazine	HYD ISO	Tomato Barley	^{3}H,CC,TLC,Synth,β-Glcase, Acet-Conj-IR,Aglycone-IR	170
Cytrolane	OXD	Cotton	^{14}C,CC,TLC,β-Glcase, Aglycone-MS	171
2,4-D	OXD	Soybean, Corn	^{14}C,TLC,β-Glcase, Aglycone-Co-TLC,PC	172

Table III. Continued

Xenobiotic	Phase I[a] Reaction	Species	Methods of Isolation[b] and Characterization	Ref
Diamidafos	HYD	Tobacco	^{14}C,TLC,PC,β-Glcase, Aglycone-Co-TLC	173
Dichlofop-Methyl	HYD OXD	Wheat, Oat	^{14}C,TLC,HPLC,H$^+$-Hyd,β-Glcase,Aglycone-Glc-Anal	32
Diphenamid	OXD	Tomato, Pepper	^{14}C,TLC,H$^+$-Hyd,β-Glcase, Glc-Anal,Acet-Conj-MS	68 69
Diuron	OXD	Cotton, Barley, Wheat	^{14}C,TLC,β-Glcase, Aglycone-Co-TLC	41
Ethanol	---	Pea	CC,TLC,H$^+$-Hyd,β-Glcase, GLC-Anal	213
Fluorodifen	GSH	Peanut	CC,β-Glcase,Acet-Conj MS,NMR	174
Hymexazol	---	Rice, Cucumber, Tomato	^{14}C,CC,TLC,β-Glcase, Glc-Anal,TMS-Conj-MS, IR,NMR	175
Isouron	OXD	K. Bean Sugar Cane	^{14}C,TLC,β-Glcase, Aglycone-Co-TLC,MS	40
Isoxathion	HYD	Cabbage, Bean	^{14}C,Synth,TLC,H$^+$-Hyd,β-Glcase,Acet-Conj-IR,NMR	49
Maleic Hydrazide	ISO	Tobacco	^{14}C,TLC,β-Glcase,Glc-Anal,Acet-Conj-MS	176
Monolinuron	OXD	Cress, Potato, Spinach	^{14}C,TLC,β-Glcase, Aglycone-Co-TLC	177
Pentachlorophenol	---	Soybean	^{14}C,HPLC,β-Glcase, Acet-Conj-MS	36
Perfluidone	OXD	Peanut	^{14}C,CC,β-Glcase,Glc-Anal, Aglycone-Co-TLC,HPLC,GC,MS	178
Permethrin	HYD	Cotton, Bean	^{14}C,Synth,TLC,β-Glcase, Aglycone-Co-TLC	179
Phenols (m- or p-Dihydroxybenzene)	---	Angiosperms Gymnosperms	PC,Diazotization	25
m-Phenoxybenzyl Alcohol	---	Cotton	^{14}C,TLC,β-Glcase, Acet-Conj-MS	30
Triton X-100	---	Pea	^{14}C,TLC,β-Glcase, Acet-Conj-MS	148

[a] Phase I reaction preceeding conjugation: OXD=oxidation, HYD= hydrolysis, GSH=glutathione conj, ISO=isomerization.

[b] Acet=Acetylated; CC or PC=column or paper chromatography; Conj= Conjugate; Glc=glucose; Glc-Anal=Glucose Analysis; β-Glcase= β-glucosidase; H$^+$-Hyd=Acid Hydrolysis; Synth=Synthesis.

glucosides by hydrolysis with β-glucosidase, synthesis and chromatographic comparison (12), or in more recent studies, by NMR analysis as described by Feil in Chapter 9. Although plants are capable
of forming galactose, glucuronic acid, and other carbohydrate conjugates with endogenous substrates such as flavones (27-29), mono-
saccharide conjugates of xenobiotics involving carbohydrate moieties
other than glucose have been extremely rare. However, as metabolism
studies are conducted with more diverse chemicals and plant species,
and with more sophisticated methods of analysis, it seems likely
that other carbohydrate moieties will be found conjugated to xenobiotics.

    In plants, β-O-glucosides of xenobiotics can be formed very
rapidly. For example, m-phenoxybenzyl alcohol was metabolized to
O-glucosides in excised cotton leaves in ca. 90% yield within 8 hr
following treatment (30). The most common reaction(s) that appear
to compete with glucoside formation in the metabolism of phenols are
those that lead to bound residue. If a phase I oxidation is
required to generate a free hydroxyl group prior to glucoside formation, the free phenol or alcohol may not be observed or the phenol
may be observed only at a very low concentration (3). It appears
that uptake or oxidation are more frequently the limiting steps.
Chlorsulfuron is metabolized in wheat by ring-hydroxylation followed
by glucoside formation (Equation 1). In chlorsulfuron-treated
wheat, 60% of the herbicide was in the form of the 5-hydroxygluco-
side 24 hr after treatment and no free chlorsulfuron was observed.
The half-life of chlorsulfuron was estimated to be only 2-3 hr (31).
Diclofop-methyl is metabolized in wheat by hydrolysis of an ester-
group followed by ring-hydroxylation and subsequent formation of a
β-O-glucoside (Equation 2). The glucoside accounted for 85% of the
herbicide 24 hr after treatment and the free ring-hydroxylated
form(s) accounted for only 3.1% (32).

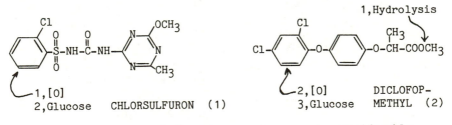

    When more than one hydroxyl group is present, considerable
variation may be observed among plant species regarding the site
that will become glucosylated. When cell cultures of eight species
were treated with o-hydroxybenzyl alcohol, six species formed the
glucoside at the benzylic hydroxyl group (Nicotiana, Datura,
Duboisia, Catharanthus, and Bupleurum, 97-100%) while two species
formed primarily the phenolic glucoside (Gardenia and Lithospermum,
69 and 81%, respectively) (33).
    Most β-O-D-glucosides are thought to be formed by glucosyl
transferase enzymes that require UDPG as the glucosyl donor.
However, a trans-glucosylation system that can utilize certain glucosides in the formation of new O-glucosides and a system that
requires ATP and CoA for the synthesis of some glucose ester conjugates have been reported (12,21). Although glucosyl transferase

enzymes that utilize phenols as substrates have been isolated from
several plant species, the conversion of xenobiotic alcohols and
phenols to β-O-D-glucosides has not been extensively studied at the
enzymatic level (12,21). A soluble UDPG:glucosyl transferase from
germinating mung bean was shown to have a broad substrate specificity for various phenols and alkyl alcohols. Substrate specificity
was based on molecular size (34). A soluble UDPG:glucosyl transferase from tomato utilized various hydroxy-cinnamic acids as
substrates and catalyzed the formation of O-glucosides at pH 8 and
glucose esters at pH 7 (35). Particulate UDPG:glucosyl transferases
that utilize sterols as substrates have also been reported (21).
Recently, a soluble UDPG:glucosyl transferase that utilized pentachlorophenol as a substrate was isolated from wheat and soybean cell
cultures (36). A non-UDPG-dependent enzyme was isolated from pea
that catalyzes the trans-glucosylation of phenols, but the role of
this enzyme system in xenobiotic metabolism has not been well
explored (37,38).

A direct relationship has been observed between the ability of
some plant species to form O-glucosides of herbicides and resistance
of those species to the herbicides; however, in most cases a phase I
reaction preceeds glucose conjugation and it is not known whether
the phase I reaction or conjugation results in herbicide detoxification. Chlorpropham is metabolized in plants by ring-hydroxylation
and subsequent conjugation with glucose (Equation 3). In vitro, the

CHLORPROPHAM   (3)   ISOURON   (4)

phenolic intermediate was at least as toxic by several criteria as
the parent herbicide, but the O-glucoside was not biologically
active (39). In contrast, isouron, a dimethylurea herbicide, was
rapidly metabolized in resistant sugarcane by oxidation and subsequent glucoside conjugation (Equation 4). In susceptible kidney
bean, metabolism occured more slowly by N-demethylation. Oxidation
of the tert-butyl group resulted in loss of activity as a photo-
synthetic inhibitor in Chlorella, but it was not determined whether
Chlorella formed an O-glucoside of the oxidized form of isouron. N-
demethylation resulted in only a partial loss of activity (40,211).
Resistance of plants to chlorsulfuron herbicide is related to metabolism. Resistant species rapidly metabolized chlorsulfuron by
ring-oxidation followed by O-glucoside formation while susceptible
species did not metabolize chlorsulfuron at an appreciable rate (31)
(Equation 1). It was not proven whether oxidation or glucose conjugation resulted in detoxification. The phenylurea herbicide,
chlortoluron, was rapidly metabolized in resistant species by oxidation and glucoside conjugation of a methyl group substituted on the
phenyl ring, but metabolism in susceptible species occured more
slowly by alternate routes such as N-demethylation (41). Diclofop-
methyl herbicide was more rapidly metabolized by ring-hydroxylation
followed by glucoside formation in a resistant species than by glucose ester formation in susceptible species (32) (Equation 2).

The stability of most xenobiotic O-glucosides in plants has not
been studied over the life of the plant. In some cases, they are
clearly stable for moderate periods of time. In other cases, they
may be metabolized rapidly to more complex glycosides as discussed
latter. A recent study on the metabolism of the herbicide chlortoluron in wheat suggested that the O-glucoside of this herbicide
was relatively stable until the plant reached maturity. At that
time, the O-glucoside appeared to be hydrolyzed and the hydroxylated
chlortoluron moiety then underwent additional oxidation (41).
Studies with m-phenoxybenzyl alcohol in cotton leaves suggested that
the β-O-glucoside of this xenobiotic undergoes rapid turn-over
within the cell (30).

N-Glucosides

Aromatic or heterocylic xenobiotics that contain a primary or secondary amino group can be metabolized in plants to N-glucosides (Table
IV). Alternate routes of metabolism of amino groups utilized by
plants include conjugation with malonic acid, or perhaps more commonly, the formation of bound residue. N-Glucoside formation is a
common mechanism of xenobiotic metabolism and has been observed in a
variety of plant species. In some cases, resistance or susceptibility to a herbicide has been correlated to the ability of a plant
species or cultivar to detoxify a herbicide by formation of an N-
glucoside. Chloramben herbicide appeared to be metabolized almost
exclusively to an N-glucoside in resistant soybean, but in susceptible barley metabolism was much slower and only 15% N-glucoside was
formed (42). Further studies with tissue sections treated for 7 hr
with chloramben verified that N-glucoside formation was a major
route of metabolism in resistant species (morning glory, 76%;
squash, 84%; snapbean, 67%; soybean, 62%), but was less signficant
in susceptible species (velvet leaf, 23%; barley, 19%; giant fox-
tail, 15%) (26). The N-glucoside of chloramben is a terminal residue in soybean grown to maturity (44). Dinoben, the nitro-analogue
of chloramben, is metabolized by reduction of the nitro-group
followed by N-glucoside formation (45). Metribuzin herbicide is
metabolized to an N-glucoside in tomato and the level of resistance
to metribuzin among different tomato cultivars was correlated to the
level of UDPG:glucosyl transferase in the young tomato seedlings
(46). The N-glucoside of metribuzin is not a terminal residue, but
is rapidly metabolized to a 6-O-malonylglucose derivative (46). The
metabolism of pyrazon herbicide to an N-glucoside is also thought to
be related to plant resistance, but evidence for this is not as good
as with the previous two herbicides (47). The primary route of
picloram metabolism in sunflower is by formation of the N-glucoside
of the primary amino group (15%, 1 day; 55%, 3 days; and 72%, 7
days). Sunflower is very susceptible to picloram and it was proposed that metabolism occured too slowly to prevent phytotoxicity
(48). The N-glucoside of picloram was also formed in French bean,
4%; barley, 2.4%; and cucumber, 20%; however, the primary route of
picloram metabolism in most species other than sunflower appears to
be by glucose ester formation (48). The N-glucoside of picloram was
not phytotoxic and appeared to be resistant to further metabolism in
French bean. Another excellent example of N-glucoside formation in

Table IV.   Xenobiotics Metabolized to N-Glucosides in Plants[a]

Xenobiotic	Phase I Reaction	Species	Methods of Isolation and Characterization	REF
Chloramben	---	Squash, Corn Soybean, etc	^{14}C, CC, TLC, H$^+$-Hydrol, Glc-Anal, Acet-Conj-MS, NMR	43
Dinoben	RED[b]	Soybean	^{14}C, TLC, Synth, H$^+$-Hydrol, Aglycone-Co-TLC	45
Halogenated Anilines	---	Soybean[c]	TLC, H$^+$-Hydrol, Color Analysis	50
Hymexazol	ISO	Rice Cucumber	^{14}C, CC, TLC, H$^+$-Hydrol, Glc-Anal, TMS-Conj-MS, IR, NMR	175
Isoxathion	HYD ISO OXD	Cabbage Bean	^{14}C, Synth, TLC, H$^+$-Hydrol, Acet-Conj-IR, NMR	49
Isoxazolinone	---	Sweet Pea, Pea[c]	^{14}C, TLC, Color Analysis	181
Metribuzin	---	Tomato	^{14}C, TLC, HPLC, H$^+$-Hydrol, Glc-Anal, Acet-Conj-MS, NMR, TMS[d]-Conj-NMR	46
Picloram	---	Sunflower Cucumber	^{14}C, CC, PC, TLC, H$^+$-Hydrol, Glc-Anal, Acet-Conj-IR, MS	48
Propanil	HYD	Rice	^{14}C, Synth, PC, TLC, H$^+$-Hydrol, Glc-Anal, TMS-Conj-GC, IR	182
Pyrazon		Red Beet	^{3}H, CC, TLC, H$^+$-Hydrol, Glc-Anal, Aglycone-Glc-GC, IR	47

[a] Abbreviations are described in Table III.
[b] Reduction (RED).
[c] Enzymatic synthesis from plant extracts.
[d] TMS=trimethyl silyl ether.

plants is the metabolism of the insecticide, isoxathion in bean, cabbage, and Chinese cabbage.   In bean, N-glucoside formation occured slowly (4.3%, 1 day; 21.6%, 6 days; and 54.1%; 15 days), but it was the major route of metabolism (49) (Equation 5).

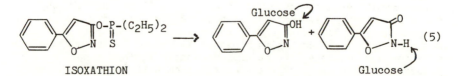

ISOXATHION   (5)

N-Glucosides can be formed by soluble UDPG:glucosyl transferase enzymes comparable to the transferases that form O-glucosides.   A soluble UDPG:glucosyl transferase with a broad substrate specificity for arylamines was isolated from soybean (50) and a similar enzyme

that catalyzed the formation of an N-glucoside of metribuzin was
detected in tomato (46). A β-configuration would be expected for N-
glucosides formed by a UDPG:glucosyl transferase system. NMR evidence supporting such a configuration was reported for the N-
glucoside of metribuzin (46). N-Glucosides appear to be resistant
to hydrolysis by β-glucosidase (12,49) and in some cases they may be
somewhat resistant to acid hydrolysis (12). Although N-glucosides
can undergo additional metabolism in plants, e.g., to malonyl-
glucosides (46), they are considered non-toxic and relatively stable
in plant tissue (12).

Glucose ester conjugates

 Xenobiotic and endogenous substrates that contain free or potential
carboxyl groups are commonly metabolized in plants to O-1 glucose
ester conjugates. At least 13 pesticides as well as a variety of
endogeneous substrates are known to be metabolized to glucose esters
in various plant species (Table V). Amino acid conjugation and formation of bound residue are two metabolic reactions involving the
carboxyl group that appear to compete with glucose ester conjugation; however, reactions not involving the carboxyl group, such as
ring hydroxylation/glucoside formation or N-glucoside formation may
also compete with glucose conjugation (32,43). Glucose ester conjugation is more common and important in plants than amino acid conjugation (6). Most xenobiotics known to form glucose ester conjugates
in plants are aromatic or heterocyclic carboxylic acid derivatives.
     Glucose ester conjugates can be formed very rapidly in plants:
m-phenoxybenzoic acid was metabolized in 81% yield to a glucose
ester conjugate within 18 hr in excised cotton leaves (51), N-
benzoyl-N-(3-chloro-4-fluorophenyl)-DL-alanine was metabolized in
85% yield to simple and complex glucose esters within 24 hr in oat
(52), and diclofop-methyl was metabolized by hydrolysis and glucose
ester conjugation in 54% yield within 24 hr in oat root (32).
     In plants, most xenobiotic glucose ester conjugates are pro-
bably formed by a UDPG:glucosyl transferase mechanism (35,53-56),
but an ATP/CoA-dependent system for the synthesis of the glucose
ester of IAA has also been reported (56). The enzymatic synthesis
of a variety of glucose esters, primarily with endogenous substrates, has been accomplished with crude and partially purified enzyme
systems isolated from several plant tissues (35,53-56). A soluble
UDPG:glucosyl transferase from tomato fruit catalyzed the synthesis
of both glucosides (pH 8) and glucose esters (pH 7) (35). A single
enzyme was thought to catalyze both reactions. The enzyme displayed
a broad substrate specificity for hydroxylated benzoic and cinnamic
acid derivatives and had Km values of 0.8 to 10 μM for both the
donor and acceptor substrates. A soluble UDPG:glucosyl transferase
enzyme from oak leaves that utilized a variety of benzoic and cinnamic acid derivatives had much higher Km values for both donor and
acceptor substrates (53). It did not appear to form simple O-
glucosides. The glucosyl transferases from both tomato fruit and
oak leaves were unstable. The glucose ester conjugation reaction
catalyzed by these enzymes was either highly reversible (53) or the
glucose esters were readily hydrolyzed by esterases present in the
enzyme preparations (35). Chloramben herbicide is metabolized to a
glucose ester in some plant species and this metabolite is hydro-

Table V. Xenobiotics Metabolized to Glucose Esters in Plants[a]

Xenobiotic	Phase I Rxn	Species	Methods of Isolation and Characterization	REF
Abscisic Acid	---	Xanthium, Spinach	^{3}H,CC,HPLC,TLC,β-Glcase, Acet-Conj-CC,MS,NMR,UV	183
Carboxin	HYD	Peanut Cell Culture	^{14}C,CC,HPLC, Acet-Conj-GC,MS	121
Chloramben	---	Corn,Barley, Barnyard-grass,etc.	^{14}C,CC,TLC,H$^+$-Hyd,Glc-Anal,Acet-Conj-MS,NMR	43
Cypermethrin	HYD	Lettuce, Cotton	^{14}C,PC,TLC,HPLC,H$^+$-Hyd, Glc-Anal,Acet-Conj-MS	73
Cyclohexanecarb-oxylic Acid	---	Bush Bean	^{14}C,TLC,Synth,Base-Hyd, Aglycone-Co-TLC	130
DIB	---	Tomato	^{14}C,TLC,H$^+$-Hyd,Glc-Anal, Acet-Conj-MS,NMR,IR	70
Diclofop-Methyl	HYD	Oat	^{14}C,TLC,HPLC,H$^+$-Hyd, Synth,Acet-Conj-MS	32
2,4-D	---	Soybean, Corn	^{14}C,TLC,β-Glcase, Aglycone-Co-TLC,PC	172
Flamprop-Isopropyl	HYD	Barley	^{14}C,TLC,β-Glcase, Aglycone-Co-TLC	184
Fluvalinate	HYD	Cabbage	^{14}C,TLC,HPLC,Acet-Conj-MS,NMR	185
Hydroxcinnamic Acids	---	Duckweed	Co-HPLC	186
IAA	---	Pine	^{14}C,TLC,β-Glcase, Aglycone-Co-TLC,Conj-NMR	187
Permethrin	HYD	Cotton, Bean	^{14}C,Synth,TLC,β-Glcase, Aglycone-Co-TLC	179
3-Phenoxybenzoic Acid	---	Soybean, Grape, Tomato	^{14}C,TLC,HPLC,H$^+$-Hyd, IR,Acet-Conj-NMR,MS	51

[a] Abbreviations are described in Table III.

lyzed readily to free chloramben in vitro and in vivo (43). The formation of glucose and myo-inositol esters of IAA in young corn seedlings was also reported to be a reversible process (56).

Several herbicides are metabolized to glucose ester conjugates in species that are susceptible to the herbicides. Diclofop-methyl is metabolized to a glucose ester in susceptible oat, but in resistant wheat it is metabolized more rapidly by ring-hydroxylation followed by glucoside formation (32). Chloramben is metabolized to a glucose ester and bound residue in susceptible barley, but it appears to be metabolized more rapidly to an N-glucoside in resis-

tant morning glory (43). It is not clear whether the phytotoxicity
in the two examples above was due to an apparent lower rate of metabolism or the reversibility of the glucose ester conjugation reaction. Glucose esters of natural plant hormones and auxin are
thought to serve as an inactive chemical reservoir of these compounds and it has been suggested that glucose esters of herbicides
may function in a similar manner (12). As with other xenobiotic
glucose conjugates, glucose ester conjugates can be further metabolized by conjugation reactions that involve the glucose moiety.
These additional metabolic processes may not significantly stabilize
the O-1 glucose ester bond from non-enzymatic hydrolysis, but they
may block enzymatic processes such as β-glucosidase hydrolysis or
the reverse reaction in the synthesis of the O-glucose ester. Thus,
the formation of these more complex metabolites may result in
detoxification of the xenobiotic.

Most xenobiotic glucose ester conjugates appear to be β-O-1
esters; however, the glucose ester of chloramben appeared to be an
α-O-1 glucose ester (43). The O-1 glucose esters can undergo transesterification under mildly basic conditions (NaHCO₃) (52); acyl
migration to yield O-2, O-4, and O-6 glucose esters (56); or they
can undergo ammonolysis to yield amide derivatives of the aglycone
(43). As a result, artifacts of glucose esters have occasionaly
been reported as metabolites (52).

N-O-Glucosides

At least two well-documented cases of plant xenobiotic metabolism to
N-O-glucosides have been reported (57,58). Oxamyl insecticide was
converted to an N-O-glucoside in tobacco, young peanut plants,
alfalfa, and the fruit of orange and tomato (57) (Equation 6). This

$$(CH_3)_2N-\overset{O}{\overset{\|}{C}}-\underset{\underset{SCH_3}{|}}{C}=N-O-\overset{O}{\overset{\|}{C}}-NHCH_3 \ ----\!\!> \ ----\!\!> \ (CH_3)_2N-\overset{O}{\overset{\|}{C}}-\underset{\underset{SCH_3}{|}}{C}=N-O-Glucose \qquad (6)$$

OXAMYL

glucoside was a major metabolite (35-90% of the residue) in orange
fruit, tobacco, young peanut plants, and alfalfa. In some tissues,
the mono-N-demethylated form of this glucoside was also observed.
These N-O-glucosides were not hydrolyzed by β-glucosidase. In apple
fruit, potato tubers, and hay from mature peanut plants, more polar
metabolites were produced which apparently yielded these glucosides
upon hydrolysis with β-glucosidase. These more polar residues were
suggested to be polyglucosides of the simple N-O-glucosides (57).
The simple glucosides were identified by chromatography, EI/MS,
GC/MS, and synthesis.

Methazole herbicide was metabolized to an N-O-glucoside (13%)
and the 6-O-malonyl ester of the glucoside (22%) in spinach plants
within 48 hr following treatment (58). As with oxamyl, the R-N-OH
group was apparently formed by hydrolysis prior to glucoside formation (Equation 7). Both the glucoside and the malonyl ester of
the glucoside were identified by a combination of techniques
including ¹H- and ¹³C-NMR, CI/MS, β-glucosidase hydrolysis, and
synthesis. In contrast to the mono-N-O-glucosides of oxamyl, the

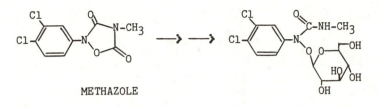

METHAZOLE

(7)

N-O-glucoside of methazole was hydrolyzed by β-glucosidase; however, the 6-O-malonyl ester of this glucoside was not hydrolyzed.

S-GLUCOSIDES

S-Glucoside conjugates have only rarely been reported as xenobiotic metabolites in higher plants; however, glucosinolates respresent a major class of natural products found in cruciferous plants (59-61). In Tropaeolum majus L., glucosinolates are synthesized from thiohydroxamates by a UDPG:glucosyl transferase that accepts a wide variety of thiohydroxamates as glucose acceptors. The products of these reactions are desulfoglucosinolates (Equation 8) which can then be converted to glucosinolates (Equation 9). The glucosyl transferase involved in this reaction can also utilize TDPG as the glucosyl donor, but this substrate is only 10% as efficient as UDPG. UDP-xylose and UDP-galactose did not serve as glycosyl donors. The enzyme was not tested with a broad range of thio-phenols or thioalcohols; therefore, it is not known whether this enzyme could function in the metabolism of a broad range of xenobiotics (62).

$$R/Aryl-\underset{\underset{N-OH}{\|}}{C}-S-glucose$$

DESULFOGLUCOSINOLATE   (8)

$$R/Aryl-\underset{\underset{N-OSO_3H}{\|}}{C}-S-glucose$$

GLUCOSINOLATE   (9)

Two xenobiotics that have been reported to form thioglucosides as metabolites in higher plants are dimethyldithiocarbamate (63) and 4-chloro-4',6-bis(isopropylamino)-6'ethylamino-di(s-triazinyl)-sulfide (64). Radioactive substrates were not used in either of these studies, but in both cases the reported S-glucoside was synthesized and compared to the isolated metabolite. The thioglucoside of dimethyldithiocarbamate was produced in cucumber, broad bean, and potato (Equation 10). It was isolated from potato by

$$(CH_3)_2N-\underset{\underset{S}{\|}}{C}-S^-Na^+ \longrightarrow (CH_3)_2N-\underset{\underset{S}{\|}}{C}-glucose \quad (10)$$

paper chromatography and was identified by chromatographic comparison to the synthetic S-glucoside. The formation of this metabolite appears analogous to the formation of glucose esters. The mechanism by which a thio-glucoside would be formed from the triazine sulfide is not clear (64).

Complex Glucose Conjugates

In higher plants, xenobiotic glucose conjugates frequently undergo
additional metabolism by conjugation of the glucose moiety with
other endogenous substrates. This occurs frequently with various
xenobiotics in many plant species; therefore, complex glucosides
should be considered in any metabolism study where glucose conjugation might be expected (Table VI).
     β-Gentiobiosides [β-(1—>6) glucosyl β-(1—>O) glucosides] were
probably the first complex glucosides of xenobiotics identified in
plants. Both o-chlorophenol and chloral hydrate were metabolized to
gentiobiose conjugates in tomato and it was subsequently shown that
chloral hydrate and closely related compounds were also metabolized
in the same manner in gladiolus and horseradish (65,66). Gentiobiose conjugate formation, illustrated in Equation 11 with o-chloro-

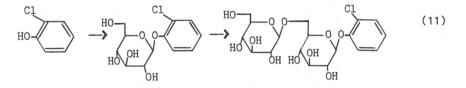

$$(11)$$

phenol, was shown to be a two-step enzymatic reaction by _in vitro_
methods that utilized two distinct UDPG:glucosyl transferase enzymes
isolated from wheat germ (67). The UDPG:glucosyl transferase that
formed the simple β-O-glucoside of the xenobiotic was separated from
the UDPG:glucosyl transferase that formed the β-O-(1—>6) glucosyl
bond (Equation 11). The second glucosyl transferase had a broad
substrate specificity for phenolic glucosides, suggesting that a
wide range of xenobiotics could be metabolized to gentiobiose conjugates. This enzyme did not catalyze the formation of gentiobiose
conjugates from free phenols, nor did it catalyze the formation of
tri- or tetra-saccharide conjugates. Diphenamid herbicide was also
shown to be metabolized to a gentiobiose conjugate in tomato and
this conjugate also appeared to be formed in a sequential manner
through a simple β-O-glucoside intermediate (68,69).
     Although the glucosyltransferase system from wheat germ did not
form tri- and tetra-saccharide conjugates, recent reports suggest
that _in vivo_ systems do form oligosaccharide conjugates of xenobiotics (57,70,71). The plant growth regulator, DIB, was metabolized to a 1—>O glucose ester (6%) as well as a di-(28%), tri-
(44%), and possibly a tetra-glucose conjugate in tomato (70)
(Equation 12). 2,4-D may be partially metabolized in cereal grains
to oligo-saccharide conjugates, as was observed with DIB in tomato
(71). Oxamyl was reported to form oligo-saccharide conjugates in the

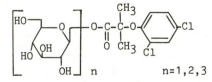

$$(12)$$

Table VI. Complex Carbohydrate Conjugates of Xenobiotics Produced
by Plants

Xenobiotic	Bond[a]	Plant Species	REF

---------------- Gentiobiose Conjugates ----------------
($\underline{\beta}$-$\underline{D}$-glucosyl 1-6 $\underline{\beta}$-$\underline{D}$-glucose)

Xenobiotic	Bond	Plant Species	REF
$\underline{o}$-Chlorophenol	1-0	tomato	66
2,2,2-Trichloroethanol	1-0	tomato, gladiolus,	65
Chloralhydrate	1-0	horseradish	
Chloralcyanohydrin	1-0		
Diphenamid	1-0	tomato, green pepper	68,69
Caviunin (natural plant flavone)	1-0	$\underline{Dalbergia}$ $\underline{sisso}$	188
Phenol[b]	1-0		
$\underline{p}$-Methoxyphenol[b]	1-0		
$\underline{m}$-Dihydroxybenzene[b]	1-0	wheat (germ)	67
$\underline{L}$-Mandelnitrile[b]	1-0		
$\underline{p}$-Dihydroxybenzene[b]	1-0		
2-Hydroxymethylphenol[b]	1-0		
Maleic hydrazide	1-0	$\underline{Malus}$, $\underline{Salix}$, $\underline{Nicotiana}$, wheat	189
DIB	1-ester	tomato	70

---------------- Polyglucoside Conjugates ----------------

Xenobiotic	Bond	Plant Species	REF
DIB	1-ester	tomato	70
2,4-D	1-ester	cereals	71
Oxamyl	1-0-N	apple, potato, peanut	57

------------ Heterodissacharide Conjugates ------------
(glucosylarabinose and glucosylxylose)

Xenobiotic	Bond	Plant Species	REF
$\underline{m}$-Phenoxybenzoic acid	1-ester	grape, cotton, and other species	51
Cypermethrin[c]	1-ester	cotton	73
$\underline{m}$-Phenoxybenzyl alcohol	1-0	cotton	30

----------- 6-0-Malonyl-$\underline{\beta}$-D-Glucose Conjugates --------

Xenobiotic	Bond	Plant Species	REF
Fluorodifen	1-0	peanut	174
Flamprop	1-ester	wheat	75
Cypermethrin[c]	1-ester	cotton	73
Methazole	1-0-N	spinach	58
Acifluorfen	1-0	soybean	94
Mobam	1-0	barley	190

Continued on next page

Table VI. Continued

Xenobiotic	Bond[a]	Plant Species	Ref.
Metribuzin	1-N	soybean	46
Fluvalinate[c]	1-ester	tomato	185
Pentachlorophenol	1-O	soybean, wheat	36

[a] Designates the bond between glucose residue and the xenobiotic
[b] These xenobiotics were tested only by *in vitro* enzymatic methods
[c] Complex pyrethroid insecticide

fruit of apple, peanut, and potato (Equation 6). It was also speculated that chlorpropham was metabolized to oligo-saccharide conjugates in alfalfa (72).
        Arabinosylglucose and xylosylglucose conjugates were reported
as metabolites of m-phenoxy benzoic acid, m-phenoxybenzyl alcohol,
and a cyclopropane carboxylic acid derivative (30,51,73) (Table VI).
In excised cotton leaves, m-phenoxybenzyl alcohol was metabolized to
the β-O-glucoside, an arabinosyl-glucoside and a xylosylglucoside
(30). The addition of the glucose and pentose residues occurred
sequentially. The glucose from these conjugates was exchangeable
with endogenous glucose, with a half-life of only several hours.
Interconversion of metabolites was demonstrated (Equation 13). It

m-phenoxybenzyl alcohol ⇌ m-phenoxybenzyl-O-glucoside
(13)
m-phenoxybenzyl-O-arabinosylglucoside

was speculated that introduction of a pentose moiety might be a
mechanism to block introduction of additional sugar residues (30).
        6-O-Malonate hemi-ester glucose conjugates of xenobiotics are
one of the most common forms of complex glucoside conjugate produced
in plants. They are formed by the action of malonyl CoA transferase
on glucose conjugates in the presence of malonyl CoA as indicated in
Equation 14. At least 9 pesticides have been isolated from 7

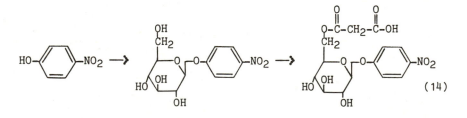

(14)

diverse plant species as malonylglucose conjugates. The xenobiotics
had widely differing structures: O-glucosides, N-glucosides, N-O-
glucosides as well as glucose esters were identified as malonyl-
glucosides. Many natural plant flavones exist as 6-O-malonylgluco-
sides and the malonyl transferase enzymes that catalyze the
formation of malonylglucosides of flavones have been isolated from
parsley and characterized (74). Recently, a malonyltransferase that
catalyzes the formation of the malonylglucoside of pentachlorophenol

was isolated from wheat and soybean (36). It appears that malonyl-
glucoside formation may be a common route of xenobiotic glucoside
metabolism in higher plants.

The malonyl hemi-ester bond is very base-labile and is also
acid-labile; therefore, great care must be taken in the isolation of
intact conjugates (70,75,76). Some mature tissues such as chickpea
and parsley contain esterase enzymes capable of hydrolyzing the 6-O
malonyl group and special precautions may be needed to isolate malonylglucosides from such tissues (74). Because of the lability of
the malonylglucose bond, it is likely that some xenobiotic and fla-
vone malonylglucosides have been incorrectly identified as simple
glucosides. The 6-O-(malonyl)-β-O-glucosides are not readily hydrolyzed by almond emulsin β-glucosidase, but some may be partially
hydrolyzed by hesperidinase (70). In plants, endogenous phenols are
commonly stored in cell vacuoles as glycosides (74). Recently, glycoside conjugates of 2,4-D were also shown to be concentrated within
the vacuole (36). In parsley, flavonoid glycosides are located
exclusively within the cell valuole, primarily as malonyl-glucosides
(74). It was suggested that the addition of the malonyl group might
aid in the transport and storage of the glycoside conjugates (74).
Certainly, compartmentation of xenobiotics as malonyl-glucosides in
the vacuole could serve as an excellent method of isolating xenobiotic metabolites from further interaction with normal metabolic
processes.

Glutathione conjugation

The conjugation of xenobiotics with glutathione (GSH) in higher
plants was first demonstrated to be a major enzymatic detoxication
mechanism with the atrazine herbicide in sorghum, corn, and other
atrazine resistant species (77-80) (Equation 15). Since this

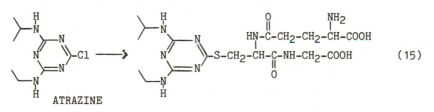

$$(15)$$

ATRAZINE

initial discovery, a variety of pesticides in numerous plant species
have been shown to be metabolized by GSH conjugation (Table VII).
GSH conjugation can be very rapid, approaching completion in 6 hr in
the metabolism of propachlor or atrazine in excised leaves of
several plant species (81,82) and in only 3 hr in the metabolism of
pentachloronitrobenzene (PCNB) in peanut cell suspension culture
(83). In some cases, such as in the metabolism of atrazine, conjugation with GSH is highly species specific, occurring in only a
few species that contain a specific glutathione S-transferase (GST)
that catalyzes the reaction (21,80,84,85). Glutathione conjugation
can thus be a mechanism of herbicide selectivity based on the presence or absence of a GST enzyme.

Propachlor and related α-chloroacetamide herbicides are metabolized by nucleophilic displacement of a chloro-group with GSH (82,

Table VII. Glutathione Conjugation of Xenobiotics in Plants

Xenobiotic	Class of Reaction	Plant Species	REF
Acifluorfen Fluorodifen	Diphenylether displacement	Soybean Peanut	94 84
EPTC	Thiocarbamate sulfoxide displacement	Corn	90
Propachlor CDAA Metolachlor Bidisine Barban 1-Chloro-2,4- dinitrobenzene[a]	Chloro-group displacement	Corn and other species	16 81 86 191 16 24 106
Captan Dichlofluanid	Trihalomethylthio conjugation	Spinach Strawberry	98 98
Atrazine Propazine Simazine Cyprazine GS-13529	Chloro-group displacement	Sorghum,Corn Sugarcane	82
Metribuzin Dimethametryn	Methylsulfoxide displacement	Soybean Rice	95 105
Tridiphane[a]	Epoxide addition	Corn	99
Methidathion Diazinon[a]	Phosphorothioate displacement	Tomato Corn	96 97
PCNB	Nitro-/chloro-group displacement	Many Species	16 102
t-Cinnamic Acid Benzo(a)pyrene[a]	Double bond addition	Pea Pea	107 107

[a] Evidence based on in vitro enzyme studies.

86,87). Propachlor is metabolized by this route in both a resistant
and a susceptible species (Equation 16). Glutathione conjugation of
propachlor occurs very rapidly in vitro in the absence of GST enzymes (78) and it is not clear whether differences in GSH and/or GST
concentrations influence the selectivity of propachlor and related
herbicides. The concentration of GSH in higher plants is estimated

PROPACHLOR   (16)   EPTC   (17)

to range from 0.10 to 0.70 mM in the cytosol and from 1 to 3.5 mM in
the chloroplast (88). If GSH conjugation occurs in the cytosol,
variations in GSH levels between species might account for some differences in selectivity. Tissue age could also be an important fac-
tor. GSH levels vary signficantly as a function of tissue age in
mammals and insects (89) and such variations might also occur in
plants. EPTC herbicide is metabolized in plants by GSH conjugation,
but conjugation is thought to occur after oxidation of the thiocarbamate group to a sulfoxide (90,91) (Equation 17). EPTC causes more
injury to weed species that contain low levels of GSH than to corn
which contains a high level of GSH (88). However, conjugation of
EPTC sulfoxide with GSH is an enzymatic process and the level or
specificity of the GST could also play a critical role in selectivity (91).

Increases in glutathione and/or GST levels can be induced in
corn by treatment with R-25788 (91), CDAA (92), and various other
compounds (24,93). The species specificity of compounds such as
R-25788 that elevate GSH levels is unknown, but an increased efflux
of GSH in tobacco suspension cultures results from treatment with
R-25788 (15); therefore, it appears that the effect is not limited
to a single species. The above chemicals decrease the phytoxicity
of certain herbicides to certain plant species and they have been
considered as herbicide antidotes or herbicide protectants (91,93).
R-25788 is very effective in decreasing the phytoxicity of EPTC to
corn and this has been attributed to an increased rate of EPTC metabolism (91). Other herbicides that are protected by R-25788 include
α-chloroacetamide and thiocarbamate herbicides that are metabolized
by GSH conjugation, as well as diclofop-methyl and chlorsulfuron
herbicides that are metabolized by glucose conjugation (93).
Therefore, factors other than or in addition to increased rates of
GSH conjugation must be involved in the protective action, or metabolism data regarding some herbicides is incomplete.

In some leguminous species such as soybean, kidney bean, lima
bean, and white clover, glutathione (γ-glutamylcysteinylglycine) is
not very abundant, and instead, homoglutathione (γ -glutamyl-
cysteinyl-β-alanine) appears to be the most abundant cysteine-
containing tripeptide (88). Recently, two xenobiotics were shown to
be metabolized to homoglutathione conjugates in soybean.
Acifluorfen herbicide was cleaved by homoglutathione to yield
S-(3-carboxy-4-nitrophenyl)homoglutathione and 2-chloro-4-trifluoro-
methylphenol (94) (Equation 18). 2-Chloro-4-trifluoromethylphenol
was further metabolized to a glucoside. Metribuzin herbicide underwent homoglutathione conjugation by displacement of an S-methyl
group, probably after oxidation to a sulfoxide (95) (Equation 19).
The metabolism of acifluorfen by homoglutathione conjugation is
similar to the enzymatic GSH conjugation of a related diphenylether

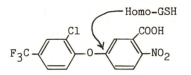

ACIFLUORFEN   (18)

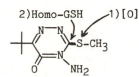

METRIBUZIN   (19)

herbicide, fluorodifen, in pea and peanut ($\underline{84},\underline{85}$); however, enzyme studies with the homoglutathione/acifluorfen system have not been reported. It is uncertain whether homoglutathione conjugation of metribuzin involves a transferase enzyme since symmetrical $\underline{S}$-methyl triazines undergo rapid nonenzymatic GSH conjugation after oxidation to the sulfoxide ($\underline{214}$).

Organophosphorothioate insecticides are partially metabolized by GSH conjugation in insects ($\underline{89}$), but methidathion is one of the few compounds in this class that has been reported to be metabolized by GSH conjugation in a higher plant. In tomato fruit, displacement of the phosphorothioate group with GSH and metabolism of the GSH conjugate to a cysteine conjugate was the major route of methidathion metabolism ($\underline{96}$) (Equation 20). In vitro, diazinon may be converted to a GSH conjugate by corn extracts, but the resulting products were not identified and there have been no reports of a GSH conjugate of diazinon produced in whole plants ($\underline{97}$) (Equation 21).

METHIDATHION   (20)   DIAZINON   (21)

Captan, folpet, and dichlofluanid react with GSH to form a GSH conjugate of trihalomethane which undergoes further reaction to yield thiophosgene. Thiophosgene then reacts with GSH or cysteine and ultimately yields the cyclic cysteine conjugate as shown in Equation 22 ($\underline{98}$).

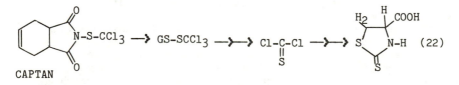

The metabolism of the herbicide antagonist, tridiphane, by a GST system from corn is one of the few examples of the conversion of an epoxide substrate to a GSH conjugate by a plant system ($\underline{99}$) (Equation 23). Based on chromatographic data, it appeared that this conversion also occurs in intact corn.

PCNB is metabolized in higher plants by GSH conjugation and by reduction of the nitro-group ($\underline{15},\underline{16},\underline{83},\underline{100},\underline{101}$). Glutathione conjugation occurs by displacement of the nitro- and/or chloro-groups, resulting in the mono- and di-glutathione and cysteine conjugates as metabolites ($\underline{102}$) (Equation 24).

Xenobiotic GSH conjugates are metabolically unstable in higher plants and rapidly undergo additional enzymatic and/or nonenzymatic transformations to a variety of products by processes that have been reviewed recently ($\underline{15},\underline{16},\underline{89}$). With several exceptions, the primary routes of metabolism of GSH conjugates in plants are similar to those observed in mammals as discussed by Bakke in Chapter 16. In sorghum and corn, the first step in the catabolism of a xenobiotic

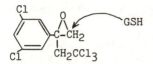

TRIDIPHANE   (23)

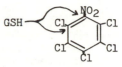

PCNB   (24)

GSH conjugate is the hydrolysis of the glycine residue rather than the glutamyl residue as occurs in mammals (15). It is not known whether this difference also occurs in other plant species.

Most xenobiotic GSH conjugates in plants are metabolized at least to cysteine conjugates and cysteine conjugates appear to be the pivotal point in metabolism. Cysteine conjugates may be end-products of metabolism, as observed in methidathion metabolism in tomato and peanut cell suspension culture, or acifluorfen metabolism in soybean and peanut cell suspension culture (16). Xenobiotic cysteine conjugates are frequently N-acylated with malonic acid as shown in Equation 25. This was demonstrated with the following

$$X-S-CH_2-\underset{\overset{|}{NH_2}}{CH}-COOH \longrightarrow X-S-CH_2-\underset{\overset{|}{HN-CO-CH_2-COOH}}{CH}-COOH \qquad (25)$$

xenobiotics in a peanut cell suspension culture: EPTC (78%), molinate cysteine conjugate (93%), fluorodifen (74%), PCNB (30%), propachlor (42%), metolachlor (24%), and butachlor (5%). Malonylcysteine conjugation was demonstrated in 8 of 11 plant species when PCNB was the substrate; therefore, the reaction appears to be general in higher plants. The sulfide bond of malonylcysteine conjugates can be oxidized, as was observed in the metabolism of propachlor in soybean (16).

Isomerization of an S-cysteine conjugate to an N-cysteine conjugate via the Smiles rearrangement has been observed in the metabolism of two triazine xenobiotics in higher plants: atrazine metabolism in sorghum (103,104) and dimethametryn metabolism in rice (105). This nonenzymatic rearrangement (Equation 26) has not

$$X-S-CH_2-\underset{\overset{|}{NH_2}}{CH}-COOH \longrightarrow X-NH-\underset{\overset{|}{COOH}}{CH}-CH_2-SH \qquad (26)$$

been reported in GSH conjugate metabolism in mammals. A lanthionine conjugate was produced by further metabolism of the N-cysteine conjugate of atrazine in sorghum. In rice, the N-cysteine conjugate of dimethametryn was degraded to an amino-triazine.

The deamination of xenobiotic cysteine conjugates can lead to thio-acetic acid conjugates, as observed with PCNB in peanut (100), or thio-lactic acid conjugates, as observed with EPTC in corn and cotton (106) or PCNB in peanut (100). Cysteine conjugates can also undergo β-lyase cleavage to thioalcohols, as observed with PCNB in peanut (100). The thioalcohols can be methylated by an S-adenosylmethionine methyltransferase system, as demonstrated with an in vitro enzyme system derived from onion root (101). Pentachloro-

phenylmethylsulfide, derived from PCNB, has been detected in several
plant species. The formation of methylsulfides as xenobiotic metabolites present in the environment is of current interest (89). The
ability of plants to form this class of metabolites from a range of
xenobiotics is uncertain. The methyl transferase enzyme system from
onion used a number of different thiophenols as substrates (15), but
the distribution of the necessary β-lyase and methyl transferase
enzymes among other higher plants is unknown.

     Several xenobiotics that are metabolized by GSH conjugation in
plants, including atrazine, propachlor, and PCNB, produce significant levels of bound residue (15). It appears that the bound residue may be formed from the GSH pathway with either a cysteine
conjugate or a thiol as a precursor (15). The chemical nature of
these bound residues has not been determined.

     Glutathione S-transferase enzymes have been detected in a
variety of different plant species (24,78,84,97,101,107-110).
However, detailed enzyme studies have been limited to corn (24,78,
109,110) and pea (85,107,110). The foliar tissue of corn contains a
GST isozyme that appears rather unique in its ability to catalyze
the conjugation of various 2-chloro-s-triazines with GSH (78). The
molecular weight and other properties of this isozyme are similar to
those reported for many mammalian GST isozymes; however, the pH
optimum of this isozyme for atrazine is somewhat lower than that
reported for many GST isozymes (24,78,109). When corn is treated
with R-25788, a second isozyme is induced (24). The induced isozyme
is very effective in catalyzing the GSH conjugation of an
α-chloroacetamide herbicide (alachlor), but it was not assayed with
atrazine (24).

     A GST enzyme from pea is very effective in catalyzing GSH con-
jugation of the herbicide fluorodifen (85). This enzyme has a pH
optimum and other properties that are comparable to mammalian GST
enzymes (85). This enzyme activity was observed in other plant species, but fluorodifen resistant species appeared to have higher
levels of this enzyme than susceptible species (85). Additional
studies with pea indicated the presence of two soluble GST isozymes,
one that utilized fluorodifen and one that utilized t-cinnamic acid
as substrates (107). These isozymes appeared to form aggregates
during purification. In addition, a microsomal GST was detected in
pea that utilized both t-cinnamic acid and benzo(a)pyrene as
substrates (107). Soybean cell suspension cultures metabolized t-
cinnamic acid in a 6% yield to a product that corresponded to the
GSH conjugate of t-cinnamic acid by paper chromatography (107).

     A transferase system capable of utilizing either cysteine or
GSH to form conjugates of isopropyl-3'-chloro-4'-hydroxy-carbanilate
was detected in etiolated oat seedlings (111,112). At least two
enzyme systems capable of catalyzing the above reactions were present (112). Excised oat tissues produced a metabolite corresponding
to the in vitro reaction product with cysteine. The product
appeared to be a cysteine conjugate, but the mechanism of the reaction and the structure of the product were not determined (111).

N-Malonyl and O-malonyl conjugates

Acetylated xenobiotic conjugates are not commonly formed in higher
plants; however, the direct or indirect conjugation of xenobiotics

with malonic acid is a common occurrence. D-Amino acids, xenobiotic
cysteine conjugates, and aromatic and heterocyclic compounds containing a primary amino group can be metabolized in plants, via
amide bond formation, to malonic acid conjugates. Xenobiotic glucose conjugates can be metabolized in plants to malonic acid conjugates via a 1-6 ester bond between malonic acid and the glucose
residue. Malonylcysteine and malonylglucose conjugates are secondary or tertiary metabolites derived from GSH- and glucose-
conjugates. These malonyl conjugates were discussed in previous
sections dealing with complex glucosides and glutathione conjugates.
     D-Amino acids are not usually natural plant constituents and
when plants are treated with D-amino acids they are generally converted to N-malonyl conjugates (113-115) (Equation 27). Naturally-

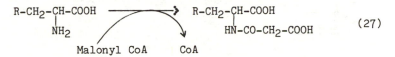

$$R-CH_2-CH-COOH \longrightarrow R-CH_2-CH-COOH \qquad (27)$$
$$\underset{NH_2}{|} \qquad\qquad \underset{HN-CO-CH_2-COOH}{|}$$

Malonyl CoA   CoA

occurring L-amino acids are not metabolized in this manner and it is
assumed that N-malonyl conjugation is a mechanism utilized by plants
to remove D-amino acids from normal metabolic pathways (113-115).
Most D-amino acids are substrates for this pathway, the N-malonyl
conjugation reaction occurs in many plant species, and the N-malonyl
conjugates appear to be terminal products of metabolism, ie., they
appear to be stable in the plant once they are formed (114). N-
Malonyl conjugation of D-amino acids appears to be similar to N-
malonyl conjugation of xenobiotic cysteine conjugates as discussed
previously. The natural ethylene precursor, aminocyclopropane carboxylic acid (ACC), is also converted to an N-malonyl conjugate in
plants (116,117). Soluble enzyme preparations that catalyze the N-
malonyl conjugation of ACC with malonylcoenzyme A as the natural co-
substrate have been isolated from mung bean (118) and peanut (119).
D-Amino acids inhibit the in vitro conjugation of ACC with malonic
acid (118) and a malonyltransferase from peanut that utilizes D-
tryptophan as a substrate for N-malonyl conjugation also utilizes
ACC as a substrate (119). These malonyltransferase enzymes have not
been assayed with xenobiotic cysteine conjugates to determine if the
same enzymes are involved in both transformations. Likewise, it is
not known if the xenobiotic cysteine conjugates undergo an isomerization from the L-to D-configuration in order for the malonylation
reaction to proceed.
     Aromatic and heterocyclic amines are usually metabolized in
higher plants by formation of N-glucosides or bound residues. In
several cases, however, aromatic and heterocyclic amines have been
partially metabolized to N-malonyl conjugates. It is not known
whether this type of N-malonyl conjugation is restricted to a few
selected amines nor is it known whether this reaction is utilized by
more than a few plant species such as peanut and soybean.
     Botran was one of the first xenobiotics reported to be metabo-
lized to an N-malonyl conjugate by this mechanism in plant tissues
(120) (Equation 28). The N-malonyl conjugate of botran was the
major metabolite in both soybean and soybean callus culture. It was
isolated by chromatographic methods and identified by synthesis and
by chemical and mass spectral methods.

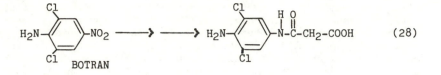

$$(28)$$

Carboxin fungicide was metabolized to malonanilic acid and the glucoside of p-hydroxymalonanilic acid in the fruit of peanut plants and in peanut cell suspension cultures (121) (Equation 29). Both

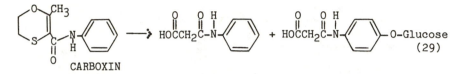

$$(29)$$

metabolites were identifed by mass spectral and chemical methods as well as by synthesis. Other major metabolites were carboxin sulfoxide (30% of the residue) and bound residue (21%). The O-glucoside of malonanilic acid is an example of an unusual dicon-jugate in which two polar groups have been conjugated at different sites on the same xenobiotic molecule.

Metribuzin is metabolized rapidly in tomato to an N-glucoside which is subsequently converted to a malonylglucoside. In soybean, metribuzin is metabolized more slowly by conjugation with homoglu-tathione and by formation of bound residue; an N-malonyl conjugate is formed in soybean as a minor product (95) (Equation 30).

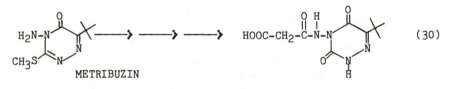

$$(30)$$

A soluble enzyme (100,000g supernatant) was obtained from peanut cell suspension cultures that catalyzed the formation of malonanilic acid from aniline and malonyl CoA in a 16% yield (121). This enzyme preparation did not utilize ACC as a substrate (122). In a detailed study of the malonyltransferases from peanut, four distinct malonyltransferase enzymes with different substrate speci-ficities were isolated from seedling hypocotyls and leaves (119). These four malonyl transferase enzymes had very distinct substrate specificity requirements, and except for one of these enzymes, no overlap in substrate specificity was observed. The four malonyl-transferases were active with the following substrates: (a) 3,5-dichloroaniline, (b) anthranilic acid, (c) 2-methoxyethanol, and (d) D-tryptophan and aminocyclopropane carboxylic acid. The malonyl-transferase from peanut leaves and hypocotyls that utilized 3,5-di-chloroaniline as a substrate is probably similar to the transferase from peanut cell culture that utilized aniline as a substrate. These malonyltransferases should be subjected to detailed substrate specificity studies to assess their potential role in xenobiotic metabolism in plants.

Although malonyl conjugation of D-amino acids was assumed to be
a detoxificiation process, some malonyl conjugates may be biologically active. A series of 25 malonanilic acids were tested for
growth-regulating properties in higher plants. Unsubstituted malonanilic acid was a potent stimulator of root-growth in cucumber
(123). It remains to be determined whether this activity was due to
malonanilic acid or to the hydrolysis product, aniline, which may
have been liberated in the roots. In peanut plants, amidase enzymes
that hydrolyze the N-malonyl conjugate of ACC are apparently produced as the plant ages (119). Therefore, caution should be exercised in making assumptions about the stability of xenobiotic
malonyl conjugates based on short-term experiments. It has been
hypothesized that N-malonyl conjugation may be a mechanism utilized
to store products in a biologically inactive state in cell vacuoles
(119).

O-Malonylconjugates of phenols or alkyl alcohols have not been
commonly reported as metabolites of xenobiotics in plants; however,
2-methoxyethanol is a substrate for a malonyl-transferase in peanut,
suggesting that this class of conjugate might ocassionaly be produced (119). It should be noted, however, that the Km value for
2-methoxyethanol was very high, greater than 100 μM.

Amino acid conjugation

Xenobiotics that contain a free or potential carboxyl group can be
metabolized by amino acid conjugation in both plants and animals.
This reaction is illustrated by the conjugation of 2,4-D with aspartic acid (Equation 31). In higher plants, amino acid conjugation is

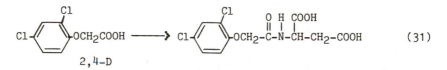

$$2,4\text{-D} \quad\quad (31)$$

relatively uncommon and it occurs usually in competition with other
reactions such as glucose ester formation or aromatic ring
hydroxylation/glucoside formation (20). It is generally restricted
to compounds that have plant growth regulator activity: 2,4-D (11,
124,125), 2,4,5-T (126,127), MCPA (11), indole-3-acetic acid (IAA)
(128,129), cyclohexane carboxylic acid (130), and naphthalene acetic
acid (131).

2,4-D was probably the first pesticide shown to be metabolized
to an amino acid conjugate in a higher plant (aspartic acid conjugate in pea) (124); however, it had previously been shown that
IAA, a natural plant auxin, was metabolized to an aspartic acid
conjugate in pea (128). Subsequently, 2,4-D was shown to be partially metabolized to amino acid conjugates in a variety of plant
species and tissues including: wheat; red and black currant;
soybean; corn; and callus tissue cultures of soybean, corn, carrot,
jackbean, sunflower, and tobacco (11,212). The metabolism of 2,4-D
in plants is highly species dependent and in most of the above species other routes of metabolism appear to be quantitatively more
important than amino acid conjugation. Glutamic and aspartic acid

conjugates are usually the most abundant amino acid conjugates,
regardless of the plant species or xenobiotic (13,20); however,
2,4-D conjugates of alanine, valine, leucine, phenylalanine, and
tryptophan have been isolated from soybean callus (132).

Some amino acid conjugates appear to be unstable in plants.
The glutamic acid conjugate of 2,4-D was metabolized to 2,4-D, 2,4-D
aspartic acid and what appeared to be glucose conjugates of 4-
hydroxy-2,5-D and 4-hydroxy-3,5-D (132). The rate of metabolism of
2,4-D glutamic acid appeared to be greater than that of 2,4-D (132).
Amino acid conjugates of 2,4-D have been reported to be hydrolyzed
by almond emulsin; consequently, some caution should be exercised in
attempting to classify conjugates as glucosides by almond emulsin
β-glucosidase hydrolysis (132). The aspartic acid conjugate of
2,4-D is acid labile and in one study the recovery of this conjugate
was only 19% (133).

Amino acid conjugation of 2,4-D appears to be more important in
callus culture than in whole plants (20), but a number of factors
can effect levels of amino acid conjugates: the concentration of
2,4-D used, age of tissue (134), length of time between treatment
and harvest, form of 2,4-D used (free acid, amine salt, or ester)
(133), and the specific source of tissue (135). Four days after
treatment of soybean plants with 2,4-D propyleneglycol butyl ester,
levels of 2,4-D glutamic and aspartic acid conjugates (75 and 57
ppm, respectively) were much higher than the level of 2,4-D acid (43
ppm). In comparison, levels of glutamic and aspartic acid conjugates were less than 4 ppm while free 2,4-D was at 47 ppm in soybean
plants treated with 2,4-D dimethylamine salt (133). Levels of 2,4-D
amino acid conjugates varied dramatically in soybean callus as a
function of time following treatment: from 1.3 ppm glutamate and 0.3
ppm aspartate at 2 days to only 0.16 ppm glutamate and 0.03 ppm
aspartate at 10 days (133). Levels of 2,4-D amino acid conjugates
also declined signficantly in whole soybean plants between 4 and 10
days following treatment. The metabolism of 2,4-D in plants is
extremely complicated and it appears that 2,4-D can be metabolized
to amino acid conjugates of the glycosides of 4-hydroxy-2,4-D and to
other highly polar unidentified products (180).

The metabolism of other xenobiotics to amino acid conjugates in
plants appears to be comparable to 2,4-D metabolism. Naphthalene
acetic acid (NAA) is a plant growth regulator used for a variety of
purposes on horticultural crops. It is metabolized to an aspartic
acid conjugate in cowpea, fruit of the mandarin orange, and in
tobacco mesophyll protoplasts (131,136). Tobacco mesophyll protoplasts that were induced to divide with NAA metabolized nearly half
of the NAA in the medium to the aspartic acid conjugate. The conjugate was identified by synthesis and negative ion CI/MS
(N_2O/CH_4)(131). Cyclohexane carboxylic acid was metabolized to a
glucose ester and an aspartic acid conjugate in bush bean leaf
discs. Metabolites were characterized by TLC comparison to
synthetic standards (130). Several endogenous compounds have been
observed in plants as peptide conjugates. Formyl- and methyl-
tetrahydrofolate form polyglutamate conjugates in carrot, potato,
turnip and beet storage tissues. Mono- and di-glutamate conjugates
were also observed, but at lower levels (137). A glycylphenylala-
nine dipeptide conjugate of ferrulic acid was isolated from barley
globulins by partial hydrolysis (138).

Some amino acid and glucose ester conjugates of 2,4-D, 2,4,5-T, and IAA are biologically active, and in some cases an amino acid conjugate has been reported to be more active than the parent compound (127,139-142). Some amino acid conjugates are unstable in plant tissues and their biological activity appears to be correlated to their ease of hydrolysis. It may be that amino acid conjugates are not biologically active, but yield the active free acid upon hydrolysis as postulated with glucose esters of some herbicides. The biological activity of the aspartic acid and alanine conjugates of IAA was directly related to the ease of hydrolysis of the amino acid conjugates in a bean stem assay. The aspartic acid conjugate was more resistant to hydrolysis and relatively inactive compared to the alanine conjugate (139). The aspartic acid conjugate of NAA was not biologically active and it was speculated that the amide bond was not hydrolyzed in vivo (131).

The in vitro synthesis of amino acid conjugates has not been demonstrated with enzyme systems isolated from plants in spite of the great importance of these enzymes to our understanding of the mode of action and metabolism of IAA, NAA, and 2,4-D (56). However, it appears that these enzymes can be induced by exogenous 2,4-D, NAA, or IAA (56). In vitro studies with enzymes from mammals suggest that xenobiotic carboxylic acids are activated by a reaction that requires ATP and CoA. The xenobiotic acyl-CoA derivative is then released from the enzyme surface. The initial activation reactions are catalyzed by different acyl-CoA synthetase enzymes with different substrate specificities. Benzoic and phenylacetic acid derivatives are activated by what appears to be a butyrl-CoA synthetase present in the mitochondrial matrix. The final reaction is catalyzed by an acyl-CoA:amino acid N-acyltransferase. Two closely related forms have been purified from bovine liver mitochondria. One is specific for benzoyl-CoA, salicyl-CoA and short chain fatty acids while the other specifically utilizes phenylacetyl-CoA or indol-3-ylacetyl-CoA (143). The substrates utilized by these mammalian enzymes are remarkably similar to the substrates metabolized to amino acid conjugates in plants.

Lipophilic Conjugates

Most xenobiotic conjugation reactions in plants and animals lead initially to the formation of polar products such as glycoside or glutathione conjugates, but several reports indicate that plants (144-148) and animals (149) may also form lipophilic conjugates.

An early indication that a widely used agricultural chemical might be metabolized to a nonpolar conjugate in plants came from an in vitro enzyme study with ^{14}C-labeled surfactants of the Triton family. A crude particulate enzyme preparation from corn shoots catalyzed the formation of fatty acid ester conjugates from the two ^{14}C-labeled polyethoxylated surfactants indicated below (Equation 32). The ester conjugates were formed primarily from palmitic and linoleic acids (≥85%). They were identified by mass spectrometry and by GLC analysis of hydrolysis products (148). In vivo, rice and

$$C_8H_{17}-\langle\!\!\!\bigcirc\!\!\!\rangle-(OCH_2CH_2)_n-OH, \; n=6 \; or \; 9 \qquad (32)$$

TRITON

barley tissue formed small amounts of ^{14}C-labeled lipophilic metabolites that were thought to be fatty acid esters of the two ^{14}C-
labeled surfactants; however, deethoxylation and glucoside conjugation were the major routes of metabolism (38). A soluble enzyme
preparation from pea catalyzed the formation of the β-glucoside conjugates of these surfactants (148).

An experimental acaricide (Ro 12-0470) was reported to be converted to fatty acid esters in apple fruit (145). One day after
treatment, 22% of the applied dose was recovered as fatty acid
methyl esters, primarily saturated fatty acids (C_{16}, C_{18}, C_{20} and
C_{22}) (Equation 33). The fatty acid esters were present both on the
surface of the apple and in extracts of washed apples.

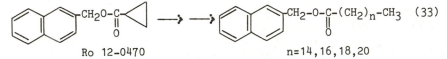

Ro 12-0470   n=14,16,18,20

These metabolites were identified by synthesis and GC/MS. The
parent compound (Ro 12-0470) and the fatty acid ester metabolites
were easily hydrolyzed (145). Additional studies should be conducted to determine if these metabolites are produced enzymatically
and/or if they might be formed as artifacts during isolation and
identification. Fatty acid esters of Ro 12-0470 were not detected
in the foliar tissues of apple.

Carbofuran is metabolized to a conjugate of angelic acid in
carrot, (146,147) (Equation 34). This metabolite was identified by

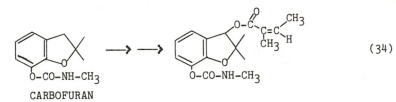

CARBOFURAN

EI/MS, ^{1}H FT-NMR, FT-IR and synthesis. It was the major residue of
carbofuran in carrot 15 days following treatment (approx. 40%), but
it was not detected in potato or radish (146,147). In potato and
radish, carbofuran was metabolized slowly to water-soluble conjugates and bound residue (146,147).

The mechanism by which xenobiotic alcohols or esters are converted to fatty acid esters has not been studied. They could be
formed by the action of lyase enzymes in the presence of fatty acid
glyceryl esters, as in the conversion of farnesol to farnesol fatty
acid esters (150). Some lipolytic acyl hydrolase enzymes from
plants readily catalyze the transfer of lipid-bound fatty acids to
low MW alcohol acceptors (150,151) and enzymes of this class could
be responsible for the occasional formation of fatty acid conjugates
of xenobiotic alcohols. Mechanisms involving fatty acid acyl CoA,
phospholipids, or direct esterification with fatty acids might also
be involved (152).

Recently, a very different class of lipophilic conjugates of
picloram and 2,4-D were isolated from radish and mustard plants

(<u>144</u>). The conjugates were formed with p-hydroxy-styryl-mustard,
vinyl-mustard, and allyl-mustard (Equation 35). In radish, picloram

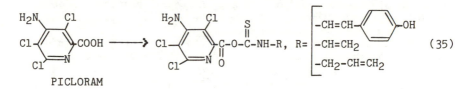

PICLORAM

and 2,4-D were metabolized to mustard oil conjugates in yields of
30% and 12%, respectively, in 72 hr. Mustard oil conjugates were
also formed in mustard plants, but the yield was lower. Mustard oil
conjugates appeared to be the only metabolites of 2,4-D in these
species (<u>144</u>). These conjugates were isolated by chromatographic
methods and identified by UV, IR, MS, and ^{1}H NMR (360 M Hz). The
metabolites were base labile, but appeared stable when introduced
into sunflower. At present, conjugation with mustard oils appears
to be a highly unusual route of metabolism, possibly restricted to
plants of the <u>Cruciferae</u> family (<u>144</u>). In other species such as
sunflower (<u>48</u>) and leafy spurge (<u>153</u>), picloram is metabolized to N-
glycoside and glucose ester conjugates and 2,4-D is metabolized to a
glucose ester, O-glucosides, and to amino acid conjugates (<u>20</u>). The
metabolism of other xenobiotic carboxylic acids should be studied in
radish and mustard to determine if this is a general pathway of
metabolism in the <u>Cruciferae</u> family.

Improved methods and instrumentation for metabolite isolation
and identification, such as capillary GC, GC/MS, HPLC, high field
NMR, FAB/MS, CI/MS, FT-IR and HPLC/MS have made the identification
of new or unusual metabolites more practical. As these advanced
techniques are employed to study xenobiotic metabolism in more
diverse species of plants, additional classes of xenobiotics will no
doubt be discovered.

Bound residues

Xenobiotics are frequently metabolized in plants by mechanisms that
lead to the incorporation or inclusion of the xenobiotic into biological polymers or tissue residues that are not soluble in commonly
used nonreactive solvents. These residues are frequently refered to
as bound, insoluble, or nonextractable residues (<u>17</u>). Bound residues in plants have most commonly been detected in plant tissues
treated with radioactively-labeled pesticides. These residues were
an important topic of a symposium held in Vail, Colo. in 1975 (<u>17</u>);
they have been discussed in many more recent papers (<u>11,154-157</u>) and
they were discussed at a symposium at the 188th ACS National
Meeting, 1984: "Non-extractable Pesticide Residues: Characteristics,
Bioavailability and Toxicological Significance".

Occasionally, xenobiotics may be extensively metabolized in
plants to CO_2 or other low MW endogenously occuring products which
can produce bound residues by reincorporation into biological polymers. Residues of this type are generally of little concern to
toxicologists and residue chemists because these residues do not
represent an unusual hazard to the biosphere. A recently proposed

definition of a bound plant residue specifically excluded residues
formed by reincorporation of such endogenously occuring compounds
(158); however, some difficulty can be encountered in determining if
recycling of this nature has occured. A general protocol to assess
the biological signficance of bound plant residues which addresses
this problem has also been suggested (158).

A partial list of xenobiotics that form bound residues in plant
tissues is presented on Table VIII. Many heterocyclic and aromatic

Table VIII. Bound Residues of Xenobiotics in Plants

Xenobiotic	Plant Species	% Bound Residue[a]	REF
Atrazine	Corn	38	192
Bentazon	Rice, Soybean	25-90	154
Benzopyrene	Soybean csc[b]	15-25	155
Buthidazole	Corn, Alfalfa	19	193
Buturon	Wheat	50	194
Carbaryl	Tobacco	40	195, 196
Carboxin	Peanut	21	121
Chloramben	10 species	4-39	43
Chloroaniline	Rice	30	197
Chlortoluron	Wheat	50	198
Cisanilide	Carrot csc	40	169
2,4-D	Several csc	3-19	155
DIB	White clover	34	162
Dichlobenil	Bean	20	199
Dieldrin	Radish	10	159
EPN	Cotton	24	200
Flamprop	Wheat	42	156
Fluorodifen	Peanut	12-26	84
Isoxathion	Bean	17	49
MCPA	Wheat	33	156
Mephosfolan	Rice	41-55	201
Metribuzin	Soybean	30	95
Nitrofen	Rice, Wheat	50	202
Oxamyl	Peanut	40	57
PCNB[c]	Peanut	13	102
PCP[d]	Soybean, Wheat	11,37	155
PCTP[e]	Peanut csc	21	15
Perfluidone	Peanut	21	178
Prometryn	Oat	20-40	203
Pronamide	Alfalfa	45	204
Propachlor	Soybean	38	178
Propanil	Rice	30	205
Propham	Alfalfa	20	206
Solan	8 species	10-20	207
Sweep	8 species	30	207
Trichlorophenol	Tomato	60	208
Triforine	Barley	75	209
Zectran	Broccoli, Bean	18,28	210

[a] Percent of radioactive residue in plant not extracted by methods
used in that study.   [b] csc=Cell suspension culture.
[c] PCNB=Pentachloronitrobenzene.   [d] PCP=Pentachlorphenol.
[e] PCTP=Pentachlorothiophenol.

compounds that contain or can be metabolized to yield functional
groups such as HO-, HOOC-, H_2N-, and HS- frequently form bound residues; but chlorinated hydrocarbons generally do not (159).
    Bound residues of xenobiotics are found incoporated or asso-
ciated with most of the biological polymers of plants, including
lignin, various carbohydrate polymers, and proteins (11). Lignin
has been implicated as the major form of bound pesticide residue in
the greatest number of cases where the residue has been studied (11,
155,158). Lignin is a highly insoluble plant cell-wall polymer of
heterogenous nature. It is probably formed by polymerization of
cinnamic acid alcohol derivatives and other endogneous substrates in
free-radical reactions catalyzed by peroxidase and laccase enzymes
(160). Bound residues of lignin are difficult to study because of
the highly insoluble nature of the material. Several methods are
frequently used to isolate lignin (158), but these methods may alter
the xenobiotic-lignin bond. There have been no truly satisfactory
methods for the structural analysis of xenobiotic-lignin residues.
In some cases, such as with 3,4-dichloroaniline, xenobiotics appear
to be incorporated into lignin by covalent bonding (155,157), but in
other examples, such as swep, buturon, and carboxin, the xenobiotic
may simply become entrapped in the cage-like matrix of the lignin
polymer (155).
    A model system for the synthesis of lignin-like polymers showed
that enols would react with the quinone-methide intermediates in
this system by a 1-6 addition (160). Utilizing this model system,
chloroanilines were copolymerized with coniferyl alcohol in the presence of horseradish peroxidase Type II enzyme, hydrogen peroxide,
vanillyl alcohol initiator and pH 7.2 buffer (157). The mechanism
of this copolymerization reaction is shown in Equation 36. The

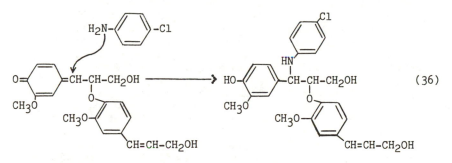

$$(36)$$

copolymers of chloroaniline and coniferyl alcohol had average MW's
of 1,000 to 1,300 and the molar ratio of chloroaniline incorporated
into the polymer was 1.19 to 1.68. When N-acetylated chloroanilines
were used in place of chloroanilines, incorporation was much lower.
This suggested the involvement of the amino group in the copolymerization reaction. The bound residue in lignin from rice treated with
3,4-dichloroaniline was similar in some respects to the synthetic
lignin. These residues were studied by ^{1}H NMR and pyrolysis/MS.
The incorporation of 3-chloroaniline and 3,4-dichloroaniline into
copolymers with coniferyl alcohol in a similar model system was confirmed in a separate laboratory where the research was expanded to
include benzo(a)pyrene quinones (155). The copolymers were studied
by ^{1}H-NMR, ^{13}C-NMR, and other techniques. The copolymers of coni-

feryl alcohol and chloroanilines appeared to contain benzylamine
bonds, but evidence was obtained in both studies that suggest that
bonds involving the aromatic rings were also formed. It would be
expected that nucleophiles such as -SH, -NH$_2$, -OH, and -COOH would
react in this manner to form lignin-like copolymers (155).

    Carbohydrate conjugation of xenobiotics is often accompanied by
the formation of bound residues. However, it is not clear whether
carbohydrate conjugation is in competition with or is an intermediate process in the formation of some bound residues. Some xenobiotics are metabolized to dissacharide and oligosaccharide
conjugates in certain tissues and it has been speculated that these
xenobiotics may be incorporated into polymeric carbohydrate fractions. Diphenamid was metabolized to a gentiobiose conjugate and
bound residue in tomato and it was speculated that the bound residue might be a polysaccharide conjugate (69). Chlorpropham was
metabolized in soybean to a glucoside, bound residue, and what
appeared to be polysaccharide conjugates. It was postulated that
the bound residue might be a more complex polysaccharide conjugate
(161). Oxamyl was metabolized to simple and complex polysaccharide
conjugates in peanut and it was speculated that the bound residue
might be carbohydrate polymers (57). In white clover, DIB was metabolized to simple, di-, and tri-glucose conjugates and a bound residue that was speculated to be a carbohydrate polymer (162). MCPA
and flamprop were metabolized in spring wheat via processes that
involved glucose conjugation and the formation of bound residue
(156). In wheat straw, 33% and 42% of the residues of MCPA and
flamprop, respectively, were bound. Characterization of these bound
residues by various enzymes including α-amylase, protease, pec-
tinase, snail digestive enzymes, hemicellulase, and cellulase was
not successful; however, fractionation of the bound residue by solubility and hydrolysis of the resulting fractions indicated that MCPA
may have been incorporated into the polysaccharide and hemicellulose
fractions. The products were separated into several different molecular weight ranges by gel permeation chromatography. Lignin
accounted for only 2% of the residue from MCPA and Flamprop in these
tissues.

    It appears that very similar xenobiotics can be metabolized to
different types of bound residue and that considerable quantitative
variation in bound residue can occur as a function of plant species
or specific source of tissue. In wheat cell suspension culture,
2,4-D, which is structurally similar to MCPA, appeared to be inco-
porated into lignin (155). This was in sharp contrast to MCPA metabolism in wheat. Carboxin (aniline-14C) metabolism in peanut cell
suspension culture and the fruit of whole peanut plants is an
example of tissue variation. In peanut cell suspension culture,
only 2.7% of the carboxin was incorporated into bound residue, but
in the fruit of whole plants, 21% incorporation into bound residue
was observed (121). The metabolism of metribuzin in tomato and
soybean is an excellent example of species variation. In tomato,
metribuzin was rapidly metabolized to N-glucosides and only 2% was
incorporated into bound residue, but in soybean, metribuzin was
metabolized slowly by homoglutathione conjugation and 20-30% of the
metribuzin was incorporated into bound residue (46,95).

    Bound residues are normally considered end-products of metabo-
lism; however, some carbohydrate components of cell walls may be

reutilized during certain stages of plant development (163).
Therefore, the incorporation of a xenobiotic residue into hemicellulose or other carbohydrate fractions can not be regarded as proof
that a residue is no longer subject to further metabolic transformations.

Conclusions

The metabolism of xenobiotics in higher plants has been studied
extensively over the last 20 years. In common plant species such as
corn, it is frequently possible to predict the conjugation reactions
that may be utilized in the initial phases of metabolism of a new
xenobiotic. In less commonly studied species, predictions are more
uncertain and exotic metabolites are occasionaly formed. In those
cases where phase I oxidative reactions are likely, it is difficult
to predict the course of metabolism because phase I oxidation reactions in plants are frequently very substrate and species specific.
Phase I oxidative reactions have a profound effect on ensuing conjugation reactions. The presence of multiple functional groups on a
xenobiotic also increases the uncertainty of the route of metabolism
likely to be followed in a particular species.
     Although these uncertainties exist in our ability to predict
initial conjugation reaction(s), some generalizations can be made.
Most phenols, anilines, and carboxylic acids are initially metabolized to glucose conjugates and/or bound residues. Occasionally,
carboxylic acids are metabolized to amino acid conjugates and anilines to malonic acid conjugates. Xenobiotics with electrophilic
sites are frequently metabolized to glutathione or homoglutathione
conjugates, depending upon the species. Lipophilic conjugates of
xenobiotic phenols and carboxylic acids have been reported, but few
generaliziations can be made with this class of conjugate. Most
bound residues appear to be associated with lignin or carbohydrate
polymers.
     In higher plants, xenobiotic conjugates frequently undergo
additional metabolic transformations. It is difficult to predict
the precise nature of these secondary or tertiary transformations.
Simple glucose conjugates may be stable or they may be metabolized
to malonylglucose-, gentiobiose-, oligosaccharide-, or hetero-
dissacharide-conjugates, or they may be incorporated into bound
residue. At present, we can not predict which of the above routes
of metabolism will be utilized in a specific case. Glutathione or
homoglutathione conjugates nearly always undergo additional metabolic transformations to cysteine-, malonylcysteine-, malonylcysteine
sulfoxide-, thiolactic acid-conjugates, etc., or to bound residue,
but the route that will be utilized in a specific case can not be
predicted. Xenobiotic amino acid conjugates and glucose ester conjugates may be transitory and further metabolized in a manner similar to the parent compound.
     Most of the enzymes that catalyze the formation of xenobiotic
conjugates in plants have not been well-studied. Since some conjugation reactions are involved in herbicide selectivity, it is
likely that research relating to these enzymes will intensify as a
result of efforts to develop herbicide resistant crops through
bioengineering. Enzymes that may be useful in bioengineering for
herbicide resistance are the GST enzymes, N-glucosyl transferases,

and perhaps malonyl transferases. Oxidative enzymes are extremely
important in regards to herbicide selectivity, but they may present
some unusually difficult problems in bioengineering studies.
Intensive research on the enzymes responsible for xenobiotic conjugation in plants will greatly increase our knowledge of the behavior
and fate of xenobiotics in plants.

Literature Cited

1. Casida, J. E.; Lykken, L. Ann. Rev. Plant Physiol. 1969, 20,
    607.
2. Matsunaka, S. In "Environmental Toxicology of Pesticides";
    Matsumura, F.; Boush, G. M.; Misato, T., Eds.; Academic: New
    York, 1972; pp. 341-364.
3. Shimabukuro, R. H.; Lamoureux, G. L.; Frear, D. S. In
    "Biodegradation of Pesticides"; Matsumura, F.; Krishma Murti,
    C. R., Eds.; Plenum: New York, 1982.
4. Baldwin, B. C. In "Drug Metabolism from Microbe to Man";
    Parke, D. V.; Smith, R. L., Eds.; Taylor and Francis Ltd.:
    London, 1977; pp. 191-217.
5. Rouchard, J.; Decallone, J. R.; Meyer, J. A. Pestic. Sci.
    1978, 9, 74.
6. Quistad, G. B.; Menn, J. J. Residue Rev. 1983, 85, 173.
7. Menzie, C. M. In "Metabolism of Pesticides Update III";
    United States Department of the Interior, Fish and Wildlife
    Service, Special Scientific Report -- Wildlife No. 232,
    Superintendent of Documents, U.S. Government Printing Office,
    Washington, D.C. 20402, 1980.
8. Frear, D. S.; Hodgson, R. H.; Shimabukuro, R. H.; Still, G. G.
    Adv. Agronomy 1972, 24, 317.
9. Kearney, P. C.; Kaufman, D. D. "Herbicide Chemistry,
    Degradation and Mode of Action"; Marcel Dekker: New York,
    1975; 2 Volumes.
10. Naylor, A. W. In "Herbicide Physiology, Biochemistry,
    Ecology"; Audus, L. J., Ed.; Academic: New York, 1976; Vol.
    I, pp. 397-426.
11. Hatzios, K. K.; Penner, D. "Metabolism of Herbicides in
    Higher Plants"; Burgess Publishing Co.: Minneapolis, 1982.
12. Frear, D. S. In "Bound and Conjugated Pesticide Residues";
    Kaufman, D. D.; Still, G. G.; Paulson, G. D.; Bandal, S. K.,
    Eds.; ACS SYMPOSIUM SERIES No. 29, American Chemical Society:
    Washington, D.C., 1976; pp. 35-54.
13. Mumma, R. O.; Hamilton, R. H. In "Bound and Conjugated
    Pesticide Residues"; Kaufman, D. D.; Still, G. G.; Paulson,
    G. D.; Bandal, S. K., Eds.; ACS SYMPOSIUM SERIES No. 29,
    American Chemical Society: Washington, D.C., 1976; pp. 68-85.
14. Hutson, D. H. In "Bound and Conjugated Pesticide Residues";
    Kaufman, D. D.; Still, G. G.; Paulson, G. D.; Suresh, K. B.,
    Eds.; ACS SYMPOSIUM SERIES No. 29, American Chemical Society:
    Washington, D.C., 1979; pp. 103-131.
15. Lamoureux, G. L.; Rusness, D. G. In "Sulfur in Pesticide
    Action and Metabolism"; Rosen, J. D.; Magee, P. S.; Casida,
    J. E., Eds.; ACS SYMPOSIUM SERIES No. 158, American Chemical
    Society: Washington, D.C., 1981; pp. 133-164.
16. Lamoureux, G. L.; Rusness, D. G.; IUPAC Pesticide Chemistry
    3; Miyamoto, J., Ed.; Pergamon: New York, 1983; pp. 295-300.

17. Kaufman, D. D.; Still, G. G.; Paulson, G. D.; Bandal, S. K.
     "Bound and Conjugated Pesticide Residues"; ACS SYMPOSIUM
     SERIES No. 29, 1976; American Chemical Society: Washington,
     D.C.
18. Shimabukuro, R. H.; Walsh, W. C. In "Xenobiotic Metabolism:
     In Vitro Methods"; Paulson, G.; Frear, D. S.; Marks, E. P.,
     Eds.; ACS SYPOSIUM SERIES No. 97, American Chemical Society:
     Washington, D.C., 1979; pp. 1-34.
19. Sanderman, H.; Diesperger, H.; Scheel, D. In "Plant Tissue
     Culture and Its Biotechnolgical Application"; Barz, W.;
     Reinhard, E.; Zenk, M. H., Eds.; Springer-Verlag: Berlin,
     1977; pp. 178-196
20. Mumma, R. O.; Hamilton, R. H. In "Xenobiotic Metabolism: In
     Vitro Methods"; Paulson, G.; Frear, D. S.; Marks, E. P.,
     Eds.; ACS SYMPOSIUM SERIES No. 97, American Chemical Society:
     Washington, D.C., 1979; pp. 35-76.
21. Lamoureux, G. L.; Frear, D. S. In "Xenobiotic Metabolism: In
     Vitro Methods"; Paulson, G. D.; Frear, D. S.; Marks, E. P.,
     Eds.; ACS SYMPOSIUM SERIES No. 97; Washington, D.C., 1979;
     pp. 77-128.
22. Sandermann, Jr., H. In "Environmental Mutogenesis,
     Carcinogenesis, and Plant Biology"; Klekowski, E. J., Jr.,
     Ed.; Praeger: New York, 1982; Vol. I, pp. 3-32.
23. Dohn, D. R.; Krieger, R. I. Drug Metabolism Reviews 1981,
     12, 119.
24. Mozer, T. J.; Tiemeier, D. C.; Jaworski, E. G. Biochemistry
     1983, 22, 1068.
25. Pridham, J. B. Phytochemistry 1964, 3, 493.
26. Frear, D. S.; Swanson, C. R.; Kadunce, R. E. Weeds 1967, 15,
     101.
27. Aritomi, M.; Kawasaki, T. Phytochemistry 1984, 23, 2043.
28. Markham, K. R.; Whiteshouse, L. A. Phytochemistry 1984, 23,
     1931.
29. Nakano, K.; Maruhashi, A.; Nohara, T.; Tomimatsu, T.;
     Imamura, N.; Kawaskai, T. Phytochemistry 1983, 22, 1249.
30. Roberts, T. R.; Wright, A. N. Pestic. Sci. 1981, 12, 161.
31. Sweetser, P. B.; Schow, G. S.; Hutchison, J. M. Pestic.
     Biochem. Physiol. 1982, 17, 18.
32. Jacobson, A.; Shimabukuro, R. H. J. Agric. Food Chem. 1984,
     32, 742.
33. Mizukami, H.; Terao, T.; Miura, H.; Ohashi, H. Phytochemistry
     1983, 22, 679.
34. Storm, D.; Hassid, W. Z. Plant Physiol. 1974, 54, 840.
35. Fleuriet, A.; Macheix, J. J.; Suen, R.; Ibrahim, R. K. Z.
     Naturforsch 1980, 35c, 967.
36. Schmitt, R.; Kaul, J.; v.d. Trenck, T.; Schaller, E.;
     Sandermann, H., Jr. Pestic. Biochem. Physiol. 1985 (in press).
37. Liu, T-y.; Castelfranco, P. Plant Physiol. 1970, 45, 424.
38. Stolzenberg, G.; Olson, P.; Tanaka, F.; Mansager, E.;
     Lamoureux, C. In "Advances in Pesticide Formulation
     Technology", Scher, H., Ed.; ACS SYMPOSIUM SERIES No. 254,
     American Chemical Society: Washington, D.C., 1984; p 207.
39. Still, G. G.; Rusness, D. G.; Mansager, E. R. In "Mechanism
     of Pesticide Action"; Kohn, G. K., Ed.; ACS SYMPOSIUM SERIES

No. 2, American Chemical Society: Washington, D.C., 1974; pp. 117-129.

40. Ishizuka, M.; Kondo, Y.; Takeuchi, Y. J. Agric. Food Chem. 1982, 30, 882.
41. Ryan, P. J.; Gross, D.; Owen, W. J.; Laanio, T. L. Pestic. Biochem. Physiol. 1981, 16, 213.
42. Colby, S. R. Science 1965, 150, 619.
43. Frear, D. S.; Swanson, H. R.; Mansager, E. R.; Wien, R. J. Agric. Food Chem. 1978, 26, 1347.
44. Swanson, C. R.; Hodgson, R. H.; Kadunce, R. E.; Swanson, H. R. Weeds 1966, 14, 323.
45. Colby, S. R. Weeds 1966, 14, 197.
46. Frear, D. S.; Mansager, E. R.; Swanson, H. R.; Tanaka, F. S. Pestic. Biochem. Physiol. 1983, 19, 270.
47. Ries, S. K.; Zabik, J. J.; Stephenson, G. R.; Chen, T. M. Weed Science 1968, 16, 40.
48. Chkanikov, D. I.; Makeev, A. M.; Pavlova, N. N.; Nazarova, T. A. Fiziol. Rast. 1983, 30, 95.
49. Ando, M; Iwasaki, Y; Nakagawa, M. Agr. Biol. Chem. 1975, 39, 2137.
50. Frear, D. S. Phytochemistry 1968, 7, 381.
51. More, J. E.; Roberts, T. R.; Wright, A. N. Pestic. Biochem. Physiol. 1978, 9, 268.
52. Pillmoor, J. B.; Roberts, T. R.; Gaunt, J. K. Pestic. Sci. 1982, 13, 129.
53. Gross, G. Phytochemistry 1983, 22, 2179.
54. Michalczuk, L.; Bandurski, R. S. Biochem. and Biophys. Research Communications 1980, 93, 588.
55. Corner, J. J.; Swain, T. Nature 1965, 207, 634.
56. Bandurski, R. S. In "The Biosynthesis and Metabolism of Plant Hormones"; Corzier, A.; Hillman, J. R., Eds.; Society for Experimental Biology Seminar Series No. 23, Cambridge Univ. Press: Cambridge, 1984; Chap. 8.
57. Harvey, J., Jr.; Han, J. C-Y.; Reiser, R. W. J. Agric. Food Chem. 1978, 26, 529.
58. Suzuki, T.; Casida, J. E. J. Agric. Food Chem. 1981, 29, 1027.
59. Truscott, R. J. W.; Johnstone, P. K.; Minchinton, I. R.; Sang, J. P. J. Agric. Food Chem. 1983, 31, 863.
60. Olsen, O.; Sorensen, H. Phytochemistry 1980, 19, 1783.
61. Carlson, D. G.; Daxenbichler, M. E.; VanEtten, C. H.; Tookey, H. L.; Williams, P. H. J. Agric. Food Chem. 1981, 29, 1235.
62. Matsuo, M.; Underhill, E. W. Phytochemistry 1971, 10, 2279.
63. Kaslander, J.; Sijpesteijn, A. K.; Van der Kerk, G. J. M. Biochim. Biophys. Acta 1961, 52, 396.
64. Mildner, P.; Mihanovic, B.; Poje, M. FEBS Letters 1972, 22, 117.
65. Miller, L. P. Contrib. Boyce Thompson Inst., 1941; 12, 15.
66. Miller, L. P. Contrib. Boyce Thompson Inst. 1941, 11, 387.
67. Yamaha, T.; Cardini, C. E. Arch. Biochem. Biophysics 1960, 86, 133.
68. Hodgson, R. H.; Frear, D. S.; Swanson, H. R.; Regan, L. A. Weed Science 1973, 21, 542.
69. Hodgson, R. H.; Hoffer, B. L. Weed Science 1977, 25, 331.
70. Schneider, B.; Schutte, H. R.; Tewes, A. Plant Physiol. 1984, 76, 989.

71. Chkanikov, D. I.; Makeev, A. M.; Pavlova, N. N.; Nazarova,
     T. A. Fiziol. Rast. 1982, 29, 542.
72. Still, G. G.; Mansager, E. R. Chromatographia 1975, 8, 129.
73. Wright, A. N.; Roberts, T. R.; Dutton, A. J.; Doig, M V.
     Pestic. Biochem. Physiol. 1980, 13, 71.
74. Matern, U.; Heller, W.; Himmelspach, K. I. Eur. J. Biochem.
     1983, 133, 439.
75. Dutton, A. J.; Roberts, T. R.; Wright, A. N. Chemosphere
     1976, 5, 195.
76. Beck, A. B.; Know, J. R. Aust. J. Chem. 1971, 24, 1509.
77. Lamoureux, G. L.; Shimabukuro, R. H.; Swanson, H. R.; Frear,
     D. S. J. Agric. Food Chem. 1970, 18, 81.
78. Frear, D. S.; Swanson, H. R. Phytochemistry 1970, 9, 2123.
79. Shimabukuro, R. H.; Frear, D. S.; Swanson, H. R.; Walsh, W. C.
     Plant Physiol. 1971, 47, 10.
80. Shimabukuro, R. H.; Lamoureux, G. L.; Frear, D. S. In
     "Chemistry and Action of Herbicide Antidotes", Pallos, F.;
     Casida, J. E., Eds.; Academic Press: New York, 1978; pp.
     133-149.
81. Lamoureux, G. L.; Stafford, L. E.; Tanaka, F. S. J. Agric.
     Food Chem. 1971, 19, 346.
82. Lamoureux, G. L.; Stafford, L. E.; Shimabukuro, R. H. J.
     Agric. Food Chem. 1972, 20, 1004.
83. Lamoureux, G. L.; Gouot, J-M.; Davis, D. G.; Rusness, D. G.
     J. Agric. Food Chem. 1981, 29, 996.
84. Shimabukuro, R. H.; Lamoureux, G. L.; Swanson, H. R.; Walsh,
     W. C.; Stafford, L. E.; Frear, D. S. Pestic. Biochem.
     Physiol. 1973, 3, 483.
85. Frear, D. S.; Swanson, H. R. Pestic. Biochem. Physiol. 1973,
     3, 473.
86. EPA, 2-chloro-N-(2-ethyl-6-methyl-phenyl)-N-(2-methoxy-1-
     methylethyl)acetamide; Metolachlor Pesticide Registration
     Standard; 1980.
87. Hussain, M.; Kapoor, I. P.; Ku, C. C.; Stouts, S. J. Agric.
     Food Chem. 1983, 31, 232.
88. Rennenberg, H. Phytochemistry 1982, 21, 2771.
89. Lamoureux, G. L.; Bakke, J. E. In "Foreign Compound
     Metabolism", Caldwell, J.; Paulson, G. D., Eds., Taylor and
     Francis: Philadelphia, PA, 1984; pp. 185-199.
90. Hubbell, J. P.; Casida, J. E. J. Agric. Food Chem. 1977, 25,
     404.
91. Lay, M.; Casida, J. E. In "Chemistry and Action of Herbicide
     Antidotes"; Pallos, F.; Casida, J. E., Eds.; Academic Press:
     New York, 1978; p. 151.
92. Ezra, G.; Rusness, D. G.; Lamaoureux, G. L.; Stephanson, G. R.
     Pestic. Biochem. Physiol. 1985, 23, 108.
93. Parker, C. Pestic. Sci. 1983, 14, 40.
94. Frear, D. S.; Swanson, H. R.; Mansager, E. R. Pestic.
     Biochem. Physiol. 1983, 20, 299.
95. Frear, D. S.; Swanson, H. R.; Mansager, E. R. Pestic.
     Biochem. Physiol. 1985, 23, 56.
96. Simoneaux, B. J.; Martin, G.; Cassidy, J. E.; Ryskiewich, D.
     P. J. Agric. Food Chem. 1980, 28, 1221.
97. Ioannou, Y. M.; Dauterman, W. C. Pestic. Biochem. Physiol.
     1979, 10, 212.

98. Schuphan, I.; Westphal, D.; Haque, A.; Eling, W. ACS
     SYMPOSIUM SERIES No. 158, 1981; p. 85
99. Lamoureux, G. L.; Rusness, D. G. Proc. 188th National ACS
     Meeting, 1984.
100. Rusness, D. G.; Lamoureux, G. L. J. Agric. Food Chem. 1980,
     28, 1070.
101. Lamoureux, G. L.; Rusness, D. G. Pestic. Biochem. Physiol.
     1980, 14, 50.
102. Lamoureux, G. L.; Rusness, D. G. J. Agric. Food Chem 1980,
     28, 1057.
103. Lamoureux, G. L.; Stafford, L. E.; Shimabukuro, R. H.;
     Zaylskie, R. H. J. Agric. Food Chem. 1973, 21, 1020.
104. Shimabukuro, R. H.; Walsh, W. C.; Lamoureux, G. L.; Stafford,
     L. E. J. Agric. Food Chem. 1973, 21, 1031.
105. Mayer, P.; Kriemler, H-P.; Laanio, T. L. Agric. Biol. Chem.
     1981, 45, 361.
106. Lamoureux, G. L.; Rusness, D. G.; Unpublished research,
     1982-1985.
107. Diesperger, H.; Sandermann, H., Jr. Planta 1979, 146, 643.
108. Balabaskaran, S.; Muniandy, N. Phytochemistry 1984, 23, 251.
109. Guddewar, M. B.; Dauterman, W. C. Phytochemistry 1979, 18,
     735.
110. Burkholder, R. R. S.; M.S. Thesis, North Dakota State Univ.,
     Fargo, ND 1977.
111. Still, G. G.; Rusness, D. G. Pestic. Biochem. Physiol. 1977,
     7, 210.
112. Rusness, D. G.; Still, G. G. Pestic. Biochem. Physiol. 1977,
     7, 220.
113. Good, N. E.; Andreae, W. A. Plant Physiol. 1957, 32, 561.
114. Rosz, N.; Neish, A. C. Can. J. Biochem. 1968, 46, 797.
115. Ladesic, B. C.; Pokorny, M.; Keglevic, D. Phytochemistry
     1970, 9, 2105.
116. Hoffman, N. E.; Fu, J-R.; Yang, F. Y. Plant Physiol. 1983,
     71, 197.
117. Amrhein, N.; Schneebeck, D.; Skorupka, H.; Tophof, S.
     Naturwissenshafften (in press).
118. Kionka, C; Amrhein, N. Planta 1984, 162, 226.
119. Matern, U.; Feser, C.; Heller, W. Arch. Biochem. Biophys.
     1984, 235, 218.
120. Kadunce, R.; Stolzenberg, G.; Davis, D.; Proc. 168th National
     ACS Meeting 1974.
121. Larson, J. D.; Lamoureux, G. L. J. Agric. Food Chem. 1984,
     32, 177.
122. Lamoureux, G. L.; Larson, J. D. Unpublished research, 1983.
123. Shindo, N.; Keto, M. "Abstracts of Papers", 5th-International
     Congress of Pesticide Chemistry, Kyoto, Japan, Aug. 1982;
     International Union of Pure and Applied Chemistry: Oxford,
     1982.
124. Andreae, W. A.; Good, N. E. Plant Physiol. 1957, 32, 566.
125. Feung, C-s.; Hamilton, R. H.; Mumma, R. O. J. Agric. Food
     Chem. 1975, 23, 373.
126. Arjmand, M.; Hamilton, R. H., Mumma, R. O. J. Agric. Food
     Chem 1978, 26, 1125.
127. Davidonis, G. H.; Arjmand, M.; Hamilton, R. H.; Mumma, R. O.
     J. Agric. Food Chem. 1979, 27, 1086.

128. Andreae, W. A.; Good, N. E. Plant Physiol. 1955, 30, 380.
129. Badenoch-Jones, J.; Summons, R. E.; Rolfe, B. G.; Letham, D. S. J. Plant Growth Regul. 1984, 3, 23.
130. Severson, J. G., Jr.; Bohm, B. A.; Seaforth, C. E. Phytochemistry 1970, 9, 107.
131. Aranda, G.; Tabet, J.-C.; Leguoy, J.-J.; Caboche, M. Phytochemistry 1984, 23, 1221.
132. Feung, C-s.; Hamilton, R. H.; Mumma, R. O. J. Agric. Food Chem. 1973, 21, 637.
133. Zama, P.; Mumma, R. O. Weed Sci. 1983, 31, 537.
134. Davidonis, G. H.; Hamilton, R. H.; Mumma, R. O. Plant Physiol. 1978, 62, 80.
135. Davidonis, G. H.; Hamilton, R. H.; Mumma, R. O. Plant Physiol. 1980, 65, 94.
136. Archer, T. E.; Stokes, J. D. J. Agric. Food Chem. 1983, 31, 286.
137. Fedec, P.; Cossins, E. A. Phytochemistry 1976, 15, 359.
138. Van Sumere, C. F.; dePotter, H.; Ali, H.; Degrauw-VanBussel, M. Phytochemistry 1973, 12, 407.
139. Bialek, K.; Meudt, W. J.; Cohen, J. D. Plant Physiol. 1983, 73, 130.
140. Aberg, B.; Popoff, T.; Theander, O. Swedish J. Agric. Res. 1982, 12, 41.
141. Davidonis, G. H.; Hamilton, R. H.; Vallejo, R. P.; Buly, R.; Mumma, R. O. Plant Physiol. 1982, 70, 357.
142. Feung, C-s.; Mumma, R. O.; Hamilton, R. H. J. Agric. Food Chem. 1974, 22, 307.
143. Hirom, P. C.; Millburn, P. In. "Foreign Compound Metabolism in Mammals, Vol. 6"; Hathway, D. E., Ed.; The Royal Soc. Chem.: Burlington House, London, 1981; pp. 111-132.
144. Chkanikov, D.; Pavlova, N.; Makeev, A.; Nazarova, T. Fiziol. Rast. 1984, 31, 321.
145. Pryde, A.; Hanni, R. J. Agric. Food Chem. 1983, 31, 564.
146. Sonobe, H.; Kamps, L.; Mazzola, E.; Roach, J. J. Agric. Food Chem. 1981, 29, 1125.
147. Sonobe, H.; Carver, R.; Krause, T.; Kamps, L. J. Agric. Food Chem. 1983, 31, 96.
148. Frear, D. S.; Swanson, H. R.; Stolzenberg, G. E. Proc. 174th ACS National Meeting, 1977.
149. Quistad, G. B.; Staiger, L. E.; Schooley, D. Nature 1982, 296, 462.
150. McMichael, K.; Overton, K.; Picken, D. Phytochemistry 1977, 16, 1290.
151. Galliard, T.; Dennis, S. Phytochemistry 1974, 13, 1731.
152. Dennis, S.; Galliard, T. Phytochemistry 1974, 13, 2469.
153. Frear, D. S.; Swanson, H. R.; Mansager, E. R. Proc. 187th National ACS Meeting, 1984.
154. Otto, S.; Buetel, P.; Drescher, N.; Huber, R.; IUPAC Advances in Pesticide Science, Zurich 1978; Pergammon: New York, 1979; pp. 551-556.
155. Sandermann, H., Jr.; Scheel, D.; v.d. Trenck, T. J. Appl. Polymer Sci; Applied Polymer Symp. 1983, 37, 407.
156. Pillmoor, J.; Gaunt, J.; Roberts, T. Pestic. Sci. 1984, 15, 375.

157. Still, G. G.; Balba, H. M.; Mansager, E. R. J. Agric. Food
      Chem. 1981, 29, 739.
158. Huber, R.; Otto, S. In "Pesticide Chemistry Human Welfare and
      the Environment Vol. 3"; Miyamoto, J.; Kearney, P. C., Eds.;
      Pergammon Press: Oxford, England, 1983; p. 357.
159. Stratton, G.; Wheeler, W. J. Agric. Food Chem. 1983, 31,
      1076.
160. Freudenberg, K.; Neish, A. C. "Constitution and Biosynthesis
      of Lignin"; Springer-Verlag: New York, 1968.
161. Still, G. G.; Mansager, E. R. Pestic. Biochem. Physiol.
      1973, 3, 87.
162. Smith, A. E. Weed Sci. 1979, 27, 392.
163. Takeuchi, Y.; Komamine, A.; Saito, T.; Watanabe, K.;
      Morikawa, N. Physiol. Plant. 1980, 48, 536.
164. Bull, D. L.; Whitten, C. J.; Ivie, G. W. J. Agric. Food
      Chem. 1976, 24, 601.
165. Mine, A.; Miyakado, M.; Matsunaka, S. Pestic. Biochem.
      Physiol. 1975, 5, 566.
166. Ogawa, K.; Tsuda, M.; Yamuchi, F.; Yamaguchi, J.; Misato, T.
      J. Pestic. Sci. 1976, 1, 219.
167. Marshall, T. C.; Dorough, H. W. J. Agric. Food Chem. 1977,
      25, 1003.
168. Miller, L. P. Contrib. Boyce Thompson Inst., 1941; 12, 167.
169. Frear, D. S.; Swanson, H. R. Pestic. Biochem. Physiol. 1975,
      5, 73.
170. Nakagawa, M.; Kawakubo, K.; Ishida, M. Agric. Biol. Chem.
      1971, 35, 764.
171. Zulalian, J.; Blinn, R. C. J. Agric. Food Chem. 1977, 25,
      1033.
172. Feung, C-s.; Loerch, S. L.; Hamilton, R. H.; Mumma, R. O. J.
      Agric. Food Chem. 1978, 26, 1065.
173. Meikle, K. W. J. Agric. Food Chem. 1977, 25, 746.
174. Shimabukuro, R. H.; Walsh, W. C.; Stolzenberg, G. E.; Olson,
      P. A. Abstr. Weed Sci. Soc. America 1975, No. 171.
175. Kamimura, S.; Nishikawa, M.; Saeki, H.; Takahi, Y.
      Phytopathology 1974, 64, 1273.
176. Frear, D. S.; Swanson, H. R. J. Agric. Food Chem. 1978, 26,
      660.
177. Schuphan, I.; Ebing, W. Pestic. Biochem. Physiol. 1978, 9,
      107.
178. Lamoureux, G. L.; Stafford, L. E. J. Agric. Food Chem. 1977,
      25, 512.
179. Gaughan, L. C.; Casida, J. E. J. Agric. Food Chem. 1978, 26,
      525.
180. Drinkwine, A. D.; Fleeker, J. R. J. Agric. Food Chem. 1981,
      29, 763.
181. Murakoshi, I.; Ikegami, F.; Kato, F.; Haginiwa, J.; Lambein,
      F.; Rompuy, L. V.; Van Parijs, R. Phytochemistry 1975, 14,
      1269.
182. Still, G. G. Science 1968, 159, 992.
183. Boyer, G. L.; Zeevaart, J. A. D. Plant Physiol. 1982, 70,
      227.
184. Roberts, T. R. Pestic. Biochem. Physiol. 1977, 7, 378.
185. Quistad, G. B.; Staiger, L. E.; Mulholland, K. M.; Schooley,
      D. A. J. Agric. Food Chem. 1982, 30, 888.

186. Krause, J.; Strack, D. J. Chromatogr. 1979, 176, 465.
187. Riov, J.; Gottlieb, H. E. Physiol. Plant. 1980, 50, 347.
188. Sharma, A.; Chibber, S. S.; Chawla, H. M. Phytochemistry 1979, 18, 1253.
189. Towers, G.; Hutchinson, A.; Good, A. Nature 1958, 181, 1535.
190. Lamoureux, G. L. Unpublished Data, 1968.
191. Pont, V; Collet, G. F. Phytochemistry 1980, 19, 1361.
192. Shimabukuro, R. H. Plant Physiol. 1967, 42, 1269.
193. Yu, C.; Atallah, Y.; Whitacre, D. J. Agric. Food Chem. 1980, 28, 1090.
194. Haque, A.; Weisgerber, J.; Klein, W. Chemosphere 1976, 3, 167.
195. Locke, R. H.; Bastone, V. B.; Baron, R. L. J. Agric. Food Chem. 1971, 19, 1205.
196. Locke, R. K.; Chen, J. Y. T.; Damico, J. N.; Dusold, L. R.; Sphon, J. A. Arch. Environ. Contamination Toxicol. 1976, 4, 60.
197. Balba, H. M.; Still, G. G.; Mansager, E. R. J. Assoc. Off. Anal. Chem. 1979, 62, 237.
198. Gross, D.; Laanio, T.; Dupuis, G.; Esser, H. Pestic. Biochem. Physiol. 1979, 10, 49.
199. Verloop, A.; Nimmo, W. Weed Res. 1969, 9, 357.
200. Chrzanowski, R.; Leitch, R. J. Agric. Food Chem. 1982, 30, 155.
201. Ku, C.; Kapoor, I.; Rosen, J. J. Agric. Food Chem. 1978, 26, 1352.
202. Honeycutt, R. C.; Adler, I. L. J. Agric. Food Chem. 1975, 23, 1097.
203. Khan, S. U. J. Agric. Food Chem. 1980, 28, 1096.
204. Yih, R.; Swithenbank, C. J. Agric. Food Chem. 1971, 19, 314.
205. Yih, R. Y.; McRea, H. D.; Wilson, I. F. Science 1968, 161, 376.
206. Still, G. G.; Mansager, E. R. Pestic. Biochem. Physiol. 1975, 5, 515.
207. Chin, W. T.; Tanovick, R. P.; Cullen, T. E.; Holsing, G. C. Weeds 1964, 12, 201.
208. Fragiadakis, A.; Sotiriou, N.; Korte, F. Chemosphere 1981, 10, 1315.
209. Rouchard, J.; Decallone, J. R.; Meyer, J. A. Pestic. Sci. 1978, 9, 74.
210. Abdel-Wahab, A. M.; Kuhr, R. J.; Casida, J. E. J. Agric. Food Chem. 1966, 14, 290.
211. Ishizuka, M.; Kondo, Y.; Ishizuka, I.; Sumimoto, S.; Takeuchi, Y.; Fifth International Congress of Pesticide Chemistry, IUPAC, Kyoto, Japan, Aug. 29 – Sept. 4, 1982 (Poster)
212. Feung, C-s.; Hamilton, R. H.; Mumma, R. O. J. Agric. Food Chem. 1973, 21, 632.
213. Liu, T-y.; Oppenheim, A.; Castelfranco, P. Plant Physiol. 1965, 40, 1261.
214. Crawford, M. J.; Hutson, D. H.; Stoydin, G. Xenobiotica 1980, 10, 169.

RECEIVED August 12, 1985

ISOLATION AND IDENTIFICATION
OF XENOBIOTIC CONJUGATES

5

Isolation of Xenobiotic Conjugates

W. Muecke

Ciba-Geigy Ltd., Basel, Switzerland

Procedures and instruments are available for the
isolation of virtually all types of low-molecular,
polar xenobiotic conjugates. Most isolation procedures involve the use of liquid chromatography
(anion exchange and partition mode), high performance liquid chromatography (partition mode), and thin
layer chromatography (adsorption mode). A dedicated
isolation strategy must be designed, after taking
into account the properties of the conjugates under
study, the nature of the matrix and the features of
available methods with regard to capacity, selectivity, and resolution.

The Vail symposium in 1975 on bound and conjugated pesticide
residues (1) summarized the state of the art on methods and instru
ments that were available for the isolation of conjugates at
that time. Although dedicated to pesticides, the presentations
reviewed the frontiers of methodology available for the
isolation of all types of xenobiotics conjugates (2,3,4,5,6,7).
When summarizing the symposium, Geissbuehler (8) stated,
"HPLC has been mentioned as a possibility for speeding up and improving separations ... and ought to be utilized without delay".
In the last decade HPLC has become the most powerful tool the
metabolism chemist has for isolating xenobiotic conjugates.
During this period , many improvements have also been made in older
traditional techniques. Noteworthy here is the positive impact
HPLC had on conventional liquid chromatography regarding equipment
and stationary phases and the significantly enhanced sensitivity of
the radioactivity detectors for monitoring column effluents.
Dramatic improvements in instrumentation and methods used for
the identification of xenobiotic conjugates (and other metabolites),
notably proton nuclear magnetic resonance and mass spectroscopy,
have significantly influenced the strategies for the isolation of
conjugates. The recently developed soft ionization techniques make
it possible to obtain useful mass spectral information for a wide
variety of intact underivatized xenobiotic conjugates. The state of

0097–6156/86/0299–0108$06.00/0
© 1986 American Chemical Society

the art NMR spectrometers give interpretable spectra from sample
amounts as low as about 10 µg. Consequently methods with limited
capacity, but high resolution can be utilized at an earlier stage of
the isolation procedure. Thus a wider variety of purification
systems can now be selected from. These developments in spectroscopy are thus changing the isolation strategy and thereby significantly improving the economy of the whole process.

During the recent years several reviews on the analytical
methodology and strategy involved in isolating xenobiotic
metabolites, including conjugates have been published,
(9,10,11,12,13). This presentation will not be a review, but will
attempt to place within a frame of the presently used techniques and
methods my very personal experiences and views. This discussion
will be limited to the "classical" polar conjugates. Very different
methodology is required to purify the non polar "lipophilic"
conjugates. Furthermore, only methods that result in the isolation
of intact conjugates in amounts adequate for multiple-method
structural identification studies will be discussed (i.e. combined
methods such as GC-MS and HPLC-MS and derivatization techniques will
not be covered).

Prerequisites

For the sake of convenience the isolation process may be arbitrarily
subdivided into several interdependent steps. The xenobiotic
conjugates present in solid matrices must be solubilized and separated from the bulk of other materials concomittantly extracted or
present in the original liquid sample (enrichment).
Thereafter the mixtures of xenobiotic conjugates must be separated into individual fractions (separation) which in turn must be
ultimately purified (purification) for spectroscopic identification.
As the investigator moves from the enrichment via separation to
purification steps the methods require less capacity and selectivity, but greater resolution. Suitable methods have to be selected
and arranged in a meaningful sequence in order to obtain an adequate, (i.e. a successful and economical) isolation strategy.

The overall strategy should take into consideration the nature
of the matrix and the properties of the metabolites present. Relevant properties, to be investigated by appropriate analytical procedures include hydrolytical stability, volatility, lipophilicity,
solvent partition properties, electrical charge and susceptibility
towards defined enzymes. A reliable analytical system providing
efficient separation of individual metabolites present, needs to be
established as early in the isolation procedure as possible. This
analytical system is used to monitor the result of each single isolation step, and - of equal importance - to recognize any artifact
formation which may occur during the execution of that step. Two-
directional TLC and reversed phase HPLC are fast and, therefore,
commonly used methods for this purpose.

Enrichment Methods

The extraction of polar conjugates from solid samples (e.g. plant
material, animal tissue, and feces) is most commonly carried out

with polar solvents, such as methanol, acetone and acetonitrile and
mixtures thereof with water. These solvents will, when adequately
applied, solubilize all low molecular conjugates, including glutathionates. After reconstituting these extracts in water, the
solubilized metabolites may be processed as aqueous samples.

Partitioning of the aqueous solution with water non-miscible
organic solvent is frequently used to remove lipophilic metabolites
and contaminants as well. The extraction of polar conjugates from
aqueous solution with organic solvents at or below their pk-values
or at their isoelectric points has been used, occasionally, but
usually does not result in a complete partitioning into the organic
solvent. Ion pair extraction of ionizable conjugates has never
aquired a widespread use for preparative isolations, because optimation of the extraction conditions with regard to the selection and
concentration of the counter ions, the pH-value, etc. is tedious.
Therefore, this method of enriching a complex mixture of metabolites
is competitive in only special cases (14).

Currently the most powerful and commonly used enrichment
technique is solid extraction applying non-ionic hydrophobic resins,
such as Amberlite XAD-2, Amberlite XAD-4, Porapak Q and their
analogs. These resins extract a wide variety of conjugates from
aqueous phases providing a hydrophobic moiety is present in the
molecule. Extractions of virtually all types of conjugates from
various matrices have been reported [e.g. O-, S- and N-glucuronides
(15-24), O- and N-glucosides (25-27), sulfates (17, 28), amino acid
conjugates including glutathionates, homoglutathionates and
taurinates (17,21,24, 29-32)].

Controlling the pH value of the aqueous solution, (to maintain
the conjugates in their non-ionized forms) and/or increasing
the ionic strength by adding salts (preferably buffer) are very
useful tools to enhance adsorption of polar xenobiotic conjugates to
hydrophobic resin columns. Usually the adsorbed metabolites are
quantitatively desorbed from the column by elution with water-
miscible organic solvents, such as methanol, acetone or acetonitrile. However, unreasonably low recoveries are sometimes obtained,
even when columns such as those prepared from Amerlite XAD are
sequentially eluted with polar, non-polar, acidified and
alkalinized solvents. We have found that this type of adsorption is
not a problem if silica bound reversed phase columns (e.g. RP 8 or
RP 18) are used instead of XAD-2 and XAD-4 resin columns. As these
materials are now available as bulk material at reasonable prices
and particle sizes allowing low pressure chromatography, they represent an attractive alternative to the forementioned copolymers
Amberlite XAD-2, Amberlite XAD-4 and Porapak Q (33). Under
favourable circumstances these reversed phase enrichment methods can
be - in a preparative scale - extended to the separation of
individual metabolites.

Other solid extractants, e.g. ion exchange resins (34,35),
charcoal (36-38), Sephadex LH 20 (39,40) were used for the recovery
of conjugates from crude aqueous extracts of biological samples.
Even though these solid extractants are not recommended for general
use, however, an excellent separation of conjugates in the urine of
rats dosed with of o-toluidine on Sephadex LH-20 was published by
Son et al. (40) and is demonstrated in Figure 1.

Liquid Column Chromatography

Conventional liquid chromatography, in its different operation
modes, still represents a versatile and powerful technique. The
development of HPLC during the last decade has also made a significant impact on the "hardware" used with traditional column chromatography (i.e. solvent delivery pumps, gradient programmers, sample
application systems, continuous flow UV and radioactivity detectors,
have been greatly improved). These advances make it possible to
carefully control the conventional liquid chromatographic conditions. Furthermore, the increasing commercial availability of synthetic stationary phases, originally developed for HPLC, but adapted
for low pressure chromatography (particle-form, -size and -distri-
bution) has widened the applicability of this traditional technique
and the technical differences between these methods are diminishing
(41,42).

Although all modes of column chromatography have been used in
the isolation of underivatized polar conjugates, ion exchange and
partition chromatography can be considered as modes of first choice.
Cellulose and dextran based weakly basic anion exchangers, such as
the DEAE-type, are the most versatile and most commonly used ion
exchange columns. Two decades ago Knaak et al. (43) demonstrated
excellent separation of various glucuronic acid and sulfate ester
conjugates in the urine of carbaryl treated rodents (Figure 2).

Gradients formed by increasing ionic strength (constant pH) are
preferable to gradients formed by changing the pH. Volatile buffers, such as formates and acetates of pyridine and ammonium are
usually used because they can be easily removed from the eluted
fractions. However, the use of non-ionic hydrophobic resins and the
silica bonded alkyl phases offer a convenient alternative to
remove also non-volatile buffers.

The separation of conjugates from the urine of rats dosed with
the fungicide pyroquilon is presented in Figure 3 (44).

Chromatography on polyglycoside based DEAE and QAE anion exchangers has also been used successfully for the isolation of
various glutathionate pathway metabolites (45-48). Figure 4 demonstrates the separation of this type of metabolites of methidathion
produced by rat liver in-vitro experiments (47).

Cation exchange chromatography has been used occasionally,
mainly for amino acid conjugates, such as glutathionates, glutamyl
cysteinates, malonyl cysteinates, N-acetyl cysteinates and
cysteinates (49-51). However, Larsen et al. (52) also used it for
isolating S- and O-glucuronides by exploiting the properties of the
corresponding exocons. In contrast to anion exchange chromatography
changing the pH is the most common way of forming a gradient for the
elution of cation exchange columns. It should be noted that strongly acidic divinylbenzene cross-linked polystyrene based resins, such
as Dowex 50, tend to generate artifacts when used to isolate acid
labile conjugates.

Recently Sjövall et al. (53) proposed the use of lipophilic
ion exchangers for group separation of conjugates. These exchangers
can be eluted with organic solvents and should be essentially devoid

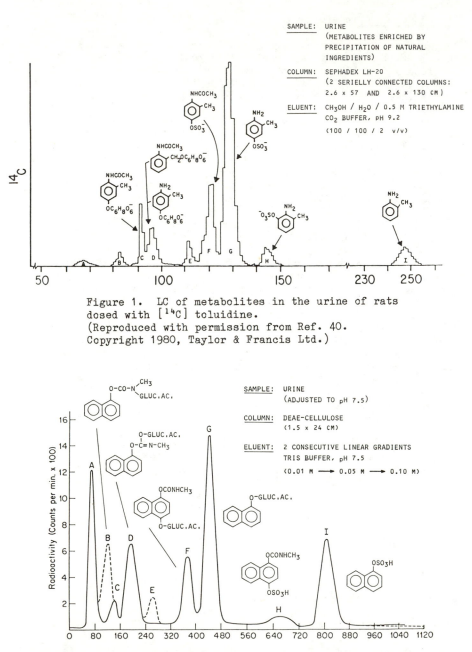

SAMPLE: URINE
         (METABOLITES ENRICHED BY
         PRECIPITATION OF NATURAL
         INGREDIENTS)

COLUMN: SEPHADEX LH-20
         (2 SERIELLY CONNECTED COLUMNS:
         2.6 x 57 AND 2.6 x 130 CM)

ELUENT: CH_3OH / H_2O / 0.5 M TRIETHYLAMINE
         CO_2 BUFFER, pH 9.2

         (100 / 100 / 2 v/v)

Figure 1. LC of metabolites in the urine of rats
dosed with [^{14}C] toluidine.
(Reproduced with permission from Ref. 40.
Copyright 1980, Taylor & Francis Ltd.)

SAMPLE: URINE
         (ADJUSTED TO pH 7.5)

COLUMN: DEAE-CELLULOSE
         (1.5 x 24 CM)

ELUENT: 2 CONSECUTIVE LINEAR GRADIENTS
         TRIS BUFFER, pH 7.5

         (0.01 M ⟶ 0.05 M ⟶ 0.10 M)

Figure 2. LC of metabolites in the urine of guinea-pigs and rats
dosed with [^{14}C] carbaryl.
(Reproduced with permission from Ref. 43.
Copyright 1965, American Chemical Society)

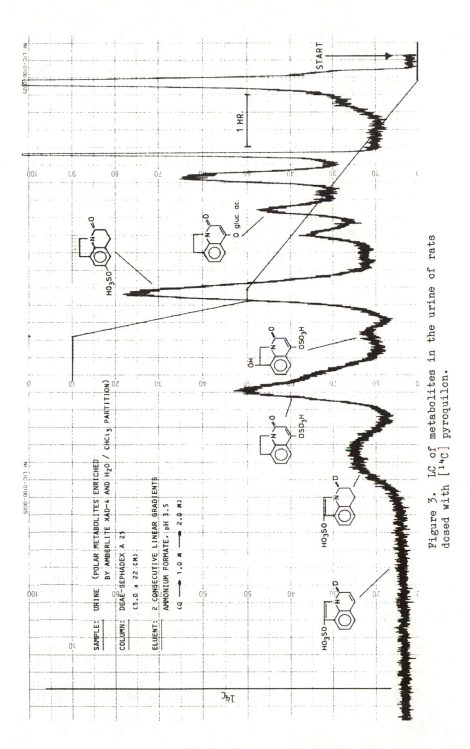

Figure 3. LC of metabolites in the urine of rats dosed with [^{14}C] pyroquilon.

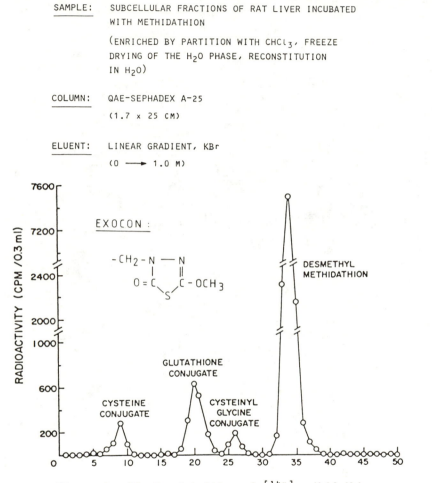

SAMPLE: SUBCELLULAR FRACTIONS OF RAT LIVER INCUBATED
          WITH METHIDATHION

          (ENRICHED BY PARTITION WITH CHCl$_3$, FREEZE
          DRYING OF THE H$_2$O PHASE, RECONSTITUTION
          IN H$_2$O)

COLUMN: QAE-SEPHADEX A-25

          (1.7 x 25 CM)

ELUENT: LINEAR GRADIENT, KBr

          (0 ⟶ 1.0 M)

Figure 4. LC of metabolites of [^{14}C] methidathion
produced by rat liver in-vitro.
(Reproduced with permission from Ref. 47.
Copyright 1981, Academic Press Inc.)

of non-ionic interactions. This very interesting approach has still to be validated under practical conditions.

Besides ion exchange, partition chromatography is the most powerful technique for isolating polar conjugates. The older technique for preparing a partition column involved coating a deactivated carrier with the stationary phase. This usually resulted in a rather labile system that was tedious to operate and is now considered to be obsolete. The formation of a reversed phase partition system in-situ involves washing an appropriate carrier such as Amberlite XAD-2 with a polar solvent containing a small amount of a lipophilic solvents. The lipophilic solvent is preferentially taken up by the resin, thus forming a stationary phase, that is stabile, and gives high resolution reversed phase chromatographic separations (54,55). Figure 5 illustrates the use of this type of stationary phase for the separation of several glucuronides including two enantiomers in the urine of man (55). Correspondingly, using a polar carrier such as silica and washing with an unpolar solvent containing small amounts of a hydrophilic solvent leads to a "normal" phase partition system.

To date the literature contains only a few examples of the use of the non-ionic resins (e.g. Amberlite XAD-2, Amberlite XAD-4) and especially the organic phases covalently bound to porous silica for the preparative separation of conjugates by conventional liquid column chromatography. Alkyl, amino, diol, nitril, phenyl moieties have an inherently high potential for the preparative separation of polar conjugates. The use of corresponding HPLC systems to aid in the optimation of the separation conditions for the low pressure chromatography is an additional and very advantageous feature of these systems. Figure 6 illustrates the use of RP 18 on a preparative scale to separate conjugates in the urine of rats orally dosed with the safening agent 4,6-dichloro-2-phenyl-pyrimidine (44).

In the past size exclusion chromatography, utilizing hydrophilic as well as lipophilic gels, such as Sephadex G-10 and Sephadex LH-20 were used quite extensively for the penultimate and ultimate purification of individual conjugates. Some excellent separations of complex mixtures have been accomplished using these types of columns and reported in the literature (39,40,56). However, the separation mode of these systems is probably best described as partition chromatography. Thus these techniques have been effectively replaced by reversed phase chromatography on lipophilic resins or silica bound phases.

High Performance Liquid Chromatography

Although HPLC was developed as an analytical tool it has proven to be the most powerful and versatile method invented during the last decade for the isolation of metabolites, including polar conjugates, in preparative amounts. The commercial availability of a large variety of stationary phases allows the application of all modes of liquid chromatography. When used in combination with adequate eluting solvents chromatographic systems for virtually all types of conjugates are provided. Increasing the size of the chromatography column and/or repetitive sample injections are used to overcome the

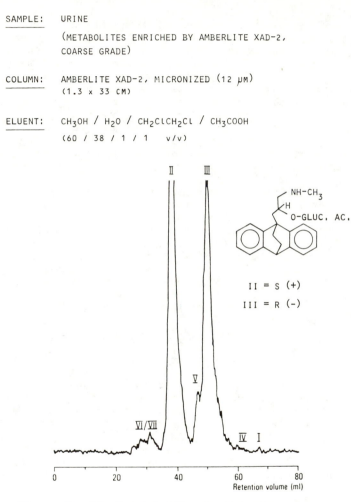

SAMPLE: URINE

           (METABOLITES ENRICHED BY AMBERLITE XAD-2,
           COARSE GRADE)

COLUMN: AMBERLITE XAD-2, MICRONIZED (12 μM)
           (1.3 x 33 CM)

ELUENT: CH₃OH / H₂O / CH₂ClCH₂Cl / CH₃COOH
           (60 / 38 / 1 / 1   v/v)

II = S (+)

III = R (-)

Retention volume (ml)

Figure 5. LC of metabolites in the urine of man
dosed with [¹⁴C] oxaprotiline.
(Reproduced with permission from Ref. 55.
Copyright 1984, Taylor & Francis Ltd.)

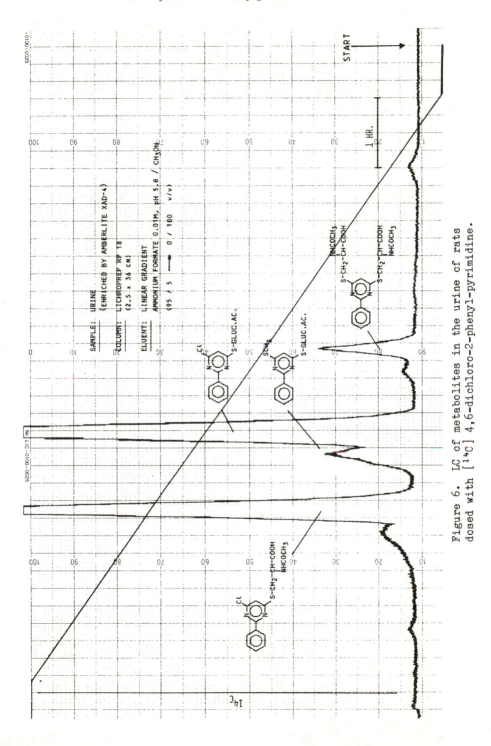

Figure 6. LC of metabolites in the urine of rats
dosed with [^{14}C] 4,6-dichloro-2-phenyl-pyrimidine.

rather low capacity of the method. High selectivity can be achieved by the proper selection of the HPLC column and high resolution is an inherent feature of this technique. Commercially available equipment is designed to allow convenient and precise control of the chromatographic parameters. Therefore, the chromatographic conditions can be quickly and efficiently optimized. Because the HPLC system is a closed system, usually operated under mild conditions at ambient temperature with exclusion of light and air, artifact formation is usually not a serious problem. Therefore, despite the high price of the equipment, the technique is highly economical. In order to maximize the utilization of these advantages, HPLC should be utilized as early as possible in the isolation process. Occasionally, lipophilic contaminants can be removed from the sample by solvent partition, and the polar metabolites (enriched by non-ionic resin extraction) can be directly applied to a HPLC column. However, in complex mixtures or in very crude samples, separation of individual metabolites or groups of metabolites utilizing high capacity methods prior to HPLC is usually advisable for the sake of economy.

The high resolution afforded by HPLC usually make it the method of choice for the ultimate purification step.

Reversed phase HPLC is the most popular chromatographic method for purification of polar conjugates. The reversed phase technique is commonly performed on silica bonded phases, such as RP 8 and RP 18.

When these systems are used control of a variety of parameters is of paramount importance. The specific properties of conjugates can be individually addressed in the design of the chromatographic procedure. The control of the pH of the eluent is of importance in maintaining ionizable conjugates (and/or the other ionizable contaminants present in the sample) in the desired status to optimize separation. Precisely controlling the ionic strength of the eluent is also important in maintaining the metabolites in the appropriate form to maximize resolution. By applying these principles in a dedicated manner, virtually all low-molecular, polar conjugates can be separated and/or purified using these reversed phase systems. This also applies to the separation of isomers, including enantiomers. Reviewing the literature on the isolation of conjugates, reveals that the advantages of reverse phase HPLC has been exploited in nearly all studies reported in recent publications.

Silica bonded phases with functional groups (e.g. CN, NH_2, Diol) which can be operated as "normal" and "reversed" phase systems have not been extensively used for preparative purposes. Some preliminary experiments in our laboratory indicated that complex conjugate mixtures can be separated using these types of columns but that the capacity and the resolution is lower compared to RP 18 columns. When operating in the "normal" phase mode, polar conjugates are eluted with essentially non-aqueous solvents. Unfortunately, polar compounds in non-aqueous solvents have a strong tendency to stick to the glass beads scintillator of the radioactivity detector causing unpleasant memory effects. On the basis of the limited information availabe no significant advantage of these "normal" phase systems is obvious, however, more experience is necessary to make a sound judgement on the usefulness of these systems for polar conjugates.

Adsorption HPLC on silica phases has been occasionally used to isolate polar conjugates (57), but it is not generally recommended, as the desorption of polar compounds requires highly polar (mostly aqueous) solvents, which tend to deteriorate the stationary phases.

Other Methods

The thin layer chromatography technique, now more than 30 years old, continues to deserve its reputation as a preparative tool for the isolation of polar conjugates. Reasons for its continuous popularity include that it is easy to use, the technique is fast, the excellent predictibility of the results from analytical runs, the low price of the equipment, and the high resolution of the method.

Preparative TLC is commonly performed using silica as the support, but other supports have been occasionally used as well (58). The method is generally used for prepurification rather than for separation of complex mixtures; generally an additional ultimate purification step is required to remove the contaminants coextracted from the support. Preparative TLC has been used for all types of polar compounds, but precautions must be taken with labile conjugates, as the method is prone to artifact formation.

Paper chromatography was occasionally used to isolate polar conjugates until recently, but the usage of this technique is now diminishing. Gas chromatography has never gained recognition as a common tool for the isolation of polar conjugates. However, one laboratory (i.e. the Metabolism and Radiation Research Laboratory, Fargo, N.D., USA) has had the patience and the eagerness to acquire the experience to successfully apply this technique to glucuronic acid, amino acid and glucose conjugates after appropriate derivatization (21,24,27,52,59). Nevertheless, looking through their recent publications, it appears that these colleagues are now tending to replace gas chromatography by more common methods.

The recently developed counter current chromatography, originating from the counter current distribution techique, is in its various technical modifications, potentially suitable for the isolation of polar xenobiotic conjugates. The fractionation of polar plant constituents, including glycosides has been demonstrated (60-62). Also, separation of a mixture of the glucuronic acid, sulfuric acid and glucose conjugate of p-nitrophenol using this approach has been reported (63). Although the high capacity of the method is a very attractive feature of counter current chromatography, especially for crude plant extracts, the comparatively low resolution of this method will probably not allow it to successfully compete against other established methods.

The Ultimate Purification

After most of the contaminating materials have been removed from a compound an ultimate purification step is required prior to spectroscopic analysis. For polar conjugates this most commonly implies HPLC in the reversed phase mode. However, even when the final chromatographic procedure yields the metabolite in highly purified form special considerations must be taken in this final step including the succeeding transfer procedure.

In addition to mass spectral analysis, proton NMR is usually
required for the unequivocal structure elucidation of most xenobiotic conjugates. Modern NMR instruments will produce interpretable spectra from samples of about 10 μg. In contrast to MS, which
may act as a separation technique by sequential volatilization of
components in a mixture, the NMR spectrum will directly reflect the
purity of the sample in the NMR tube. This means, that impurities
in the sample will jeopardize the assignment of a structure to the
xenobiotic conjugate.

Based on my experience, there are often more impurities brought
into the sample during the ultimate purification and transfer procedure, than concomittantly isolated from the matrix. Having this in
mind, I will devote the remainder of this presentation to a discussion of procedures used in our laboratory to minimize contamination
of the sample during these final procedures. The manufacturers of
the high quality solvents commonly used in HPLC guarantee $\leqslant 10^{-3}$ %
of non-volatile contaminants. This means, that up to 10 μg/ml may
be present, which is about in the same order of magnitude as the
metabolite to be elucidated. Sticking to a reliable brand of a
solvent and monitoring its quality by "blank" NMR spectroscopy helps
to circumvent this potential problem. When for the sake of
separation efficiency the metabolite is eluted as a broad peak, the
purified metabolite is rechromatographed using elution conditions
that yield a lower K'-value. This modification leads to a narrow
peak (i.e. to a smaller elution volume) and consequently to less
solvent-borne contaminants in the final sample. The same procedure
is used, when the metabolite was obtained after multiple
injections.

The eluted fractions are manually collected in small conical
flasks, which are carefully visually inspected. All vessels having
scratches or other signs of potential contamination sources are
rejected. Immediately before use, they are rinsed - in and
outside - successively with water, methanol, chloroform and ether of
highest quality. Thereafter, the flasks are only handled with
forceps. During collection of the metabolite, care is taken to
assure that the eluate contacts the smallest inner surface area of
the flask possible. All instruments, tools, devices, etc. and even
the site of preparation are exclusively used for this purpose to
minimize any risk of sample contamination. The eluate is evaporated
under nitrogen protection in a miniature rotary evaporator.
Thereafter, the flask is put in a beaker and placed in a dessicator,
operated at about 100 millibar in the dark using dust-free silica as
a dessicating agent. The sample flask, the beaker and the silica
are covered with alumina foil in which holes are punched as to allow
the movement of vapor, but to reduce the risk of contamination by
particles upon aeration of the dessicator.

After drying, usually overnight, the sample is transferred with
the deuterated solvent (preferably acetone or methanol) into the
micro NMR tube. The qualities of the NMR solvents often fluctuate
(even from the same manufacturer). Therefore, each batch of solvent
is checked by NMR spectroscopy for contaminants prior to use with an
unknown compound.

The micro NMR tube is rinsed with the same sequence of solvents
as mentioned before and kept under vacuum at 120 °C during the night

before use. A syringe to be used for the transfer of the deuterated solvent is kept under the same conditions. With this syringe three aliquots of the solvent are added sequentially to the flask containing the sample. The three aliquots of sample solution are carefully transferred into the NMR tube with a pasteur pipette (which is drawn out in a flame immediately before use). The tube is then closed with a cap, and stored in the refrigerator under nitrogen and dessicating conditions until the spectra are taken. The cap and the rubber bulb used to operate the pasteur pipette are used only once and than discarded to avoid contamination. Only after the NMR spectrum has been recorded and interpreted are aliquots taken for MS investigations. These procedures are used for all types of metabolites. Even when ammonium formate buffers (up to 10^{-2} M) were used as eluents in the ultimate purification step, there was never any significant effect on the quality of the spectra, because the buffer components completely volatilize under these conditions.

Literature Cited

1. Kaufman, D.D.; Still, G.G.; Paulson, G.D.; Bandal, S.K., Eds. "Bound and Conjugated Pesticide Residues"; ACS SYMPOSIUM SERIES No. 29, American Chemical Society: Washington, D.C., 1976.
2. Frear, D.S. In (1); 1976; pp. 35-54.
3. Bakke, J.E. In (1); 1976; pp. 55-67.
4. Mumma, R.O.; Hamilton, R.H. In (1); 1976; pp. 68-85.
5. Paulson, G.D. In (1); 1976; pp. 86-102.
6. Hutson, D.H. In (1); 1976; pp. 103-131.
7. Iwan, J. In (1); 1976; 132-152.
8. Geissbuehler, H. In (1); 1976; pp. 368-377.
9. Tomasic, J. In "Drug Fate and Metabolism, Method and Techniques"; Garret, E.R.; Hirtz, J.L., Eds.; Marcel Dekker: New York and Basle, 1978; Vol 2, pp. 281-335.
10. Coutts, R.T.; Jones, G.R. In "Concepts in Drug Metabolism"; Jenner, P.; Testa, B., Eds.; Marcel Dekker: New York and Basle, 1980; Part A, pp. 1-51.
11. Martin, L.E.; Reid, E. In "Progress in Drug Metabolism"; Bridges, J.W.; Chasseaud, L.F., Eds.; John Wiley: Chichester, New York, Brisbane, Toronto, 1981; Vol. 6, pp. 197-248.
12. Dieterle, W.; Faigle, J.W. "Preparative Liquid Chromatographic Methods in Drug Metabolism Studies" presented at the 4th International Bioanalytical Forum, University of Surrey, Sept. 7-10, 1981.
13. Muecke, W. In "Progress in Pesticide Biochemistry and Toxicology"; Hutson, D.H.; Roberts, T.R., Eds.; John Wiley: Chichester, New York, Brisbane, Toronto, Singapore, 1983; Vol. 3, pp. 279-366.
14. Baldwin, M.K.; Hutson, D.H. Xenobiotics 1980, 10, 135-144.
15. Lynn, R.K.; Smith, R.G.; Thompson, R.M; Deinzer, M.L.; Griffin, D.; Gerber, N. Drug Metab. Dispos. 1978, 6, 494-501.
16. Schwartz, D.E.; Jordan, J.C.; Vetter, W.; Oesterhelt, G. Xenobiotica 1979, 9, 571-581.

17. Stierlin, H.; Faigle, J.W.; Sallmann, A.; Kueng, W.; Richter,
     W.J.; Kriemler, H.P.; Alt, K.O.; Winkler, T. Xenobiotica 1979,
     9, 601-610.
18. Kojima, S.; Muruyama, K.; Babasaki, T. Drug Metab. Dispos.
     1980, 8, 463-466.
19. Mihara, K.; Ohkawa, H.; Miyamoto, J. J. Pestic. Sci. 1981, 6,
     211-222.
20. Hignite, Ch. E.; Tschanz, Ch.; Lemons, S.; Wiese, H.; Azarnoff,
     D.L.; Huffmann, D.H. Life Sci. 1981, 28, 2077-2081.
21. Bakke, J.E.; Price, C.E. J. Environ. Sci. Health 1979, 14,
     427-441.
22. Pekas, J.C. Pestic. Biochem. Physiol. 1979, 11, 166-175.
23. Bakke, J.E.; Rafter, J.J.; Lindeskog, P.; Feil, V.J.;
     Gustafsson, J.A.; Gustafsson, B.E. Biochem. Pharmacol. 1981,
     30, 1839-1844.
24. Larsen, G.L.; Bakke, J.E. Xenobiotica 1981, 11, 473-480.
25. Tang, B.K.; Kalow, W.; Grey, A.A. Drug. Metab. Dispos. 1979, 7,
     315-318.
26. Paulson, G.D.; Giddings, J.M.; Lamoureux, C.H.; Mansager, E.R.;
     Struble, C.B. Drug Metab. Dispos. 1981, 9, 142-146.
27. Frear, D.S.; Swanson, H.R. J. Agric. Food Chem. 1978, 26,
     660-666.
28. Maurer, G.; Donatsch, P.; Galliker, H.; Kiechel, J.R.; Meier, J.
     Xenobiotica 1981, 11, 33-41.
29. Bakke, J.E.; Gustafsson, J.A.; Gustafsson, B.E. Science 1980,
     210, 433-435.
30. Wolf, D.E.; Vandenheuvel, W.J.A.; Tyler, T.R.; Walker, R.E.;
     Koninszy, F.R.; Gruber, V.; Arison, B.H.; Rosegay, A.; Jacob,
     T.A.; Wolf, F.J. Drug Metab. Dispos. 1980, 8, 131-136.
31. Aschbacher, P.W.; Mitchell, A.D. Chemosphere 1983, 12,
     961-965.
32. Frear, D.S.; Swanson, H.R.; Mansager, E.R. Pestic. Biochem.
     Physiol. 1983, 20, 299-310.
33. Frear, D.S., Swanson, H.R.; Mansager, E.R. Pestic. Biochem.
     Physiol. 1985, 23, 56-65.
34. Lamoureux, G.L.; Davison, K.L. Pestic. Biochem. Physiol. 1975,
     5, 497-506.
35. Axness, M.E.; Fleeker, J.R. Pestic. Biochem. Physiol. 1979,
     11, 1-12.
36. Kripalani, K.J.; Dreyfuss, J.; Nemec, J.; Cohen, A.I.; Meeker,
     F.; Egli, P. Xenobiotica 1981, 11, 481-488.
37. Ruelins, H.W.; Tio, C.O.; Knowles, J.A.; McHugh, S.L.;
     Schillings, R.T.; Sisenwine, S.F. Drug Metab. Dispos. 1979, 7,
     40-43.
38. Dieterle, W.; Faigle, J.W.; Mory, H.; Richter, W.J.; Theobald,
     W. Europ. J. Clin. Pharmacol. 1975, 9, 135-145.
39. Bakke, J.E.; Feil, V.J., Price, C.E. Biomed. Mass Spectrom.
     1976, 3, 226-229.
40. Son, O.S.; Everett, D.W.; Fiala, E.S. Xenobiotica 1980, 10,
     457-468.
41. Verzele, M.; Geevaert, E. J. Chromatogr. Sci. 1980, 18,
     559-570.
42. Nettleton, D.E., Jr. J. Liq. Chromatogr. 1981, 4, 141-173.

43. Knaak, J.B.; Tallant, M.J.; Bartley, W.J.; Sullivan, L.J.
    J. Agric. Food Chem. 1965, 13, 537-543.
44. Muecke, W. unpublished data.
45. Simoneaux, B.J.; Martin, G.; Cassidy, J.E.; Ryskiewich, D.P.
    J. Agric. Food Chem. 1980, 28, 1221-1224.
46. Chopade, H.M.; Dauterman W.C.; Simoneaux, B.J. Pestic. Sci.
    1981, 12, 17-26.
47. Chopade, H.M.; Dauterman, W.C. Pestic. Biochem. Physiol 1981,
    15, 105-119.
48. Bakke, J.E.; Aschbacher, P.W.; Feil, V.J.; Gustafsson, B.E.
    Xenobiotica 1981, 11, 173-178.
49. Lamoureux, G.L.; Shimabukuro, R.H.; Swanson, H.R.; Frear, D.S.
    J. Agric. Food Chem. 1970, 18, 81-86.
50. Lamoureux, G.L.; Stafford, L.E.; Shimabukuro, R.H. J. Agric.
    Food Chem. 1972, 20, 1004-1010.
51. Lamoureux, G.L; Rusness, D.G. J. Agric. Food Chem. 1980, 28,
    1057-1070.
52. Larsen, G.L.; Bakke, J.E.; Feil, V.J. Biomed. Mass Spectrom.
    1978, 5, 382-390.
53. Sjövall, J.; Rafter, J.; Larsen, G.; Egestad, B. J. Chromatogr.
    1983, 276, 150-156.
54. Dieterle, W.; Faigle, J.W.; Mory, H. J. Chromatogr. 1979, 168,
    27-34.
55. Dieterle, W.; Faigle, J.W.; Kriemler, H.P.; Winkler, T.
    Xenobiotica 1984, 14, 311-319.
56. Bakke, J.E.; Rafter, J.; Larsen, G.L.; Gustafsson, J.A.;
    Gustafsson, B.E. Drug Metab. Dispos. 1981, 9, 525-528.
57. Dixon, R.; Evans, R.; Crews, Th. J. Liq. Chromatogr. 1984,
    7, 177-190.
58. Roberts, T.R.; Wright, A.N. Pestic. Sci. 1981, 12, 161-169.
59. Rusness, D.G.; Lamoureux, G.L. J. Agric. Food Chem. 1980, 28,
    1070-1077.
60. Hostettmann, K.; Hostettmann-Kaldas, M.; Nakanishi, K.
    J. Chromatogr. 1979, 170, 355-361.
61. Putman, L.J.; Butler, L.G. J. Chromatogr. 1985, 318, 85-93.
62. Zhang, T.Y. J. Chromatogr. 1984, 315, 287-297.
63. Conway, W.D.; Ito, Y. J. Liq. Chromatogr. 1984, 7, 275-289.

RECEIVED August 12, 1985

6

Synthesis of Xenobiotic Conjugates

Åke Bergman

Section of Organic Chemistry, Wallenberg Laboratory, University of Stockholm,
S-106 91 Stockholm, Sweden

Chemical methods for the synthesis of the major
types of conjugates known to be formed as metabolites of xenobiotics are reviewed. The synthesis
of glucuronic acid, sulfate, glutathione, cysteine,
N-acetyl-cysteine, glucose, amino acid, bile acid,
fatty acid and methylthiolated conjugates can be
accomplished by a variety of reactions. These
synthetic methods are described, exemplified and the
utility of these in macro and micro scale preparations is discussed in relation to synthesis of
radiolabeled and unlabeled conjugates. Preparative
scale enzymatic systems that have been used for the
synthesis of conjugates are included.

Most xenobiotic conjugation reactions result in products that have
very different physical and chemical properties than the unconjugated xenobiotic. Many xenobiotic conjugates have less biological
activity than the parent compound; however some conjugates are
reactive species with greater biological activity than the xenobiotic from which they were derived (1). Some xenobiotic conjugates have lipophilic characteristics quite similar to the parent
xenobiotic (2,3) and may be bioaccumulated in animal tissues with
or without enterohepatic circulation. Xenobiotic conjugates may be
further metabolized to either more or less polar products through
degradation or additional conjugation reactions. Xenobiotic conjugates must be synthesized or isolated from biological matter in a
preparative scale in order to prove the structure of these
conjugates, and to describe their physical and chemical characteristics, their biological activity and bioaccumulation properties.
The chemical synthesis of xenobiotic conjugates may also elucidate
the often complex sterochemistry involved in the formation of
these compounds.

The present review on synthesis of xenobiotic conjugates will
concentrate on methods for synthesis of the carbohydrate
(glucuronic acid, glucose), sulfate, and mercapturic acid pathway
conjugates, the latter including glutathione, cysteine, N-acetyl-
-cysteine, methyl sulfide and sulfone containing compounds.

0097-6156/86/0299-0124$09.50/0
© 1986 American Chemical Society

The synthesis of model compounds to verify the structures of
xenobiotic amino acid adducts (products of nonenzymatic reactions
of electrophilic xenobiotics with amino acids in proteins) are
included. Methods for synthesis of some rarely observed conjugate types will be briefly introduced. The major organic methods
for synthesis are exemplified and xenobiotic conjugates prepared
via these reactions are tabulated with references to the original
publications.

Synthesis of Glucuronic acid conjugates

In vivo D-glucuronic acid conjugates of xenobiotics are formed
from UDP-glucuronic acid and xenobiotics. The reactions are catalyzed by UDP-glucuronyltransferase enzymes; O-, S-, N- and
C-glucuronides can be formed in this manner. The biochemical
formation of glucuronic acid conjugates and the chemical properties of both enzyme and conjugates have been thoroughly studied
(4,5,6).
    Chemical methods for the synthesis of glucuronic acid conju-
gates of O-, S-, N- and C-aglycones were most recently reviewed by
Keglevic (7). Both direct and indirect methods are available for
synthesis of glucuronic acid conjugates.

Koenigs–Knorr synthesis. This method, based on the method described by Koenigs and Knorr (8) for the synthesis of glycopyrano-
sides is the procedure most often used for synthesis of glucuronic
acid conjugates. The Koenigs–Knorr reaction with its modifications
(cf. synthesis of glucose conjugates) has been reviewed by
Igarashi (9). The reaction is typically carried out as described
by Bollenback (10): methyl (tri-O-acetyl-α-D-glucopyranosyl
bromide) uronate is stirred with a ten times molar excess of a
phenol in the presence of silver carbonate (Figure 1). Solvents,
such as diethyl ether, methylene chloride, acetone, benzene,
toluene, 1,2-dichloroethane, methanol and dimethylformamide (DMF),
suitable to dissolve the aglycone, may be used. However, the
yields are improved by dry reaction conditions; the water can be
trapped by magnesium sulfate, calcium sulfate, or molecular sieves
(3Å) added to the reaction mixture or can be removed by continous
azeotropic destillation of the solvent. The reaction proceeds with
inversion of configuration at C-1 of the glucuronic acid.
    The Koenigs–Knorr reaction, in its original form, requires
reaction at room temperature over night. The appropriate molar
ratio of the reactants can further improve the yields as shown in
the synthesis of certain steroid glucuronides (11,12). Promotors
other than silver salts have been used as shown in Table I.
Cadmium carbonate may be especially useful in this regard (12);
the cadmium bromide formed in this reaction likely acts as a
catalyst. Silver perchlorate (0.2 mol/mol aglycone) has been
reported to be an effective catalyst when used together with
silver carbonate (13). Phase transfer catalysis (PTC) conditions
were used for synthesis of methyl (2-benzyloxy-4-formylphenyl
2,3,4-tri-O-acetyl-β-D-glucopyranoside) uronate (Figure 2) but
only a moderate yield (28%) was obtained (14). Other protective
groups for the carbohydrate hydroxyls can also be used, thus
methyl (2,3,4-tri-O-pivaloyl-α-D-glucopyranosyl bromide) uronate

was sucessfully synthesized and further used for the preparation
of primary, secondary and tertiary alcohol glucuronides in high
yields (15).

Synthesis of S-, N-, N-hydroxy- and C-glucuronides can also
be performed by reactions analogous to the one shown in Figure 1.
S-Nucleophiles may induce partial hydrolysis of the acetyl protective groups but addition of an acetylating agent prior to isolation of the S-glucuronide derivatives will improve the yields.
Some N-glucuronides can be formed by rearrangement of certain
O-glucuronides as illustrated in Figure 3 (16). A C-glucuronide of
equilenine was obtained as a byproduct from the synthesis of
methylequilenine (tri-O-acetyl-β-D-glucopyranoside) uronate (12).

The methyl (tri-O-acetyl-α-D-glucopyranosyl bromide) uronate
is readily prepared (10,17) but it is also commercially available.
The derivatized conjugate is commonly purified by chromatographic
procedures before it is deprotected by alkaline hydrolysis at +4°C
or room temperature (4,7,10,12). The synthesis of a selected
number of derivatized and deprotected glucuronic acid conjugates
of xenobiotics are summarized in Table I.

Fusion type synthesis. Methyl (tetra-O-acetyl-β-D-glucopyranose)
uronate reacted with an excess of phenol in the presence of zinc
chloride at reduced pressure and 110-120°C (10). The product,
methyl (phenyl-tri-O-acetyl-β-D-glucupyranoside) uronate, was
isolated in 8% yield. If p-toluenesulfonic acid was used instead
of zinc chloride a 55% yield of the glucuronide was obtained.
Penta-O-acetyl-α-L-idopyranose was fused under similar conditions
with phenol to give the anomeric phenyl tetra-O-acetyl-L-ido-
pyranosides (18). These compounds were oxidized to the corresponding glucuronic acid conjugates by oxygen over a platinium
oxide catalyst. Diethylstilboestrol-β-D-glucuronic acid was
obtained in 40% yield after condensation using a catalytic amount
of p-toluenesulfonic acid and subsequent alkaline hydrolysis (19).
Another catalyst, stannic chloride, has shown advantageous
properties for carrying out this type of condensation reaction
under mild conditions (20).

There is only limited information on the synthesis of
C-glucuronides of xenobiotics probably due to the fact that
relatively few conjugates of this type have been reported. The
C-glucuronide of Δ^6-tetrahydrocannabinol (Δ^6-THC) was prepared
from Δ^6-THC and methyl (tetra-O-acetyl- β-D-glucopyranose)
uronate using boron trifluoride etherate as a catalyst at 0°C for
4 h (21). If the reaction was catalyzed by p-toluenesulfonic acid
under reflux an equal amount of the Δ^6-THC 4'-C-glucuronide and
Δ^6-THC O-glucuronide and a minor amount of Δ^6-THC 6'-C-gluc-
uronide were obtained.

Oxidative reactions. Oxidation of the primary alcohol group of
partially protected glucopyranosides can be performed with
oxidizing agents such as potassium permanganate, hydrogen peroxide
or hypohalogenite (7). Ruthenium tetraoxide, present in catalytic
amount together with molar amounts of periodate effectively
oxidizes alcohols such as the primary hydroxyl group of glucosides
if the remaining hydroxyl groups are protected (35). Oxidation of
glucosides with oxygen and a platinum catalyst is an important

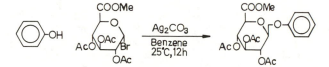

Figure 1. Synthesis of methyl (phenyl-2,3,4-tri-O-acetyl--
β-D-glucopyranoside) uronate by the Koenigs-Knorr method.

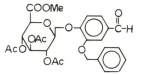

Figure 2. Structure of methyl (2-benzyloxy-4-formylphenyl-
2,3,4-tri-O-acetyl-β-D-glucopyranoside) uronate.

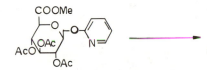

Figure 3. O-N rearrangement of methyl (2-pyridyl-2,3,4-tri-O-
-acetyl-β-D-glucopyranosid) uronate.

Table I. Xenobiotics or xenobiotic metabolites (aglycones) used in Koenigs–Knorr type synthesis or fusion reaction.

Aglycone	Promotor	Ref.
Phenol*[1]	Ag_2CO_3	10
2- and 4-Halophenols*	Ag_2CO_3	10
2-,3- and 4-Methylphenol*	Ag_2CO_3	10
Catechol*	Ag_2CO_3	10
1- and 2-Naphthol*	Ag_2CO_3	10
Methyl gentisate*	Ag_2CO_3	10
N-Acetyl-N-phenylhydroxylamine	Ag_2CO_3	22
4-Methoxy-2-methyl-6-sulfanilamidopyrimidine[2]	–	23
2-Pyridinol	–	16
Cortisone	Ag_2CO_3	24
Cortisol	Ag_2CO_3	24
Corticosterone	Ag_2CO_3	24
Potassium methyl and ethyl xanthate[3]	–	25
Potassium benzyl xanthate[3]	–	25
18β-Glycyrrhetic acid	Ag_2CO_3	11
Estrone	$CdCO_3$	12
17β-Estradiol	$CdCO_3$	12
Equilin	$CdCO_3$	12
Equilenin[4]	$CdCO_3$	12
α,α'-Diethyl-4,4'-stilbenediol*	–/$CdCO_3$	26/27
2-Benzyloxy-4-hydroxybenzaldehyde	–	14
Phenolphthalein methyl ester	$CdCO_3$	28
Hexoestrol	$CdCO_3$	27
Procaterol	–	29
Methaqualone	Ag_2CO_3	30
1-Phenylethanol	Ag_2CO_3	31
N[2]-Methyl-9-hydroxyellipticinium acetate	–	32
4-Chloro-3-hydroxybenzotrifluoride	–	33
tert-Butanol	Ag_2CO_3	15
2,2,2-Trichloroethanol	Ag_2CO_3	15
2-Adamantanol	Ag_2CO_3	15
5-Ethoxy-2-nitrophenol	–	34

1. * = Aglycones used also in synthesis of glucuronides via the fusion reaction; 2. N-glucuronide; 3. S-glucuronide; 4. C-glucuronide.

method for the preparation of glucuronic acid conjugates (36). For example, (2-ethoxy-5-nitrophenyl)-β-D-glucopyranoside uronic acid was obtained by a Pt catalyzed oxidation of the corresponding glucoside at 80°C (37). Since the starting material for the oxidation must be the glucose conjugate, or a derivative thereof, the method is valuable if both glucose and glucuronic acid conjugates are needed. The synthesis of glucose conjugates will be discussed below. However, oxidation procedures can not always be used because other functional groups in the aglycone moiety may also be oxidized.

Enzymatic methods. In vitro synthesis of glucuronides by use of
isolated UDP-glucuronyltransferase and an aglycone may be performed (e.g. 38) but only very small amounts of aglycones can be
processed by this method. Immobilized UDP-glucuronyltransferase,
covalenty bound to cyanogen bromide activated agarose (39) or
Sepharose (40), has been used for the preparation of a variety of
glucuronide conjugates. The purified and immobilized enzyme from
rabbit liver was more stable than the enzyme obtained from rat or
calf liver. Immobilization stabilizes the enzyme which may be used
repeatedly. Small amounts of ethanol, dimethylsulfoxide (DMSO),
and propylene glycol can be added to the aqueous reaction medium
without a major loss of enzymatic activity. However, the additon
of only 2% of dioxane resulted in 100% inhibition (39,40). The
originally reported 50-70% yield of 4-nitrophenol glucuronide was
improved by additions of Mg^{2+}($\cong$3.0 mM) and Ca^{2+}(0.8-3.0 mM)
salts (41), but higher concentrations of these ions were in-
hibitory. The reaction has been carried out in a 5-30 mg scale of
aglycone and the products were purified by TLC methods (39,41). If
larger amounts of cosolvents must be added in order to dissolve
more nonpolar xenobiotics lower yields are obtained. Sometimes
this problem can be overcome by extending the incubation time.
Several glucuronide conjugates of xenobiotics, exemplified in
Table II, have been synthesized by the immobilized UDP-glucuronyltransferase enzyme method (40-47).

Glucuronide ester synthesis. The chemical synthesis of glucuronic
acid conjugates of xenobiotic containing a carboxyl group can be
accomplished by reacting the xenobiotic and a protected glucuronic
acid in the presence of molar amounts of e.g. dicyclohexylcarbodiimide (DCC) (48,49), or 1,1'-carbonyldiimidazole as illustrated
in Figure 4A (50). The diimide activated xenobiotic acid reacts
with the protected glucuronic acid to give ester conjugates in
fair yield ($\geqq$50%). The reactions may be carried out in organic
solvents such as DMF, methylene chloride and tetrahydrofuran. The
DCC mediated procedure cannot be used to prepare the glucuronic
acid conjugates of aroyl acids. Instead the corrsponding aroyl
acid chlorides can be used as shown in Figure 4B (51). The
Koenigs-Knorr reaction may also be used for the synthesis of
glucuronic acid conjugates. A more readily deprotected glucuronic
acid derivative, 2,2,2-trichloroethyl (2,3,4-tri-O-(2,2,2-tri-
chloro-ethyloxycarbonyl)-D-glucopyranoside) uronate, has been used
in glucuronide ester synthesis in order to minimize hydrolysis of
the 1-ester bond (49).

Miscellaneous methods and comments. The N^4-glucuronic acid
conjugate of metoclopramide was obtained in low yield by simply
reacting this compound with a molar amount of glucuronic acid at
room temperature overnight (52). Trimethylsilyl trifluoromethane-
sulfonate has been used to activate methyl (2,3,4-tri-O-acetyl-
glucopyranoside) uronate for reaction with hydroxy-containing
aromatic and aliphatic aglycons at low temperatures (53). Both
α- and β-anomeric products were formed.
    Chemical synthesis of glucuronic acid conjugates of poly-
functional xenobiotics may yield undesirable side products;
enzymatic methods may be preferable for glucuronic acid

Table II. Xenobiotics or xenobiotic metabolites (aglycones) used
in synthesis of glucuronic acid conjugates by oxidative (A),
enzymatic (B) and esterification (C) methods.

Aglycone	Method	Ref.
Methanol	A	54
Triphenylmethanol	A	35
2-Ethoxy-5-nitrophenol	A	37
4-Nitrophenol	A,B	38-41
Borneol	B	41
Meprobamate	B	41
4-Nitrothiophenol	B	41
Δ^8-Tetrahydrocannabinol	B	42
Cannabinol	B	43
Isoborneol	B	44
N-Hydroxyphenacetin	B	46
N-Hydroxy-4-chloroacetanilide	B	46
N-Hydroxy-2-acetylaminonaphthalene	B	46
N-Hydroxy-2-acetylaminofluorene	B	46
Flurbiprofen	B	47
Gemfibrozil	B	47
Prodolic acid	B	47
Flufenamic acid	B	47
Tripelennamine	B	45
Cyproheptadine	B	45
Indomethacin	C	49
Ananilino acid[1]	C	55
3-Phenoxybenzoic acid	C	48
4-Chlorobenzoic acid	C	56

1. Fluvalinate metabolite.

conjugation of compounds of this type and if only limited amounts
of the aglycone is available. However, the Koenigs-Knorr reaction
can also be used for micro scale synthesis. Oxidative methods, on
the other hand, require somewhat larger amounts of reactants.

The synthesis of radiolabeled xenobiotic glucuronides seem to
be carried out less often. It is usually preferable to label the
xenobiotic moiety but the glucuronic acid can certainly be labeled
as well. Labeled methyl (2,3,4-triacetyl-α-D-glucopyranosyl
bromide) uronate is at present not commercially available.

Synthesis of Glucose conjugates

Glucopyranoside conjugates of exogenous compounds are commonly
found in plants (57) and insects (58). Steroid glycosides were
also recently considered as potential prodrugs with specific
tissue distribution (59). The synthesis of α- and β-glucosides of
numerous exogenous or endogenous compounds has been reported over
the years, mainly in journals specialized in carbohydrate chemistry.
In addition to O-, N-, S-glucosides (Table III) and glucoside
esters a number of related conjugates such as gentiobiosides and
6-O-malonyl-glucosides (57), have also been reported; however the

Table III. The synthesis of selected xenobiotic O-, N- and
S-glucoside conjugates.

Aglycone	Conjugate	Ref.
Phenobarbital	D[1]	68,69
Sulfamethazine	G[2]	67,70
Phenol	D	61,66,71
3-Nitrophenol	D	61,72
4-Nitrophenol	D	61,71,73
2,4-Dinitrophenol	D	61
2,4,6-Triiodo-3-acetamido-5-		
-N-methylcarboxamidophenol	D,G	62
4-Methylphenol	D,G	66
Aniline	D,G	66
4-Methylaniline	D,G	66
3-Nitroaniline	D,G	66
2-Aminonaphthalene	D,G	66
Theophylline	D,G	66
2-Methylphenol	D,G	71
3-Chlorophenol	D,G	71
1-Naphthol	D,G	71
Morphine	D	71
8-Hydroxyquinoline	D	71
Thiophenol	D,G	71

1. D = The final synthetic product is a protected glucoside
conjugate; 2. G = The final product is the free β-D-glucoside
conjugate.

synthesis of these will not be discussed here.

<u>Koenigs Knorr method</u>. The most commonly used synthetic method for
preparation of glucoside conjugates is based on the Koenigs-Knorr
reaction (8) that has been thorougly reviewed by Igarashi (9).
Improved conditions for carring out this reaction with aliphatic
and aromatic aglycones respectively, are shown in Figure 5. Silver
triflate mediated glycosylation with the benzoylated
α-D-glucopyranosyl bromide was carried out in nitromethane or
toluene at -78°C (60). The reactions were completed in less than
15 min and 70-90% yields of products were reported. 4-Nitrophenyl
2,3,4,6-tetra-O-acetyl-β-D-glucopyranoside was obtained in high
yield (88%) from 2,3,4,6-tetra-O-acetyl-α-D-glucopyranosyl bromide
and the aromatic aglycone in presence of silver imidazolate and
zinc chloride (61). This reaction was completed in 48 h in the
dark at 40°C and the solvent was kept dry by molecular sieves
(3Å). It is notable that a fair yield of 2,4,6-triiodo-3-acet-
amido-5-N-methylcarboxamidophenyl-D-tetraacetyl glucopyranoside
was obtained in a synthesis carried out in a water/acetone
solution (62).
      Crown ethers have been used as phase transfer catalysts in
the formation of O-glucosides (63,64). A number of xenobiotic
glucoside conjugates were also prepared under PTC conditions
(30-70% yield) from tetra-O-benzyl-α-D-glucopyranosyl bromide,
aglycone and triethylbenzylammonium chloride in a two-phase

solvent system (water/methylene chloride) (65). However, the
corresponding tetraacetyl glucopyranosyl bromide gave only low
yields of conjugates.

Miscellaneous methods and comments. Several xenobiotics (Table
III) were successfully condensed with penta-O-acetyl-β-D-gluco-
pyranose in the presence of polyphosphoric acid or ethyl poly-
phosphate (66). The O-N rearrangement, discussed above for glucuronic acid conjugates (Figure 3), is also a potential synthetic
method for the preparation of certain N-glucoside conjugates (65).
Sulfamethazine N^4-glucoside was obtained by reaction of sulfamethazine with D-glucose in DMF and ethylene glycol and acetic
acid (67).
     Even though the number of glycoside conjugates of xenobiotics
that have been synthesized so far is limited, the methods described here are convenient both for large and small scale operations. No specific restrictions in synthesis of isotopically
labeled conjugates should be encountered due to the synthetic
methods.

Synthesis of Sulfate conjugates

Conjugation of O-, N- and S-containing functional groups with
sulfate is very common in animal systems. Sulfate is transferred
from adenosine 3'-phosphate-5'-sulfatophosphate (PAPS) by a
variety of transfer enzymes. The mechanisms for sulfation of
xenobiotics (74,75) and the chemistry of sulfate esters and
related compounds and the synthesis of these have been reviewed
(75-77). Syntheses of O-, N- and S-sulfate conjugates of xenobiotics may be carried out by the methods described below.
Sulfated xenobiotics that have been synthesized by chemical
methods are exemplified in Table IV.

Reactions with sulfur trioxide. Sulfur trioxide (SO3) may react
violently with nucleophiles unless the conditions are well controlled and this route is not commmonly used for syntheses of
xenobiotic sulfate conjugates. However a method suitable for both
macro and micro scale synthesis of ascorbic acid 2-[^{35}S]sulfate
(Figure 6A) was recently reported where the SO3 was introduced
to the substrate dissolved in DMF at -180°C (78). The reaction was
completed at -15°C in 30 min and gave a 74% yield of the
2-O-sulfate.
     By use of adduct deactivated SO3 more readily controlled
reaction of SO3 can be performed. Sulfur trioxide forms adducts
with Lewis bases such as DMF, pyridine, trialkylamines and
dioxane. These adducts are commercially available or conveniently
synthesized (79). The reactivity of the SO3 complex increases
with decreasing base strength. Thus, the derivatized ascorbic acid
2-sulfate described above could also be synthesized from
5,6-O-isopropylidene-2-ascorbic acid (as above) and SO3-pyridine
in DMF at 25°C for 7 h (Figure 6B) (80). The yield of the
2-O-sulfate reported in this case was 49%. The reaction conditions
choosen are dependent on the reactivity of the SO3 adduct used
but slightly elevated temperatures are often necessary. Amine
containing xenobiotics are more reactive and consequently the

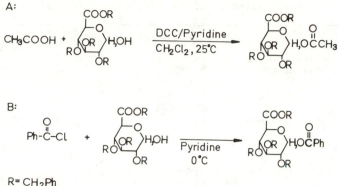

R= CH₂Ph

Figure 4. A: Dicyclohexylcarbodiimide mediated esterification
of a carboxylic acid (acetic acid) with benzyl (2,3,4-tri-0-
-benzyl-D-glucopyranose) uronate.
B: Esterification of an aroyl acid via reaction of the
corresponding acid chloride and the protected glucuronic acid.

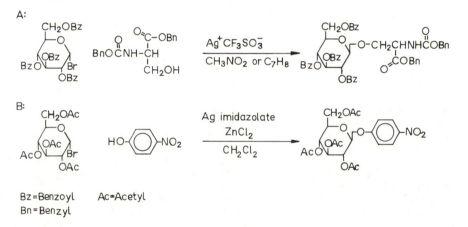

Bz=Benzoyl   Ac=Acetyl
Bn=Benzyl

Figure 5.   0-Glucosidation at a hydroxy-alkyl group (A) and a
phenol group (B) via improved Koenigs-Knorr reactions.

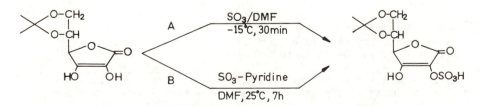

Figure 6.   Synthesis of 5,6-0-propylidene ascorbic acid
2-sulfate.

Table IV. Chemical synthesis of xenobiotic sulfate conjugates

Substrate	Reagents	Sulfate type	Ref.
Anthraquinone[1]	Et_3N-SO_3	O-,diester	100
1,2-Phthaloylcarbazole	Et_3N-SO_3	O-,diester	100
Phenylhydroxyl amine[2]	$Pyr-SO_3$	N-	101
Harmol	Me_3N-SO_3	O-	102
2-Octanol	$Pyr-SO_3(*)$[3]	O-	81
Sec. alcohol[4]	Et_3N-SO_3	O-	82
3,4-Dihydroxybenzylamine	$Pyr-SO_3$	3-O-;4-O-	103
Tiaramide	$Pyr-SO_3$	N-	104
Azetidinone	$DMF-SO_3$	N-	105
Phenol	$ClSO_3H$	O-	106,107
Methylphenol (2-,3-,4-)	$ClSO_3H$	O-	107
4-Bromophenol	$ClSO_3H$	O-	107
Aniline	$ClSO_3H$	N-	108
4-Aminobiphenyl	$ClSO_3H$	N-	108
1- and 2-Aminonaphthalene	$ClSO_3H$	N-	108
4-Nitrocatechol	$ClSO_3H(*)$	O-	109
4-Octylphenol	$ClSO_3H$	O-	106
2,4-Dichloro-1-naphthol	$ClSO_3H$	O-	106
4-Biphenylol	$Et_3N-SO_3(*)$	O-	83
2- and 4-Cyclohexylphenol	$ClSO_3H(*)$	O-	83
2-Naphthol	$ClSO_3H(*)$	O-	83
4-Chlorophenol	$ClSO_3H$	O-	110
Morphine	$ClSO_3H$	3-O-;6-O-	111
5-Chloro-7-iodo-8-quinolinol	$ClSO_3H$	O-	112,113
8-Hydroxyquinoline	$ClSO_3H$	O-	114
Methaqualone	$ClSO_3H$	O-	30
10-Undecen-1-ol	$ClSO_3H(*)$	O-	115
Hydroquinone monoacetate	$ClSO_3H$	O-	116
Octadecanol	$H_2SO_4(*)$	O-	84
Cyclohexanol	$H_2SO_4(*)$	O-	85,88
Phenol	$H_2SO_4(*)$	O-	85,88
1- and 2-Naphthol	$H_2SO_4(*)$	O-	85,88
Octanethiol	$H_2SO_4(*)$	S-	85,88
Benzenethiol	$H_2SO_4(*)$	S-	85,88
Cyclohexylamine	$H_2SO_4(*)$	N-	85,88
Aniline	$H_2SO_4(*)$	N-	85,88
Acetophenone oxime	$H_2SO_4(*)$	N-	85,88
1-,3-,7- and 9-Hydroxy-Benzo(a)pyrenes	H_2SO_4	O-	117
Phenol	$K_2S_2O_7$	O-	92
Morphine	$K_2S_2O_7$	3-O-	93
3-Nitrophenol[5]	$K_2S_2O_8$	O-	91
Methyl 4-nitrophenylsulfate	$MeO^-/MeOH$	O-	97
Ethyl 4-bromophenylsulfate	$MeO^-/MeOH$	O-	98
Methyl 4-methylphenylsulfate	$MeO^-/MeOH$	O-	98

1. Reduced prior to the sulfating reaction; 2. N-Sulfates of several substituted phenyl hydroxylamines were also synthesized; 3. (*) Indicates that ^{35}S-labeled sulfate conjugates were prepared; 4. Secondary pentanol, octanol, nonanol, decanol, tetra-decanol; 5. Product: 4-hydroxy-2-nitrophenylsulfate.

reaction temperatures can be lowered. A cosolvent in addition to
the solvent of the SO_3-adduct may be used. Improved yields were
reported when carbon tetrachloride was substituted for
1,2-dichloroethane in the synthesis of octyl-2-sulfate (81).
   ^{35}S-Labeled sulfate conjugates have been synthesized from
both [35]SO_3 and [35]SO_3-adducts. However [35]SO_3-adducts are
preferable for small scale reactions. SO_3-Triethylamine was used
in synthesis of several secondary alcohol conjugates due to its
greater stability compared to SO_3-pyridine upon storage (82).
The lower reactivity of the SO_3-triethylamine was compensated
for by use of the substrate alkoxides instead. No formation of
positional isomers were obtained as was the case when sulfation
was performed with sulfuric acid or chlorosulfonic acid.

Reactions with chlorosulfonic acid. The preparation of sulfate
conjugates from chlorosulfonic acid ($ClSO_3H$) and solid substrates must be performed in a solvent, e.g. chloroform, carbon
tetrachloride, carbon disulfide or N,N-dimethylaniline. Hydrogen
chloride formed as a biproduct of this reaction can result in
undesired products if the xenobiotic contains other labile groups.
[35]S-Labeled $ClSO_3H$ is commercially available and can be used
as such for sulfation or for in situ preparation of [35]SO_3-
-amine adducts (83). The hydrogen chloride formed in this reaction
may be removed prior to the addition of the substrate.
Sulfamations are also efficiently carried out by use of $ClSO_3H$
(77). A diethyl ether-$ClSO_3H$ complex can be used as a milder
sulfating agent than $ClSO_3H$ alone.

Carbodiimide mediated reactions with sulfuric acid. In a series of
papers Mumma and co-workers described the development of an
alternative sulfate conjugate synthetic method using sulfuric acid
(H_2SO_4) as the sulfating agent, Figure 7, (84-90). DCC was
initially employed in the reaction and is probably the most
commonly used carbodiimide even though several others have been
studied (90). In addition to DCC, 1-cyclohexyl-3-(2-morpholino-
ethyl)carbodiimide p-toluene sulfonate was shown to give fair
yields of sulfate conjugates. An important feature of the latter
carbodiimide is that the urea formed upon reaction is soluble. The
reaction is carried out in DMF and gives O-sulfate conjugates of
alcohols (86). More concentrated reagent solutions are necessary
for preparation of O-, N- and S-sulfate conjugates of phenols,
alkyl- and aryl-thiols, amines and oximes. It is thus possible to
selctively sulfate hydroxyalkyl groups in the presence of other
functional groups. The use of solvents other than DMF has been
investigated. The use of tetrahydrofuran and dioxane resulted in
the formation of several biproducts and the reactions were
uncontrollable (90).
   By carefully controlling the molar ratio of the reactants
micro and macro scale syntheses can be performed. DCC in DMF is
typically mixed with the nucleophile dissolved in DMF prior to the
addition of H_2SO_4. The reaction is performed at $0°C$ and is
finished in less than 30 min. Since small volumes of solutions are
mixed the method is applicable for synthesis of radiolabeled
compounds (86).

Reactions with persulfate and pyrosulfate. The persulfate (peroxydisulfate ion, $S_2O_8^{2-}$) reaction, also known as the Elbs
persulfate oxidation, has been important in synthesis of hydroxylated phenols. The method has occasionally been used for synthesis
of O-sulfate conjugates. For example, 4-hydroxy-2-nitrophenyl
sulfate was obtained when 3-nitrophenolate was stirred with
potassium peroxydisulfate at room temperature for 2 days (91). The
persulfate reaction has been used for the sulfation of various
phenols and aromatic amines; however the yields are usually low to
moderate (77). The sulfate group is preferentially introduced in
the 4-position of phenols and in the 2-position of aromatic amines
but if these positions are blocked substitution at the 2- and
4-positions will occur, respectively.
        Pyrosulfate (disulfate, $S_2O_7^{2-}$) has been used for
synthesis of aryl sulfate esters from the corresponding phenols
(92). A high yield of phenyl sulfate was obtained when phenol was
heated with potassium disulfate in dimethyl- or diethylaniline.
Morphine-3-sulfate ester was isolated from an aqueous mixture of
morphine, potassium hydroxide and potassium disulfate after 8 h at
room temperature (93).
        The two methods described here are very different since in
reaction with peroxydisulfate the whole sulfate group is introduced into the substrate molecule, while in the pyrosulfate reaction a hydroxyl group of the substrate molecule is sulfated.

Miscellaneous methods. Sulfamic acid (NH_2SO_3H) has been used
mainly in synthesis of sulfate conjugates of steroids (94,95).
However the reaction, catalyzed by pyridine and performed at
80-100°C for an hour, has also been used for phenolic subtrates
(77). Pyridinium sulfate and acetic anhydride has been used for
synthesis of estrone sulfate (96).
        A different approach to the synthesis of sulfate esters was
discussed by Buncel and co-workers (97). The sulfate monoester was
obtained after neutral or alkaline methanolysis of methyl 4-nitro-
phenyl sulfate. The methoxide ion attacks only at the alkyl carbon
with the formation of dimethyl ether and 4-nitrophenyl sulfate ion
as the leaving group. Other substituted methyl phenyl sulfates
have been converted to phenyl sulfates in the same manner (98).
Correspondingly 1,2:5,6-di-O-isopropylidine-α-D-glucofuranose was
reacted with phenyl chlorosulfate to give the carbohydrate
3-phenyl sulfate (99). The carbohydrate monosulfate ester was
obtained after hydrogenation over platinum oxide. The phenyl group
was hydrogenated to form a cyclohexyl group and was removed as
cyclohexanol. These indirect methods have not been used
extensively so far but the properties of the neutral compounds
(with possibly better properties for isolation) should be
considered.

Synthesis of Glutathione, Cysteinylglycine, Cysteine and
N-Acetyl-Cysteine conjugates

The conjugation of glutathione (GSH) to various types of xenobiotics is well known. The formation of GSH conjugates is usually
catalyzed by a group of enzymes known as glutathione S-transferases (118), although nonenzymatic reactions between a strong
electrophile and the strong nucleophile GSH may occur. The GSH

conjugate may be enzymatically hydrolyzed to the corresponding
cysteinylglycine and cysteine conjugate, and the latter acetylated
to form the mercapturic acid. The mercapturic acid conjugates had
been regarded as the final products of this sequence of reactions
known as the mercapturic acid pathway (MAP). However, several
xenobiotics that are known to be metabolized via MAP are further
degraded to thiols, methyl sulfides, methyl sulfoxides and
sulfones. The synthesis of these metabolites will be discussed
separately below. The reaction of GSH and an arene oxide gives
GSH-dihydro-hydroxy type compounds, that are often further trans-
formed into several other types of compounds and are here referred
to as the premercapturic acid pathway (PMAP) metabolites.

Except in the case of N-acetyl-cysteine, the chemical
problems encountered in synthesis of the conjugates summarized
under this heading are similar due to the amphoteric character of
the nucleophiles. In general the reactions with thiol compounds
must be carried out in an inert atmosphere in order to minimize
oxidation. The starting material, for synthesis of some of the
xenobiotic conjugates below, require multistep syntheses, but the
procedures for their preparations will not be discussed here.

Nucleophilic reactions with haloalkanes, haloalkenes and similar
alkyl compounds. A number of conjugates of simple alkyl compounds
have been synthesized directly from the haloalkane and cysteine or
N-acetyl-cysteine (Table V). The nucleophilicity of the thiolate
anion is utilized in the direct displacement of the leaving group
that takes place with inversion at the substituted carbon atom.
Polar solvents, commonly ethanol, but also other alcohols, liquid
ammonia and even acetic acid, have been used mostly with bases
(trialkylamines, hydroxides or alkoxides) for dissolving the
compound and forming the thiolate anion. The displacement of a
halogen on primary carbon atoms is rapid and the reaction often
goes to completion at room temperature. Even a tertiary haloalkane, 1-bromoadamantane (Figure 8) upon reaction with cysteine
hydrochloride gave a 74% yield of the corresponding conjugate
(119). The same product was also obtained from 1-adamantanol and
cysteine in the presence of boron trifluoride.

Both optically active 1-chloroethylbenzene, obtained from
(R)- and (S)-mandelic acid, and the racemic mixture were
substituted with GSH as the nucleophile (120). The reaction was
catalyzed by a 18-crown-6 and the optically active (R)- and
(S)-S-(1-phenylethyl)-glutathione compounds were isolated in a 60%
yield. Hexachlorocyclohexane (lindane) has been used as starting
material for synthesis of glutathione conjugates of dichloro-
benzenes. Three products were isolated and characterized as the
2,5-, 3,5- and 2,3-dichlorophenyl glutathione conjugates (121-123).
Since haloalkanes and haloalkanes are readily available,
these compounds are frequently used as starting material for the
reaction method discussed but other leaving groups can be used as
well. The sulfonamide group in several 5-substituted
1,3,4-thiadiazole- and 1,3,4-thiadiazoline-2-sulfonamides was
displaced by cysteine and GSH but the rate of displacement
decreased with increasing pK_a of the sulfonamides (124). No
substitution was observed when benzenesulfonamides were used.

The syntheses of several alkyl mercapturic acids were per-

Table V. Glutathione, cysteine and N-acetyl-cysteine conjugates
synthesized by nucleophilic displacement of an appropriate leaving
group.

Substrate	Nucleophile	Ref.
Trityl chloride	Cys[1],N-AcCys	127
Iodopropane	Cys	128
3-Bromopropanol	Cys	129
Benzylchloride [2]	GSH	130
1-Bromoadamantane	Cys	119
Ordram sulfone	GSH,Cys	131
Lindane	GSH	122
3-Chloro-1,2-propanediol	CYS	132
1-Amino-3-chloropropane-2-ol	Cys	133
epi-Bromohydrin[3]	Cys	134
1-Chloro-1-phenylethane	GSH	120
2-Bromo-1-phenylethane	N-AcCys	135
2-Bromo-2-phenylethanol	GSH	136
Phosgene	GSH	137
Propachlor	Cys	138,139
Tetrafluoroethylen	GSH,Cys	140
Allylbromide	N-AcCys	141
1-Bromo-2-butyne	N-AcCys	141
4-Nitrobenzylbromide	GSH	142
Chlorotrifluorethylene	GSH,Cys	143
Hexachloro-1,3-butadiene	GSH,N-AcCys	144
4-Cyano-N-(chloroacetyl)aniline	N-AcCys	145
Clofibric acid chloride	N-AcCys	146
α-Bromo-o-xylene	N-AcCys	147
5-Bromopentanoic acid	N-AcCys	148

1. L-cysteine was used in all cases; 2. Several substituted
benzylchlorides were also used; 3. Mono- and bis-cysteine
conjugates were formed.

formed in high yields by the same general outline as above. The
N-acetyl-cysteine di-anion was generated in methanol or DMSO with
methoxide or hydride respectively, as the base (125). Due to the
large number of GSH, cysteine and N-acetylcysteine conjugates
prepared by this method only selected examples from recent reports
are given in Table V. Occasionally it may be preferable to
synthesize the methyl N-acetyl-cysteine ester conjugate of
xenobiotics instead, a reaction that proceeds by use of e.g.
methyl N-acetyl-cysteine ester as the nucleophile (125,126).

Nucleophilic reactions with substituted aromatic compounds.
Aromatic nucleophilic substitutions are not favored unless the
nucleous is activated (e.g. nitro groups or halogen atoms). The
reaction may be rapid and specific, as between GSH and
2,4-dinitrochlorobenzene or 2,4-dinitrofluorobenzene, reactions
that both give S-(2,4-dinitrophenyl) glutathione in 90% yield
(149,150). However nucleophilic displacement reactions with poly-
functional or polysubstituted substrates may lead to product
mixtures. For example, pentachloronitrobenzene (PCNB) upon

reaction with GSH gave one major and three minor products (151).
The major product S-(pentachlorophenyl) glutathione was formed by
displacement of the nitro group and a minor product, S-(tetra-
chloronitrophenyl) glutathione, was formed by displacement of a
chlorine atom. The other two products obtained from the reaction
of PCNB and GSH were isomers containing two GSH moieties. Mixtures
of glutathione or cysteine conjugates obtained from nucleophilic
aromatic substitution reactions are usually purified by chromatographic methods. If the radiolabelled aromatic substrates can be
obtained the displacement reaction is well suited for small scale
preparation of the corresponding conjugates.

Nucleophilic reactions with diazotized arylamines. Conjugates of
aromatic xenobiotics may also be synthesized by coupling of an
appropriate thiol with a diazotized arylamine (Figure 9). Improved
yields were obtained when the cuprous mercaptides of cysteine
(152), GSH (153,154) and N-acetyl-cysteine (155) were used in the
coupling reaction. In order to obtain a rapid decomposition of the
initially formed and labile diazosulfide a catalytic amount of a
nickel salt can be added to the thiol solution, prior to the
addition of the diazonium salt (155). This method is readily performed on a semi-micro scale with substrates that are not labile
to the acidic conditions during diazotisation. In contrast to the
multiple products obtained from the reaction of GSH, cysteine or
N-acetyl-cysteine with arene oxides or polysubstituted aromatic
compounds only one fully aromatic conjugate of a known structure
is formed by the present method. Some conjugates synthesized by
this procedure are shown in Table VI.

Reactions with aliphatic and aromatic epoxides. The metabolism of
many alkenes and arenes is known to proceed via epoxides formed by
phase I reactions (166). Chemical methods for the preparation of
epoxides will not be discussed here but valuable references (or
references cited therein) for their synthesis are found in
Table VII. The electrophilic character of the epoxides has been
utilized in chemical synthesis of several PMAP metabolites.
Differences in the reactivites of arene oxides are probably less
important in chemical synthesis of their conjugates but may
influence the ratio of positional and diasteromers formed. The
reaction of GSH with a xenobiotic epoxide gives the 1,2-dihydro-
-1-hydroxy-2-glutathionyl-xenobiotic conjugate and with cysteine
and N-acetyl-cysteine the corresponding dihydro-hydroxy
conjugates. The nucleophile (GSH, cysteine or N-acetyl-cysteine)
is usually mixed with the arene oxide in a polar solvent such as
acetone, dioxane, acetonitrile, DMSO or methanol but the reaction
may also be run in aqueous buffer solutions. Even though the
addition of base increases the nucleophilicity of the sulfur, the
pH should not be increased too much since that will lead to more
complex reaction product mixtures (167). A PTC mediated reaction
with tetrabutylammonium bisulfate has been suggested for the
synthesis of N-acetyl-cysteine conjugates of epoxides (167). The
recent synthesis of some premercapturic acid type conjugates from
epoxides is summerized in Table VII.

In order to determine the stereochemistry of metabolites
formed and to reduce the complexity of reaction products formed

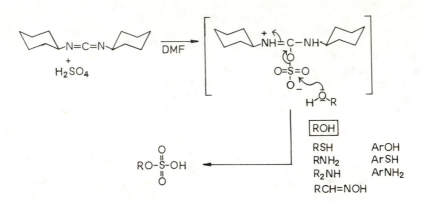

Figure 7. Mechanism for dicyclohexylcarbodiimide (DCC) mediated
sulfation of xenobiotics with sulfuric acid.

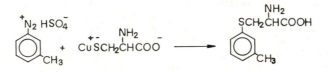

Figure 8. Nucleophilic substitution of 1-bromoadamantane with
cysteine.

Figure 9. Synthesis of S-(3-methylphenyl) cysteine.

Table VI. Aryl glutathione, cysteine and N-acetyl-cysteine
conjugates synthesized by nucleophilic displacement of halogen or
some other leaving group including the one obtained by
diazotisation of arylamines.

Substrate	Nucleophile	Ref.
2,4-Dinitrochlorobenzene	GSH	149
2,4-Dinitrofluorobenzene	GSH,Cys	150
1-Methyl-4-nitro-5-chloroimidazole	GSH	156
1-Fluoro-4-nitrobenzene	GSH,Cys	157
4-Fluoro-5-nitro-2-(4-thiazolyl)-benzimidazole[1]	Cys	158
Pentachloronitrobenzene	GSH,Cys,N-AcCys	151,159
Methyl 2-nitro-5-fluorobenzoate[2]	GSH,Cys	160
4-Bromoaniline	Cys	152
Aniline	Cys	161,162
4-Chloroaniline	Cys	162
2- and 3-Haloanilines[3]	Cys	163
1- and 2-Aminonaphthalene	Cys,N-AcCys	163,164
2-, 3- and 4-Methylanilines	Cys	163
2- and 4-Methoxyaniline	Cys	163
1-Aminonaphthalene	GSH,Cys	153
2-Chloro-4-nitroaniline	GSH,Cys	154
2,4-Dichloroaniline	GSH	121
1-Amino-4-nitrophenol	Cys	165
Amino-2,4',5-trichlorobiphenyl	N-AcCys	155

1. Cambendazole metabolism study; 2. Acifluorfen metabolism study;
3. Chloro- and bromoanilines.

Table VII. Epoxides of aromatic and alkene compounds used for
synthesis of S-alkyl dihydro-hydroxy-xenobiotics.

Epoxide	Nucleophile	Ref.
Phenanthrene 9,10-oxide	GSH,Cys,N-AcCys	170
	N-AcCys	167
Styrene oxide	N-AcCys	171,172
		135,173
	GSH	136,174
		175
	Cys,N-AcCys,	
	Cys-Gly,GSH	176
Benzo(a)pyrene 4,5-oxide	GSH,N-AcCys	177
	GSH	167,178
Pyrene oxide	GSH	167
Naphthalene 1,2-oxide	N-AcCys	179
2,2',5,5'-Tetrachlorobiphenyl 3,4-oxide	Cys,GSH	180
3-(4-Nitrophenoxy)propen 1,2-oxide	GSH,N-AcCys	181
Cyclohexene oxide	N-AcCys	182
2-,3- and 4-Methylstyrene oxides	N-AcCys	183
2,3-Epoxypropanol	GSH	132

synthetically, it is neccessary to use optically pure starting
materials. Additional information on the stereochemistry may be
obtained by use of halohydrins as described by Yagan et al (136).
A reaction of 2-bromo-2-phenylethanol and GSH yielded primarily a
mixture of C-2 enantiomers but minor amounts of the C-1
enantiomers were also formed.

The stability of PMAP conjugates must be considered as a
potential problem; severe decomposition of S-alkyl dihydrohydroxy-
phenanthrene, including the cysteine conjugate, was recently
reported (168). When the pure S-ethyl conjugate was stored in
neutral methanol it decomposed to phenanthrene, 9-phenanthrenol,
9,10-phenanthraquinone, 9,10-dihydroxy- and 9-ethylthio-
10-hydroxy-phenanthrene. When a molar amount of hydrochloric acid
was added to 9-[S-(L-cysteinyl)]9,10-dihydro-10-hydroxy-
phenanthrene in methanol, phenanthrene, 9-[S-(L-cysteinyl)]-
9,10-dihydro-10-methoxyphenanthrene and cysteine were formed
(168). Thus it is possible that these types of degradation of PMAP
type conjugates also occur in vivo (2,169).

Reactions with quinones. The formation of quinones in vivo has
important toxicological significance because these compounds may
become covalently bound to macromolecules (184). The addition of
GSH, cysteine or N-acetyl-cysteine to quinones or iminoquinones
has proved to be a valuable synthetic method for preparation of
catechol and diaminoaryl conjugates (185-192). The mechanism for
their formation has been discussed but not completely elucidated
(187). This type of reaction has been used only occasionally for
synthesis of conjugates of xenobiotics. GSH, cysteine and
N-acetyl-cysteine conjugates of 4-hydroxyacetanilide were prepared
from N,4-dihydroxyacetanilide and the appropriate nucleophile in
30-50% yield (186). Metabolites of 4'-(9-acridinylamino)methane-
sulfone-3-anisidine (188,191) and N^2-methyl-9-hydroxy-
ellipticinium acetate (190) have also been synthesized by this
procedure.

Mannich type condensation. Primary and secondary amines when
reacted with formaldehyde and the appropriate RSH nucleophile
yield a mixture of mono- and bis-substituted conjugates and a
mono-substituted conjugate, respectively as illustrated in Figure
10. GSH conjugates of 4-aminoazobenzene (193) and N,N-dimethyl-
-4-aminoazobenzene sulfate (194) have been synthesized by this
procedure.

Addition reactions. Additions of GSH and similar nucleophiles to
activated double- or triple-bonds have been reported (Table VIII).
The reaction proceeds under mild conditions, in most cases with an
alcohol or water as the solvent and in the presence of base.
However, the addition of N-acetyl-cysteine to two hydroxybutene
compounds was accomplished using UV-irradiation for radical
induction (195).

In a similar reaction, chlorinated thiophenols were added to
α-acetamidoacrylic acid for the synthesis of mercapturic acids of
polychlorinated benzenes (196). This method produces a DL-mixture
of the N-acetyl-cysteine conjugate and it is limited to the
synthesis of mercapturic acids and cysteine conjugates, the latter
obtained after deacetylation.

Table VIII. Substrates for synthesis of mercapturic acid pathway
metabolites by miscellaneous methods[1].

Substrate	Conjugate	Method	Ref.
Benzyl isothiocyanate	GSH,Cys-gly Cys,N-AcCys	A	203
Tetramethrin	GSH,N-AcCys	A	204
1,2,3,4-Tetrahydrophthalimide	GSH,N-AcCys	A	204
Alkyl isothiocyanate[2]	N-AcCys	A	205
3- and 4-Hydroxy-1-butene	N-AcCys	A	195
1-Octyn-3-one	N-AcCys	A	206
Pentachlorothiophenol[3]	N-AcCys	A	196
2,3,5,6- and 2,3,4,6-Tetrachloro- thiophenol[3]	N-AcCys	A	196
4-Chlorothiophenol[3]	N-AcCys	A	33
2,4-Dinitrochlorobenzene	GSH	E	200
Captopril	GSH	E	200
Styrene oxide	GSH	E	200
Iminocyclophosphamid	GSH	E	200
Ethacrynic acid	GSH	E	200
Indomethacin glucuronide	GSH	E	200
Nitrosobenzene	GSH	M	207
1-(3-Mercapto-2-D-methyl-1-oxo- propyl)-2-proline (SQ-14225)	Cys	M	208
2-Nitrosofluorene	GSH	M	209
N-Acetoxy-N-acetyl-2-aminofluorene	GSH	M	210
Ronidazole[4]	Cys	M	211

1. A: addition, E: enzymatic, M: miscellaneous reactions;
2. Alkyl = methyl, ethyl and butyl; 3. Addition to α-acetamidoacrylic acid; 4. 1-methyl-5-nitroimidazole-2-methanol carbamate.

Enzymatic methods. Preparation of the GSH conjugates of xenobiotics can be performed by use of microsomal preparations (e.g.
197,198). A disadvantage of this technique is that only small
amount of substrate can be conjugated in each preparation. Larger
amounts of conjugates may be obtained from reactions between
purified glutathione S-transferases and substrate (199). The use
of a purified single GSH S-transferase also gives information on
biochemical features of the enzyme such as substrate specificity.
     Recently Pallante et al. (200) reported that GSH S-trans-
ferase had been immobilized on a cyanogen bromide activated
Sepharose gel, by a procedure similar to that discussed above for
immobilization of UDP-glucuronyltransferase (39), and used
successfully for synthesis of GSH conjugates. A variety of substrates (Table VIII) were conjugated with this system and the
products were characterized by chromatographic methods and FAB
mass spectrometry.

Miscellaneous reactions and comments. The conjugate of a few
xenobiotic substrates have been prepared by procedures other than
those described above (Table VIII). 3-Chloroalanine (L- or DL-) is

commercially available and is potentially valuable for synthesis
of cysteine and N-acetyl-cysteine conjugates from nucleophilic
sulfur-containing substrates. For example a derivatized 3-chloro-
alanine (2-benzyloxycarbonylamino-3-chloropropionic acid) was used
in synthesis of a phenacetin cysteine conjugate (201).

Mercapturic acids are most easily obtained by acetylation of
the corresponding cysteine conjugates. Acetylating agents such as
acetic anhydride (161) and ketene (183) may be used.

The reversibility of the Smiles rearrangement (Figure 11) may
occasionally be of use for synthesis of cysteine conjugates (202).
The S-N reaction is known to occur rapidly under both alkaline and
neutral conditions in DMF and DMSO and therefore it must be considered as a possible source of rearrengement products during the
isolation of cysteine conjugates involving the use of these
solvents.

Mercapturic acids are known to be more labile in alkali than
the corresponding cysteine conjugates (162). A fully aromatic GSH
conjugate of 2,2',5,5'-tetrachlorobiphenyl 3,4-oxide was unstable
upon storage in water and methanol (180). This GSH-conjugate was
suggested to undergo the cyclization reactions shown in Figure 12.

Synthesis of Methyl Sulfide, Sulfoxide and Sulfone Metabolites of Xenobiotics

The further degradation of MAP metabolites to methyl sulfides and
their oxidized products have been reported for many exogenous
compounds (212). Animal tissues and gut microflora contain C-S
lyase activity that is responsible for the formation of thiols
from GSH and related conjugates (213,214). These thiols may be
methylated and subsequently oxidized to the corresponding
sulfoxides and sulfones.

Synthesis of these simple type of thiol derivatives will be
just briefly discussed here because comprehensive reviews on
thiols and their derivatives are available (215,216). The methyl
sulfide compounds are most readily obtained by methylation of the
corresponding thiol. Nucleophilic substitutions with methane
thiolate on alkyl halides and occasionally with aryl halides lead
directly to the methyl sulfide product. The syntheses of methyl
sulfide and sulfone metabolites of xenobiotics are summarized in
Table IX.

Methylation of aryl thiols. PTC methylation of ^{14}C-aryl thiols
with methyl iodide (217) is very useful for microscale synthesis
of ^{14}C-labeled aryl methyl sulfides (218). The reaction, using
tetrabutylammonium hydrogensulfate as catalyst and sodium
hydroxide as base in methylene chloride and water, proceeds at
ambient temperature in 50-100% yield.

Nucleophilic substitution reactions. Nucleophilic displacement of
a halogen in compounds such as 2-acetamido-4-chloromethylthiazole
under reflux of an alkaline ethanol solution is a straight
foreward reaction (219). Nucleophilic aromatic substitution
reactions are more complex but under certain conditions they can
be used for single step synthesis of aryl methyl sulfides. A
number of 4-methylthio-polychlorobiphenyls were synthesized from

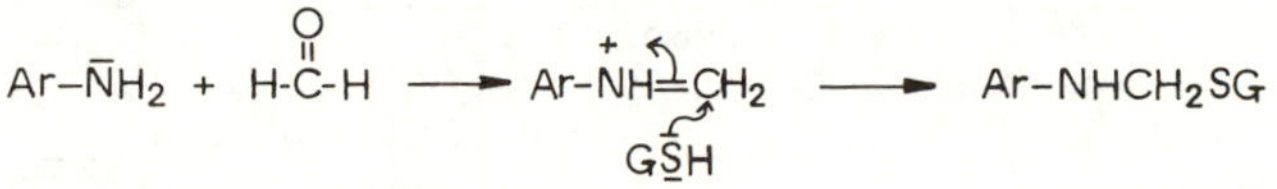

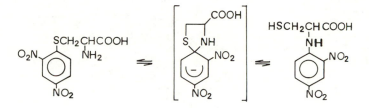

Figure 10. Mannich type synthesis of glutathione conjugates of arylamines.

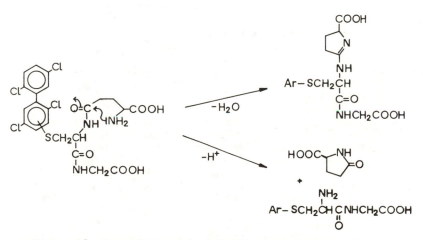

Figure 11. Smiles rearrangement of S-(2,4-dinitrophenyl) cysteine.

Figure 12. Degradation of glutathione conjugates of 2,2',5,5'-tetrachlorobiphenyl in water and methanol as suggested by Preston et al. (180).

Table IX. Synthesis of methyl sulfide, sulfoxide and sulfone
xenobiotic metabolites.

Xenobiotic	Substituent	Ref.
2-Acetamido-4-chloromethylthiazole	SMe	219
Pentachloronitrobenzene	SMe,SOMe,SO_2Me	159,227
Hexachlorobenzene	SMe,SOMe,SO_2Me	159,228
1,4-Dichlorobenzene	SMe,SOMe,SO_2Me	229
Propachlor	SMe,SO_2Me	139,230
4-Hydroxyacetanilide	SMe	201,231
Styrene oxide	SMe	232,173
PCBs[1]	SMe,SO_2Me	220,233-236
DDE[2]	SO_2Me	226
Phenanthrene	SMe	237
Naphthalene	SMe	179
2-Acetylaminofluorene	SMe	238
Biphenyl	SMe	239
3-(5-Nitro-2-furyl)-2-(2-furyl)-acrylamide	SMe	240

1. Several congeners shown to give methyl sulfides and sulfones.
2. Major DDT metabolite.

different polychlorinated biphenyls (PCB) and methane thiolate in
methanol using this method (220). Under these conditions one major
product was obtained (Figure 13) but traces of isomeric PCB methyl
sulfides were observed. The reactivity of halobenzenes with
alkanethiols has been thoroughly studied (221). The yields of
sulfides were dramatically improved by use of the polar solvent,
hexamethylphosphoric triamide (HMPA). Furthermore, yields of alkyl
aryl sulfides were improved by synthesis carried out under PTC
conditions(222,223).

Miscellaneous reactions. The sulfides are oxidized by an excess of
e.g. hydrogen peroxide or 3-chloroperoxybenzoic acid to the
corresponding sulfones. A controlled addition of peroxide yields a
mixture of sulfoxide and sulfone products. The use of nitronium
hexafluorophosphate or tetrafluoroborate to oxidize dialkyl, aryl
alkyl or diaryl sulfides yields only the corresponding sulfoxide
under extremely mild conditions (224). Sulfoxides may be further
oxidized to the corresponding sulfones in good yield by the
nitronium salts as well (225).
    In a few cases the electrophilic Friedel-Craft type acylation
can be used for the one step synthesis of aryl methyl sulfones, as
exemplified by the reaction of DDE with methanesulfonic anhydride
(226).

Synthesis of Amino acid conjugates

The largest group of amino acid conjugates studied, except for
cysteine conjugates, consists of a diversity of carboxylic acids
bound to different amino acids via an amide bond. The enzymatic
formation of this type of amino acid conjugates, examples of
carboxylic acid compounds and amino acids determined to undergo

Table X. Synthesis of amino acid conjugates of exogenous
carboxylic acids.

Carboxylic acid	Amino acid	Ref.
4-Chlorophenoxyacetic acid	gly	244
2,4-Dichlorophenoxyacetic acid	see ref.[1,2]	244
		258,247
2,4,5-Trichlorophenoxyacetic acid	gly, val	244
3,5-Di-ter.butyl-4-hydroxybenzoic acid	gly	259
Indol-3-ylacetic acid	glu,tau	245
Ferulic acid	gly	246
3-(2,2-Dichlorovinyl-2,2-dimethylcyclo-propane carboxylic acid[3]	ala,gly,ser,glu	260
3-Phenoxybenzoic acid[3]	ala,gly,ser,glu	260
1- and 2-Naphthylacetic acid	gly,gln,tau	261,262
Hydratropic acid	gly,gln,tau	263
3-Phenoxybenzoic acid[4]	tau	264
Benzoic acid[5]	gly	265
Clofibric acid	gly	266
Ananilino acid[6]	gly,thr,val	267
2-Chlorobenzoic acid	gly,glu,ala	268

1. 14 different amino acids (244); 2. 20 amino acids (258);
3. Permethrin metabolites; 4. Cypermethrin metabolites;
5. Conjugate known as Hippuric acid; 6. Fluvalinate metabolite.

conjugate formation in various animal species, have been
summarized (241-243). The amino acids most commonly conjugated to
carboxylic acids are glycine, glutamine, ornithine and taurine but
several other amino acids have also been reported.

The formation of an amide bond by organic chemical methods is
usually accomplished by reacting the acid halogenide and the amino
acid unless the acid substrate contains other reactive groups. The
acid chloride is prepared directly with thionyl chloride (244) or
indirectly via the formation of a mixed anhydride by use of
another acid halogenide (245,246). Isotopically labeled
[(2,4-Dichlorophenoxy)acetyl]valine was synthesized by a DCC
catalyzed reaction between the acid and valine (247). A few
examples of synthetic amino acid conjugates of xenobiotics are
shown in Table X. Several taurine conjugates of both alkanoic and
benzylic acids have been synthesized and their physical properties
determined by Idle et al. (248).

Reactive xenobiotic intermediates may be covalently bound to
amino acids, free or in proteins. For example the reactivity of
epoxides have been studied in relation to formation of covalent
bound products with the hemoglobin amino acids; histidine, valine
and to some extent also cysteine in dose monitor experiments
(249,250). The hemoglobin must be hydrolyzed prior to analysis of
histidine adducts while that is not neccessary in order to
characterize adducts to the N-terminal valine (251). A few
reference cysteine, histidine and valine xenobiotic adducts have
been synthesized (250,252-254). The primary amino group of
histidine is occupied in the peptide bond, therefore this group
must be protected in order to favor reaction at the other nucleo-

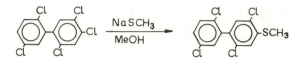

Figure 13. Synthesis of 4-methylthio-2,2',5,5'-tetrachloro-
biphenyl via nucleophilic aromatic substitution.

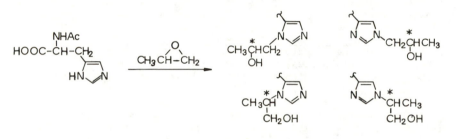

Figure 14. Formation of regioisomers and enantiomers at propene
oxide reaction with N^2-acetyl-histidine.

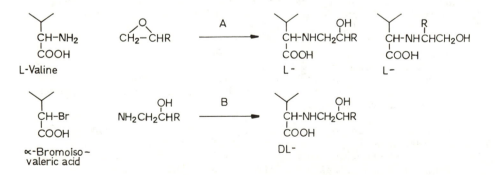

Figure 15. Synthesis of alkylated valine via A: valine and
epoxide and B: α-bromoisovaleric acid and alkylamine.

philic centers of the amino acid. N^T- and N^π-ethyl- ($\underline{254}$),
2-hydroxyethyl- ($\underline{252}$), 2-hydroxypropyl- ($\underline{253}$) and 2-hydroxy-
-2-phenylethyl- ($\underline{255}$) substituted histidines have been prepared by
use of the N^2-acetyl-L-histidine or the corresponding histidine
methyl ester and ethyl bromide, ethene oxide, propene oxide and
styrene oxide, respectively. Both positional isomers and
enantiomers are formed as shown in Figure 14 in the reaction of
propene oxide with histidine ($\underline{253}$). N^T-(2-Hydroxyethyl)-histidine
was synthesized in higher yield by reacting N^2-benzoyl-histidine
methyl ester and iodoethyl acetate followed by hydrolytic cleavage
of the ester bond ($\underline{256}$).

N-alkylated valine may be similarily synthesized from an
epoxide and valine, Figure 15A. Synthesis of DL-mixtures of
N-(2-hydroxyethyl)- and N-(2-hydroxypropyl)-valine have been
performed using α-bromo-isovaleric acid and the appropriate
alkylamine, Figure 15B ($\underline{250}$). The reactions of propylene oxide and
N-acetyl-cysteine methyl ester, valine methyl ester and
N-benzoyl-histidine methyl ester respectively were recently
thoroughly investigated and the stereochemistry discussed ($\underline{257}$).

In a separate study, N^2-protected histidine and lysine were
reacted with allyl bromide, 1-bromobutyne and 2-bromoacetophenone
to prepare the corresponding N-alkylated amino acids ($\underline{141}$).

Miscellaneous Conjugations

In addition to the classical conjugation reactions a growing
number of novel conjugates are reported and were recently listed
by Caldwell ($\underline{269}$) and in other chapters in this book. Synthesis of
all of these types of conjugates will not be discussed here but
reference to a few methods is appropriate. A bile acid conjugate
of a major fluvalinate metabolite was synthesized in high yield by
a DCC mediated esterification of the appropriate acid and methyl-
taurochemodeoxycholic ester ($\underline{270}$). Mono- and diglyceride conjugates of the same fluvalinate metabolite have also been
synthesized ($\underline{268}$). Two dipalmitoyl glycerol conjugates were
prepared from 3-phenoxybenzoyl chloride and the appropriate
glycerol ($\underline{271}$). Fatty acid conjugates of a DDT metabolite, DDOH,
and of 11-hydroxy-tetrahydrocannabinols (11-OH-THC) were
synthesized by reacting DDOH or 11-OH-THC with palmitoyl chloride
dissolved in pyridine at room temperature ($\underline{272}$,$\underline{273}$).

Acknowledgments

The inspiring help given by Drs T. Norberg, G.D. Paulson and C.A.
Wachtmeister throughout this work is gratefully acknowledged.
This work was financially supported by the Research Committee
at the National Swedish Environmental Protection Board.

Literature Cited

1. Reichert, D. Angew. Chem. Int. Ed. Engl. 1981, 20, 135.
2. Bakke, J.; Struble, C.; Gustafsson, J.-A.; Gustafsson, B.
    Proc. Natl. Acad. Sci. USA 1985, 82, 668.
3. Bakke, J.E.; Bergman, Å.; Brandt, I.; Darnerud, P.O.;
    Struble, C. Xenobiotica 1983, 13, 597.

4.   Dutton, G.J. Ed., "Glucuronic acid Free and Combined";
     Academic press, New York; 1966.
5.   Goodwin, B.L. "Handbook of Intermediary Metabolism of
     Aromatic Compounds"; Chapman and Hall, London; 1976.
6.   Dutton, G.J. "Glucuronidation of Drugs and Other Compounds";
     CRC Press Inc.; 1980.
7.   Keglevic, D. Adv. Carbohydr. Chem. Biochem. 1979, 36, 57.
8.   Koenigs, W; Knorr, E. Ber. Dtsch. Chem. Ges. 1901, 34, 957.
9.   Igarashi, K. Adv. Carbohydr. Chem. Biochem. 1977, 34, 243.
10. Bollenback, G.N.; Long, J.W.; Benjamin, D.G.; Lindqvist, J.A.
     J. Am. Chem. Soc. 1955, 77, 3310.
11. Iveson, P.; Parke, D.V. J. Chem. Soc. (C). 1970, 2038.
12. Conrow, R.B.; Bernstein, S. J. Org. Chem. 1971, 36, 863.
13. Igarashi, K.; Irisawa, J.; Honma, T. Carbohydr. Res. 1975,
     39, 213.
14. Hansson, C.; Rosengren, E. Acta Chem. Scand. 1976, B30, 871.
15. Vlahov, J.; Snatzke, G. Liebigs Ann. Chem. 1983, 42, 570.
16. Kishikawa, T.; Oikawa, Y.; Takitani, S. Chem. Pharm. Bull.
     1969, 17, 699.
17. Aboul-Enein, H.Y. J. Carbohydr. Nucleosides Nucleotides 1977,
     4, 77.
18. Friedman, R.B.; Weissmann, B. Carbohydr. Res. 1972, 24, 123.
19. Mahrwald, R.; Krüger, W. Hoppe-Seyler´s Z. Physiol. Chem.
     1978, 359, 1803.
20. Honma, K.; Nakazima, K.; Uematsu, T.; Hamada, A. Chem. Pharm.
     Bull. 1976, 24, 394.
21. Yagen, B.; Levy, S.; Mechoulam, R.; Ben-Zvi, Z. J. Am. Chem.
     Soc. 1977, 99, 6444.
22. Kato, K.; Ide, H.; Hirohata, I.; Fishman, W.H. Biochem. J.
     1967, 103, 647.
23. Bridges, J.W.; Walker, S.R.; Williams, R.T. Biochem. J. 1969,
     111, 173.
24. Mattox, V.R.; Goodrich, J.E.; Vrieze, W.D. Biochemistry 1969,
     8, 1188.
25. Sakata, M.; Haga, M.; Tejima, S. Carbohydr. Res. 1970, 13,
     379.
26. Paulson, G.D.; Zaylskie, R.G.; Dockter, M.M. Anal. Chem.
     1973, 45, 21.
27. Krohn, K. Hoppe-Seyler´s Z. Physiol. Chem. 1977, 358, 1551.
28. Nambara, T.; Takizawa, N.; Goto, J.; Shimada, K. Chem. Pharm.
     Bull. 1976, 24, 2869.
29. Shimizu, T.; Mori, H.; Tabusa, E.; Morita, S.; Miyamoto, G.;
     Yasuda, Y.; Nakagawa, K. Xenobiotica, 1978, 8, 349.
30. Ericsson, Ö. Acta Pharm. Suec. 1978, 15, 81.
31. Climie, I.J.G.; Hutson, D.H.; Stoydin, G. Xenobiotica. 1983,
     13, 611.
32. Maftouh, M.; Menuier, G.; Dugue, B.; Monsarrat, B.; Meuner,
     B.; Paoletti, C. Xenobiotica. 1983, 13, 303.
33. Quistad, G.B.; Mulholland, K.M. J. Agric. Food Chem. 1983,
     31, 585.
34. Kiss, J.; Burkhardt, F. Carbohydr. Res. 1970, 12, 115.
35. Smejkal, J.; Kalvoda, L. Czechoslov. Chem. Commun. 1973, 38,
     1981.

36. Heyns, K.; Paulsen, H. In "Advances in Carbohydrate Chemistry"; Wolfrom, M.L., Tipson, R.S., Eds; Academic press, New York, 1962, 17, p. 169.

37. Kiss, J.; Noack, K.; D´Souza, R. Helv. Chim. Acta. 1975, 58, 301.

38. Puhakainen, E.; Lang, M.; Ilvonen, A.; Hänninen, O. Acta Chem. Scand. 1976, B30, 685.

39. Parikh, I.; MacGlashan, D.W.; Fenselau, C. J. Med. Chem. 1976, 19, 296.

40. Fenselau, C.; Pallante, S.; Batzinger, R.P. Science. 1977, 198, 625.

41. Fenselau, C.; Pallante, S.; Parikh, I. J. Med. Chem. 1976, 19, 679.

42. Lyle, M.A.; Pallante, S.; Head, K.; Fenselau, C. Biomed. Mass Spectrom. 1977, 4, 190.

43. Johnson, L.P.; Fenselau, C. Drug Metab. Dispos. 1978, 6, 677.

44. Pallante, S.; Lyle, M.A.; Fenselau, C. Drug Metab. Dispos. 1978, 6, 389.

45. Lehman, J.P.; Fenselau, C. Drug Metab. Dispos. 1982, 10, 446.

46. Feng, P.C.C.; Fenselau, C.; Colvin, M.E.; Hinson, J.A. Drug Metab. Dispos. 1983, 11, 103.

47. van Breemen, R.B.; Tabet, J.-C.; Cotter, R.J. Biomed. Mass Spectrom. 1984, 11, 278.

48. Staiger, L.E.; Quistad, G.B. J. Agric. Food Chem. 1984, 32, 1130.

49. Bugianesi, R.; Shen, T.Y. Carbohydr. Res. 1971, 19, 179.

50. Thompson, R.P.H.; Hofmann, A.F. Biochim. Biophys. Acta 1976, 451, 267.

51. Keglevic, D.; Pravdic, N.; Tomasic, J. J. Chem. Soc. (C). 1968, 511.

52. Segura, J.; Bakke, O.M.; Huizing, G.; Beckett, A.H. Drug Metab. Dispos. 1980, 8, 87.

53. Fischer, B.; Nudelman, A.; Ruse, M.; Herzig, J.; Keinan, E. Abstr. XIIIth Int Symp on Med. Chem. 1984, p 66.

54. Pravdic, N.; Keglevic, D. Tetrahedron 1965, 21, 1897.

55. Quistad, G.B.; Staiger, L.E.; Jamieson, G.C.; Schooley, D.A. J. Agric. Food Chem. 1982, 30, 895.

56. Quistad, G.B.; Mulholland, K.M.; Skiles, G.; Jamieson, G.C. J. Agric. Food Chem. 1985, 33, 95.

57. Frear, D.S. In "Bound and Conjugates Pesticide Residues" Kaufman, D.D.; Still, G.G.; Paulson, G.D.; Baudal, S.K. Eds. ACS SYMPOSIUM SERIES No. 29, American Chemical Society; Washington D.C., 1976, Chpt 6.

58. Smith, J.N. In "Drug Metabolism from Microbe to Man" Parke, D.V.; Smith, R.L. Eds.; Taylor & Francis Ltd; London 1977; p 219.

59. Friend, D.R.; Chang, G.W. J. Med. Chem. 1985, 28, 51.

60. Garegg, P.J.; Norberg, T. Acta Chem. Scand. 1979, B33, 116.

61. Garegg, P.J.; Hultberg, H.; Ortega, C.; Samuelsson, B. Acta Chem. Scand. 1982, B36, 513.

62. Weitl, F.L.; Sovak, M.; Ohno, M. J. Med. Chem. 1976, 19, 353.

63. Knöchel, A.; Rudolph, G.; Thiem, J. Tetrahedron Lett. 1974, 551.

64. Knöchel, A.; Rudolph, G. Tetrahedron Lett. 1974, 3739.

65. Schmidt, G.; Farkas, J. Tetrahedron Lett. 1967, 4251.

66. Onodera, K.; Hirano, S.; Fukumi, H. Agr. Biol. Chem. 1964, 28, 173.
67. Paulson, G.D.; Giddings, J.M.; Lamoureux, C.H.; Mansager, E.R.; Struble, C.B. Drug Metab. Dispos. 1981, 9, 142.
68. Tang, B.K.; Kalow, W.; Grey, A.A. Drug Metab. Dispos. 1979, 7, 315.
69. Soine, W.H.; Bhargava, V.O.; Garrettson, L.K. Drug Metab. Dispos. 1984, 12, 792.
70. Giera, D.D.; Abdulla, R.F.; Occolowitz, J.L.; Dorman, D.E.; Mertz, J.L.; Sieck, R.F. J. Agric. Food Chem. 1982, 30, 260.
71. Brewster, K.; Harrison, J.M.; Inch, T.D. Tetrahedron Lett. 1979, 5051.
72. Ngah, W.Z.W.; Smith, J.N. Xenobiotica 1983, 13, 383.
73. Nambara, T.; Kawarada, Y. Chem. Pharm. Bull. 1976, 24, 421.
74. Mulder, G.J. "Sulfation of Drugs and Related Compounds"; CRC Press Inc.; 1981.
75. Paulson, G.D. In "Bound and Conjugated Pesticide Residues" Kaufman, D.D.; Still, G.G.; Paulson, G.D; Bandal, S.K. Eds. ACS SYMPOSIUM SERIES No. 29, American Chemical Society; Washington D.C., 1976; Chpt. 6.
76. Roy, A.B. In "Sulfation of Drugs and Related Compounds"; Mulder, G.J., Ed.; CRC Press Inc.; 1981, Chap. 2.
77. Gilbert, E.E. "Sulfonation and Related Reactions"; John Wiley & Sons, New York; 1965; Chap. 6-7.
78. Muccino, R.R.; Markezich, R.; Vernice, G.G.; Perry, C.W.; Liebman, A.A. Carbohydr. Res. 1976, 47, 172.
79. Fieser, L.F.; Fieser, M. "Reagents for Organic Synthesis"; John Wiely and Sons, New York; 1967; p. 1125.
80. Quadri, S.F.; Seib, P.A.; Deyoe, C.W. Carbohydr. Res. 1973, 29, 259.
81. Maggs, J.L.; Olavesen, A.H. J. Labelled Compd. Radiopharm. 1979, 17, 793.
82. White, G.F.; Lillis, V.; Shaw, D.J. Biochem. J. 1980, 187, 191.
83. Hearse, D.J.; Olavesen, A.H.; Powell, G.M. Biochem. Pharm. 1969, 18, 173.
84. Mumma, R.O. Lipids. 1966, 1, 221.
85. Mumma, R.O.; Fujitani, K.; Hoiberg, C.P. J. Chem. Eng. Data 1970, 15, 358.
86. Hoiberg, C.P.; Mumma, R.O. J. Am. Chem. Soc. 1969, 91, 4273.
87. Mumma, R.O.; Hoiberg, C.P.; Weber, W.W. Steroids. 1969, 14, 67.
88. Hoiberg, C.P.; Mumma, R.O. Biochim. Biophys. Acta. 1969, 177, 149.
89. Mumma, R.O.; Hoiberg, C.P.; Simpson, R. Carbohydr. Res. 1970, 14, 119.
90. Mumma, R.O.; Hoiberg, C.P. J. Chem. Eng. Data. 1971, 16, 492.
91. Jeffrey, H.J.; Roy, A.B. Anal. Biochem. 1977, 77, 478.
92. Burkhardt, G.N.; Lapworth, A. J. Chem. Soc. 1926, 684.
93. Fujimoto, J.M.; Haarstad, V.B. J. Pharmacol. Exp. Ther. 1969, 165, 46.
94. Joseph, J.P.; Dusza, J.P.; Bernstein, S. Steroids. 1966, 7, 577.
95. Nagubandi, S.; Londowski, J.M.; Bollman, S.; Tietz, P.; Kumar, R. J. Biol. Chem. 1981, 256, 5536.

96. Levitz, M. Steroids 1963, 1, 117.
97. Buncel, E.; Raoult, A.; Wiltshire, J.F.; J. Am. Chem. Soc. 1973, 95, 799.
98. Buncel, E.; Chuaqui, C. J. Org. Chem. 1980, 45, 2825.
99. Penny, C.L.; Perlin, A.S. Carbohydr. Res. 1981, 93, 241.
100. Scalera, M.; Hardy, W.B.; Hardy, E.M.; Joyce, A.W. J. Am. Chem. Soc. 1951, 73, 3094.
101. Boyland, E.; Nery, R. J. Chem. Soc. 1962, 5217.
102. Jorritsma, J.; Meerman, J.H.N.; Vonk, R.J.; Mulder, G.J. Xenobiotica 1979, 9, 247.
103. Arakawa, Y.; Imai, K.; Tamura, Z. Anal. Biochem. 1983, 132, 389.
104. Iwasaki, K; Shiraga, T.; Noda, K.; Tada, K.; Noguchi, H. Xenobiotica 1983, 13, 273.
105. Bevilacqua, P.F.; Keith, D.D.; Roberts, J.L. J. Org Chem. 1984, 49, 1430.
106. Baxter, T.H.; Kostenbauder, H.B. J. Pharm. Sci. 1969, 58, 33.
107. Barton, A.D.; Young, L. J. Am. Chem. Soc. 1943, 65, 294.
108. Boyland, E.; Manson, D.; Orr, S.F.D. Biochem. J. 1957, 65, 417.
109. Flynn, T.G.; Rose, F.A.; Tudball, N. Biochem. J. 1967, 105, 1010.
110. Paulson, G.D.; Zehr, M.V. J. Agric. Food Chem. 1971, 19, 471.
111. Mori, M.-a; Oguri, K.; Yoshimura, H.; Shimomura, K.; Kamata, O.; Ueki, S. Life Sci. 1972, 11, 525.
112. Chen, C.-T.; Samejima, K.; Tamura, Z. Chem. Pharm. Bull. 1973, 21, 911.
113. Nagasawa, K.; Yoshidome, H. Chem. Pharm. Bull. 1973, 21, 903.
114. Nagasawa, K.; Yoshidome, H. J. Org. Chem. 1974, 39, 1681.
115. Burke, B.; Olavesen, A.H.; Curtis, C.G.; Powell, G.M. Xenobiotica, 1978, 8, 145.
116. Hoffmann, K.-J.; Arfwidsson, A.; Borg, K.O. Drug Metab. Dispos. 1982, 10, 173.
117. Johnson, D.B.; Thissen, R.M. J. Chem. Soc. Chem. Commun. 1980, 598.
118. Mannervik, B.; Jensson, H. J. Biol. Chem. 1982, 257, 9909.
119. Paul, B.; Korytnyk, W. J. Med. Chem. 1976, 19, 1002.
120. Mangold, J.B.; Abdel-Monem, M.M. Biochem. Biophys. Res. Commun. 1980, 96, 333.
121. Clark, A.G.; Hitchcock, M.; Smith, J.N. Nature 1966, 209, 103.
122. Kurihara, N.; Tanaka, K.; Nakajima, M. Pestic. Biochem. Physiol. 1979, 10, 137.
123. Portig, J.; Kraus, P.; Stein, K.; Koransky, W.; Noack, G.; Gross, B.; Sodomann, S. Xenobiotica 1979, 9, 353.
124. Conroy, C.W.; Schwam, H.; Maren, T.H. Drug Metab. Dispos. 1984, 12, 614.
125. Van Bladeren, P.J.; Buys, W.; Breimer, D.D.; van der Gen, A. Eur. J. Med. Chem. Chim. Ther. 1980, 15, 495.
126. Onkenhout, W.; Guijt, G.J.; De Jong, H.J.; Vermeulen, N.P.E. J. Chromatogr. 1982, 229, ...
127. Theodoropoulos, D. Acta Chem. Scand. 1959, 13, 383.
128. Grenby, T.H.; Young, L. Biochem. J. 1960, 75, 28.
129. Barnsley, E.A. Biochem. J. 1966, 100, 362.
130. Vince, R.; Daluge, S.; Wadd, W.B. J. Med. Chem. 1971, 14, 402.

131. DeBaun, J.R.; Bova, D.L.; Tseng, C.K.; Menn, J.J. J. Agric.
     Food Chem. 1978, 26, 1098.
132. Jones, A.R. Xenobiotica 1975, 5, 155.
133. Jones, A.R.; Mashford, P.M.; Murcott, C. Xenobiotica 1979, 9,
     253.
134. Jones, A.R.; Fakhouri, G. Xenobiotica 1979, 9, 595.
135. Seutter-Berlage, F.; Delbressine, L.P.C.; Smeets, F.L.M.;
     Ketelaars, H.C.J. Xenobiotica 1978, 8, 413.
136. Yagen, B.; Hernandez, O.; Bend, J.R.; Cox, R.H. Bioorg. Chem.
     1981, 10, 299.
137. Pohl, L.R.; Branchflower, R.V.; Highet, R.J.; Martin, J.L.;
     Nunn, D.S.; Monks, T.J.; George, J.W.; Hinson, J.A. Drug
     Metab. Dispos. 1981, 9, 334.
138. Lamoureux, G.L.; Davison, K.I. Pestic. Biochem. Physiol.
     1975, 5, 497.
139. Bakke, J.E.; Rafter, J.; Larsen, G.L.; Gustafsson, J.Å.;
     Gustafsson, B.E. Drug Metab. Dispos. 1981, 9, 525.
140. Odum, J.; Green, T. Toxicol. Appl. Pharmacol. 1984, 76, 306.
141. Jones, J.B.; Hysert, D.W. Can. J. Chem. 1971, 49, 3012.
142. DeBethizy, J.D.; Rickert, D.E. Drug Metab. Dispos. 1984, 12,
     45.
143. Hassall, C.D.; Gandolfi, A.J.; Duhamel, R.C.; Brendel, K.
     Chem.-Biol. Interact. 1984, 49, 283.
144. Nash, J.A.; King, L.J.; Loch, E.A.; Green, T. Toxicol. Appl.
     Pharmacol. 1984, 73, 124.
145. Hutson, D.H.; Logan, C.J.; Regan, P.D. Drug Metab. Dispos.
     1984, 12, 523.
146. Stogniew, M.; Feuselau, C. Drug Metab. Dispos. 1982, 10, 609.
147. Van Doorn, R.; Bos, R.P.; Brouns, R.M.E.; Leijdekkers,
     Ch.-M.; Henderson, P.Th. Arch. Toxicol. 1980, 43, 293.
148. James, S.P.; Needham, D. Xenobiotica 1973, 3, 207.
149. Saunders, B.C. Biochem. J. 1934, 28, 1977.
150. Patchornik, A.; Sokolvsky, M. J. Am. Chem. Soc. 1964, 86,
     1215.
151. Lamoureux, G.L.; Rusness, D.G. J. Agric. Food Chem. 1980, 28,
     1057.
152. Du Vigneaud, V.; Wood, J.L.; Binkley, F. J. Biol. Chem. 1941,
     138, 369.
153. Booth, J.; Boyland, E.; Sims, P. Biochem. J. 1960, 74, 117.
154. Booth, J.; Boyland, E.; Sims, P. Biochem. J. 1961, 79, 516.
155. Bergman, Å.; Bakke, J.E.; Feil, V.J. J. Labelled Compd.
     Radiopharm. 1983, 20, 961.
156. DeMiranda, P.; Beacham, L.H.; Creagh, T.H.; Elion, G.B. J.
     Pharmacol. Exp. Ther. 1973, 187, 588.
157. Hollingworth, R.M.; Alstott, R.L.; Litzenberg, R.D. Life Sci.
     1973, 13, 191.
158. Wolf, D.E.; VandenHeuvel, W.J.A.; Tyler, T.R.; Walker, R.W.;
     Koniuszy, F.R.; Gruber, V.; Arison, B.H.; Rosegay, A.; Jacob,
     T.A.; Wolf, F.J. Drug Metab. Dispos. 1980, 8, 131.
159. Renner, G.; Nguyen, P.-T. Chemosphere 1981, 10, 1215.
160. Frear, D.S.; Swanson, H.R.; Mansager, E.R. Pestic. Biochem.
     Physiol. 1983, 20, 299.
161. Zbarsky, S.H.; Young, L. J. Biol. Chem. 1943, 151, 211.
162. Parke, D.V.; Williams, R.T. Biochem. J. 1951, 48, 624.
163. West, H.D.; Mathura, G.R. J. Biol. Chem. 1954, 208, 315.

164. Ing, H.R.; Bourne, M.C.; Young, L. Biochem. J. 1934, 28, 809.
165. Chrastil, J.; Wilson, J.T. J. Pharmacol. Exp. Ther. 1975, 193, 631.
166. Wolf, C.R. In "Metabolic Basis of Detoxication, Metabolism of Functional Groups"; Jakoby, W.B.; Bend, J.R.; Caldwell, J. Eds; Academic Press, 1982; Chpt. 1.
167. Hernandez, O.; Yagen, B.; Cox, R.H.; Smith, B.R.; Foureman, G.L.; Bend, J.R.; McKinney, J.D. in "Environmental Health Chemistry - The Chemistry of Environmental Agents as Potential Human Hazards", McKinney, J.D. Ed; Ann Arbor Science Publishers Inc.; 1980, p. 425.
168. Feil, V.J.; Huwe, J.K.; Bakke, J.E. Proc. 9th. Symp. on Polynucl. Arom. Hydrocabons, In press.
169. Struble, C.B.; Larsen, G.L.; Feil, V.J.; Bakke, J.E. Proc. 9th. Symp. on Polynucl. Arom. Hydrocabons, In press.
170. Boyland, E.; Sims, P. Biochem. J. 1965, 95, 788.
171. James, S.P.; White, D.A. Biochem. J. 1967, 104, 914.
172. Yagen, B.; Hernandez, O.; Bend, J.R.; Cox, R.H. Chem.-Biol. Interact. 1981, 34, 57.
173. Nakatsu, K.; Hugenroth, S.; Sheng, L.-S.; Horning, E.C.; Horning, M.G. Drug Metab. Dispos. 1983, 11, 463.
174. Ryan, A.J.; Bend, J.R. Drug Metab. Dispos. 1977, 5, 363.
175. Watabe, T.; Ozawa, N.; Hiratsuka, A. Biochem. Pharmacol. 1983, 32, 777.
176. Yagen, B.; Foureman, G.L.; Ben-Zvi, Z.; Ryan, A.J.; Hernandez, O.; Cox, R.H.; Bend, J.R. Drug Metab. Dispos. 1984, 12, 389.
177. Hylarides, M.D.; Lyle, T.A.; Daub, G.H.; VanderJagt, D.L. J. Org. Chem. 1979, 44, 4652.
178. Hernandez, O.; Walker, M.; Cox, R.H.; Foureman, G.L.; Smith, B.R.; Bend, J.R. Biochem. Biophys. Res. Commun. 1980, 96, 1494.
179. Stillwell, W.G.; Horning, M.G.; Griffin, G.W.; Tsang, W.-S. Drug Metab. Dispos. 1982, 10, 624.
180. Preston, B.D.; Miller, J.A.; Miller, E.C. Chem.-Biol. Interact. 1984, 50, 289.
181. James, S.P.; Pheasant, A.E.; Solheim, E. Xenobiotica, 1978, 8, 219.
182. van Bladeren, P.J.; Briemer, D.D.; Seghers, C.J.R.; Vermeulen, N.P.E.; van der Gen, A.; Cauvet, J. Drug Metab. Dispos. 1981, 9, 207.
183. Kühler, T. Xenobiotica 1984, 14, 417.
184. Lown, J.W. Mol. Cell Biochem. 1983, 55, 17.
185. Jellinck, P.H.; Elce, J.S. Steroids. 1969, 13, 711.
186. Gemborys, M.W.; Mudge, G.N. Drug Metab. Dispos. 1981, 9, 340.
187. Nkpa, N.N.; Chedekel, M.R. J. Org. Chem. 1981, 46, 213.
188. Shoemaker, D.D.; Cysyk, R.L.; Padmanabhan, S.; Bhat, H.B.; Matspeis, L. Drug Metab. Dispos. 1982, 10, 35.
189. Maggs, J.L.; Grabowski, P.S.; Park, B.K. Xenobiotica 1983, 13, 619.
190. Maftouh, M.; Meunier, G.; Dugue, B.; Monsarrat, B.; Meunier, B.; Paoletti, C. Xenobiotica 1983, 13, 303.
191. Gaudich, K.; Przybylski, M. Biomed. Mass Spectrom. 1983, 10, 292.

192. Vermeulen, N.P.E.; Onkenhout, W.; van Bladeren, P.J. Chromatogr. Mass. Spectrom. Biomed. Sci. 1983, 2, 39.
193. Ketterer, B.; Srai, S.K.S.; Waynforth, B.; Tullis, D.L.; Evans, F.E.; Kadlubar, F.F. Chem.-Biol. Interact. 1982, 38, 287.
194. Coles, B.; Srai, S.K.S.; Ketterer, B.; Waynforth, B.; Kadlubar, F.F. Chem.-Biol. Interact. 1983, 43, 123.
195. Suzuki, T.; Sasaki, K.; Takeda, M.; Uchiyama, M. J. Agric. Food Chem. 1984, 32, 1278.
196. Renner, G.; Richter, E.; Schuster, K.P. Chemhosphere 1978, 7, 669.
197. Buckpitt, A.R.; Rollins, D.E.; Nelson, S.D.; Franklin, R.B.; Mitchell, J.R. Anal. Biochem. 1977, 83, 168.
198. Monks, T.J.; Lau, S.S.; Highet, R.J. Drug Metab. Dispos. 1984, 12, 432.
199. Jakobsson, I.; Askelöf, P.; Warholm, M.; Mannervik, B. Eur. J. Biochem. 1977, 77, 253.
200. Pallante, S.L.; Lisek, A.; Dulik, D.; Fenselau, C. Abstr. 32nd Annual Conf. on Mass Spectrometry and Allied Topics. San Antonio, TX, 1984, p. 451.
201. Focella, A.; Heslin, P.; Teitel, S. Can. J. Chem. 1972, 50, 2025.
202. Kondo, H.; Moriuchi, F.; Sunamoto, J. J. Org. Chem. 1981, 46, 1333.
203. Brüsewitz, G.; Cameron, B.D.; Chasseaud, L.F.; Görler, K.; Hawkins, D.R.; Koch, H.; Mennicke, W.H. Biochem. J. 1977, 162, 99.
204. Smith, J.H.; Wood, E.J.; Casida, J.E. J. Agric. Food Chem. 1982, 30, 598.
205. Mennicke, W.H.; Görler, K.; Krumbiegel, G. Xenobiotica 1983, 13, 203.
206. White, I.N.H.; Campbell, J.B.; Farmer, P.B.; Bailey, E.; Nam, N.-H.; Thang, D.-C. Biochem. J. 1984, 220, 85.
207. Eyer, P.; Lierheimer, E. Xenobiotica 1980, 10, 517.
208. Ikeda, T.; Komai, T.; Kawai, K.; Shindo, H. Chem. Pharm. Bull. 1981, 29, 1416.
209. Mulder, G.J.; Unruh, L.E.; Evans, F.E.; Ketterer, B.; Kadlubar, F.F. Chem.-Biol. Interact. 1982, 39, 111.
210. Meerman, J.H.N.; Beland, F.A.; Ketterer, B.; Srai, S.K.S.; Bruins, A.P.; Mulder, G.J. Chem.-Biol. Interact. 1982, 39, 149.
211. Wislocki, P.G.; Bagan, E.S.; VandenHeuvel, J.A.; Walker, R.W.; Alvaro, R.F.; Arison, B.H.; Lu, A.Y.H.; Wolf, F.J. Chem.-Biol. Interact. 1984, 49, 13.
212. Bakke, J.E.; Gustafsson, J.-Å. Trends Pharmacol. Sci. 1984, 5, 517.
213. Tateishi, M. Drug Metab. Rev. 1983, 14, 1207.
214. Jakoby, W.B.; Stevens, J. Biochem. Soc. Trans. 1984, 12, 33.
215. Patai, S. Ed. "The Chemistry of the Thiol Group", John Wiley & Sons; 1974; Part 1 and 2.
216. Oae, S. Ed.; "Organic Chemistry of Sulfur"; Plenum press; 1977; Cap. 4, 6, 8, 10.
217. Starks, C.M.; Liotta, C. "Phase Transfer Catalysis, Principles and Techniques"; Academic Press; 1978; p. 138.
218. Bergman, Å. unpublished.

219. Chatfield, D.H.; Hunter, W.H. Biochem. J. 1973, 134, 879.
220. Bergman, Å.; Wachtmeister, C.A. Chemosphere 1978, 7, 949.
221. Chianelli, D.; Testaferri, L.; Tiecco, M.; Tingoli, M. Synthesis 1982, 475.
222. Landini, D.; Montanari, F.; Rolla, F. J. Org. Chem. 1983, 48, 604.
223. Brunelle, D.J. Chemosphere 1983, 12, 167.
224. Olah, G.A.; Gupta, B.G.B.; Narang, S.C., J. Am. Chem. Soc. 1979, 101, 5317.
225. Olah, G.A.; Gupta, B.G.B., J. Org. Chem. 1983, 48, 3585.
226. Bergman, Å.; Wachtmeister, C.A. Acta Chem. Scand. 1977, B31, 90.
227. Renner, G. Xenobiotica 1980, 10, 537.
228. Jansson, B.; Bergman, Å. Chemosphere 1978, 7, 257.
229. Kimura, R.; Hayashi, T.; Sato, M.; Aimoto, T.; Murata, T. J. Pharmacobio. Dyn. 1979, 2, 237.
230. Bakke, J.E.; Feil, V.J.; Price, C.E. Biomed. Mass. Spectrom. 1976, 3, 226.
231. Hart, S.J.; Healey, K.; Smail, M.C.; Calder, J.C. Xenobiotica 1982, 12, 381.
232. Sheng, L.-S.; Horning, E.C.; Horning, M.G., Drug Metab. Dispos. 1984, 12, 297.
233. Mizutani, T., Bull. Environm. Contam. Toxicol. 1978, 20, 219.
234. Mizutani, T.; Yamamoto, K.; Tajima, K. J. Agric. Food Chem. 1978, 26, 862.
235. Klasson-Wehler, E.; Bergman, Å.; Wachtmeister, C.A. J. Labelled Compd. Radiopharm. 1983, 20, 1407.
236. Bakke, J.E.; Feil, V.J.; Bergman, Å. Xenobiotica 1983, 13, 555.
237. Lertratanangkoon, K.; Horning, M.G.; Middleditch, B.S.; Tsang, W.-S.; Griffin, G.W. Drug Metab. Dispos. 1982, 10, 614.
238. Lotlikar, P.D.; Scribner, J.D.; Miller, J.A.; Miller, E.C. Life Sci. 1966, 5, 1263.
239. Halpaap, K.; Horning, M.G.; Horning, E.C. J. Chromatogr. 1978, 166, 479.
240. Ou, T.; Tatsumi, K.; Yoshimura, H. Biochem. Biophys. Res. Commun. 1977, 75, 401.
241. Caldwell, J. In "Conjugation Reactions in Drug Biotransformation"; Aitio, A., Ed; Elsevier, 1978; p. 111.
242. Caldwell, J.; Idle, J.R.; Smith, R.L. In "Extrahepatic Metabolism of Drugs and Other Foreign Comounds"; Gram, T.E., Ed; MTP press Ltd. 1978; Chap. 14.
243. Huckle, K.R.; Millburn, P. In "Progress in Pesticide Biochemistry"; Hutson, D.H.; Roberts, T.R., Eds; John Wiley & Sons Ltd.; 1982; vol 2, chap. 4.
244. Wood, J.W.; Fontaine, T.D. J. Org. Chem. 1952, 17, 891.
245. Bridges, J.W.; Evans, M.E.; Idle, J.R.; Millburn, P.; Osiyemi, F.O.; Smith, R.L.; Williams, R.T. Xenobiotica 1974, 4, 645.
246. De Pooter, H.; Pé, I.; Van Sumere, C.F. J. Labelled Compd. Radiopharm. 1974, 10, 135.
247. Buly, R.L.; Mumma, R.O. J. Agric. Food Chem. 1984, 32, 571.
248. Idle, J.R.; Millburn, P.; Williams, R.T. Xenobiotica 1978, 8, 253.
249. Neumann, H.-G. Arch. Toxicol. 1984, 56, 1.

250. Calleman, C.J. Ph.D. Thesis, University of Stockholm, Stockholm, 1984.
251. Törnqvist, M.; Mowrer, J.; Jensen, S.; Ehrenberg, L. Manuscript.
252. Calleman, C.J.; Wachtmeister, C.A. Acta Chem. Scand. 1979, B33, 277.
253. Svensson, K.; Osterman-Golkar, S. Toxicol. Appl. Pharmacol. 1984, 73., 363.
254. Murthy, M.S.S.; Calleman, C.J.; Osterman-Golkar, S.; Segerbäck, D.; Svensson, K. Mutation Res. 1984, 127, 1.
255. Calleman, C.J.; Kutiainen, A., personal communication.
256. Campbell, J.B. J. Chem. Soc. Perkin Trans. II. 1983, 1213.
257. Ellis, M.K.; Golding, B.T.; Watson, W.P. J. Chem. Soc. Perkin Trans. II. 1984, 1737.
258. Feung, C.-S.; Hamilton, R.H.; Mumma, R.O. J. Agric. Food Chem. 1973, 21, 632.
259. Wright, A.S.; Akintonwa, D.A.A.; Crowne, R.S.; Hathway, D.E. Biochem. J. 1965, 97, 303.
260. Unai, T.; Casida, J.E. J. Agric. Food Chem. 1977, 25, 979.
261. Dixon, P.A.F.; Caldwell, J.; Smith, R.L. Xenobiotica 1977, 7, 695.
262. Emudianughe, T.S.; Caldwell, J.; Dixon, P.A.F.; Smith, R.L. Xenobiotica 1978, 8, 525.
263. Dixon, P.A.F.; Caldwell, J.; Smith, R.L. Xenobiotica 1977, 7, 707.
264. Hutson, D.H.; Casida, J.E. Xenobiotica 1978, 8, 565.
265. Jones, A.R. Xenobiotica 1982, 12, 387.
266. Emudianughe, T.S.; Caldwell, J.; Sinclair, K.A.; Smith, R.L. Drug Metab. Dispos. 1983, 11, 97.
267. Quistad, G.B.; Staiger, L.E.; Jamieson, G.C.; Schooley, D.A. J. Agric. Food Chem. 1983, 31, 589.
268. Quistad, G.B.; Mulholland, K.M.; Jamieson, G.C. J. Agric. Food Chem. 1983, 31, 1158.
269. Caldwell, J. Drug Metab. Rev. 1982, 13, 745.
270. Stagier, L.E.; Quistad, G.B.; Duddy, S.K.; Schooley, D.A. J. Agric. Food Chem. 1982, 30, 901.
271. Crayford, J.V.; Hutson, D.H. Xenobiotica 1980, 10, 349.
272. Leighty, E.G.; Fentiman, A.F.; Thompson, R.M. Toxicology 1980, 15, 77.
273. Leighty, E.G.; Fentiman, A.F.; Foltz, R.L. Res. Commun. Chem. Pathol. Pharmacol. 1976, 14, 13.

RECEIVED June 22, 1985

Glucuronic Acid, Sulfate Ester, and Glutathione Xenobiotic Conjugates

Analysis by Mass Spectrometry

Catherine Fenselau and Lauren Yellet[1]

Department of Pharmacology, Johns Hopkins University School of Medicine, Baltimore, MD 21205

The potential of the newer desorption ionization methods for analysis of glucuronic acid, sulfate ester, and glutathione xenobiotic conjugates by mass spectrometry is discussed. Synthesis and analyses of six new glucuronides conjugated through oxime or carbinolamine functional groups are presented by way of demonstrating the strengths and limitations of fast atom bombardment mass spectrometry.

Historically mass spectrometric analysis has required that samples be volatile. This has limited the application of this important analytical technique in structure studies of conjugated metabolites to volatile derivatives. The widely realized potential of electron impact ionization, chemical ionization, and gas chromatography mass spectrometry for the analysis of glucuronides derivatized as acetates, methyl ethers and trimethylsilyl ethers has been reviewed (1). Derivatized sulfate esters (2,3) and glutathione conjugates (4,5) have been analyzed by these techniques only rarely. Attempts to analyse underivatized sulfate and glutathione conjugates by electron impact or chemical ionization, (including so-called direct chemical ionization) have resulted in analysis of pyrolysis products.

[1]Current address: A. D. Little, 15 Acorn Park, Cambridge, MA 02138

In recent years, several techniques have been developed for mass spectrometry, whereby samples are ionized and analysed from a condensed phase, without prior volatilization. These desorption techniques have permitted the extension of mass spectrometric analyses to sulfate and glutathione conjugates, as well as to underivatized and labile glucuronic acid conjugates. Primary among these techniques are field desorption (6), plasma desorption (7), laser desorption (8), fast atom bombardment (or secondary ion mass spectrometry with a liquid sample matrix) (9) and thermospray ionization (10). The latter can also serve to couple high pressure liquid chromatography and mass spectrometry for analysis of involatile and thermally labile samples.

A number of authors have pointed out that spectra acquired using various desorption techniques have many features in common (11-14). Generally the ions detected are even electron ions: molecular ions formed by protonation or addition of NH_4^+, Na^+, etc., and fragment ions formed by elimination of neutral molecules. It is not clear to what extent pyrolysis and solvolysis reactions augment the contributions of unimolecular gas phase decompositions to spectra obtained using the various desorption techniques.

Glucuronic Acid Conjugate Desorption

Features common to the spectra of glucuronic acid conjugates analysed by FAB, laser and field desorption were summarized several years ago (15). These appear to hold as well as for plasma desorption and thermospray spectra more recently examined. The situation with thermospray is somewhat more complicated as will be discussed later. Generally speaking, positive ion spectra contain protonated, natriated or analogous molecular ions species, and usually $(M+H-176)^+$ ions formed by the elimination of neutral dehydroglucuronic acid.

The $(M-H)^-$ anions detected in negative ion desorption spectra also eliminate dehydroglucuronic acid to from $(M-H-176)^-$ or $(M-177)^-$ ions. Anions comprising the glucuronic acid moiety (mass 193) are also commonly observed. Examples of laser desorption (16) and plasma desorption spectra of undervatized glucuronic acid conjugates are shown in Figures 1 and 2.

The nature of the anions and cations observed in FAB spectra formed by cleavages in the acetal or glycosidic functional group have been found to correlate with the nature of the aglycon bond conjugated (15). These correlations are proposed in scheme 1.

Despite outstanding FD measurements in laboratories at Mainz and a few other places, analysis of glucuronic acid conjugates by FAB is generally more reliable than FD analyses. However, FAB is also not completely reliable. We have worked with some glucuronic acid conjugates which provided neither an anion or cation spectrum from a variety of FAB matrices. In one of these cases the analysis has been obtained successfully by thermospray. In several other instances laser desorption has been straightforward. Plasma desorption using fission fragments from Cf-252 has also proven successful in some instances where FAB has failed. In the instances where laser and plasma desorption were more successful, it is possible that the FAB analysis was

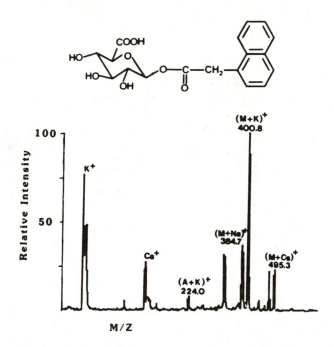

Figure 1. Laser desorption mass spectrum of the glucuronic acid conjugate of 1-naphthylacetic acid (66). The spectrum is provided by Dr. Robert Cotter.

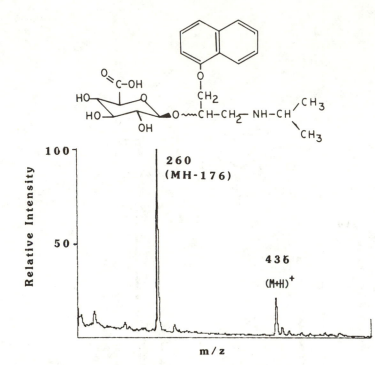

Figure 2. Plasma desorption mass spectrum of glucuronic acid conjugates of propranolol glucuronides. The spectrum is provided by Dr. Robert Cotter.

Fast atom bombardment

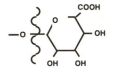

	$\oplus$	$\ominus$	
			Alcohols
			Phenols
	MH-176	M-177	Carboxy Acids
			Quaternary Amines

	MH-192	m/z 193	Carbinolamines

BOTH LOSSES	MH-176	M-177	Oximes
	MH-192	M-193,m/z 193	N-hydroxy Amides

Scheme 1

confounded by solution phenomena, probably the absence of sufficient sample molecules on the surface of the liquid matrix.

In the second half of this chapter we discuss in more detail the high potential of fast atom bombardment, demonstrating applications to analysis of some new glucuronic acid conjugates recently synthesized in our laboratory.

Sulfate Desorption

Early desorption studies of sulfate esters (17) and the several field desorption papers (18-22) confirm that cationized molecular ions can provide molecular weight information, but also suggest that multiply charged ions, salt clusters, pyrolysis and irreproduceability confound analysis of desorbed cations. Fast atom bombardment has been reported to be more effective (23-26) although cationization by alkali earth metals as well as protonation produces redundant molecular ion species in positive ion spectra. One group of authors (23) describe positive ion spectra as poorly reproduceable and of low intensity, and agreement seems to exist that negative ion spectra are simpler, independent of the presence of mixtures of alkali earth counter ions and more reproduceable. Background subtraction techniques have been applied to increase sensitivity (23,24). Derivatization has been suggested as a way to increase the molecular weight of the sulfate and move it away from glycerol matrix ions (23). Sample preparation has been found to be important to control signal suppression by impurities in achieving full scans on 15 ng samples and quantitation with stable isotope labelled internal standards (23). The FAB spectrum of the mixed anhydride sulfate ester conjugate of indomethacin is shown in Figure 3.

Glutathione Desorption

Field desorption has provided satisfactory analyses of a number of glutathione conjugates (22, 27-34) including a diglutathione formed in the metabolism of cholorform, (28). Abundant molecular ion species are observed, varying among M^+, $(M+H)^+$ and $(M+Na)^+$. Class characteristic fragmentation occurs on one side or the other of the thio ether linkage, or both, generating some or all of the cations shown in scheme 2.

Onkenhout and colleagues have pointed out that these same decomposition reactions also occur during pyrolysis (33). Certainly the possibility of pyrolytic contributions can not be excluded in most cases. However the observation of at least one of these retro-Michael fragmentations induced by collisional activations (22) also argues for their occurence independent of pyrolysis.

Frequently cleavage in the Cys-Glu bond produces cations of mass 130 and/or $(MH-130)^+$, which can be viewed as confirming the presence of a glutathione moiety. One report suggests 200 ng as a reasonable sample size (30).

Glutathione conjugates are also analysed readily by fast atom bombardment and by plasma desorption, as either anions or cations (35-39). In addition to abundant molecular ions (Fig. 4) fast atom

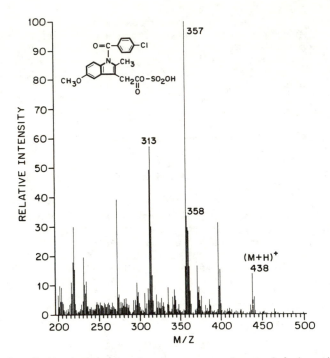

Figure 3. Fast atom bombardment mass spectrum of indomethacin acyl hydrogen sulfate (67).

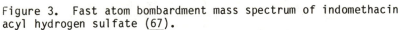

Scheme 2

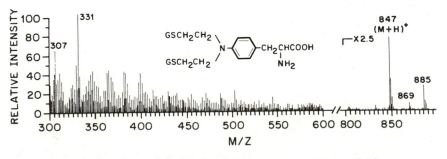

Figure 4. Fast atom bombardment mass spectrum of the diglutathione conjugate of phenylalanine mustard (68).

bombardment promotes some fragmentation analogous to field
desorption, notably formation of class characteristic (MH-306)
cations (38). The complementary cations of mass 308 are sometimes
formed as well, and in negative ion spectra anions of mass 306 are
observed, reflecting the presence of the glutathione moiety.
A comparison of the efficacy of several liquid matrices has led the
authors to recommend the analysis of positive ions from
thioglycerol as a reliable general method.

Fast atom bombardment has also been used successfully with
metabolites related to or derived from glutathione conjugates, e.g.
a homoglutathione soybean metabolite (40) and mammalian
mercapturates and cysteinyl conjugates (24, 41-44).

Ion Evaporation and LCMS

Among the several types of interfaces commercially available for
coupling high pressure liquid chromatography and mass spectrometry,
the moving belts (45) direct liquid injection (46) and thermospray
(ion evaporation) (14,47) systems have been used successfully to
analyze underivatized glucuronic acid conjugates. Since moving
belt and direct injection interfaces require at least marginally
volatile samples, they have been less readily applicable to sulfate
and glutathione conjugates. Glutathione and its conjugates have
been ionized by thermospray (38,48) and field induced ion
evaporation (49). Thermospray (10), electrospray (50) and field
induced ion evaporation (49), variants on the same mechanism
(51,52), appear to offer a true desorption technique, where ions
are formed in the condensed phase and subsequently evaporated into
the gas phase.

A separation of diastereomeric propranolol glucuronides by
reversed phase HPLC is shown in Figure 5. The chromatogram
recorded by UV detection can be compared with chromatograms
recorded by a thermospray interfaced mass spectrometer. The
potential uses of total ion current chromatograms, mass
chromatograms and selected ion monitoring parallels their
established utility in GCMS (53). Both positive and negative ions
are produced. Glucuronides were used to illustrate the point that
the sensitivity of thermospray (and likely, electrospray and field
induced ion evaporation as well) varies from compound to compound
(47). Subsequently sensitivity for positive ions has been
correlated with proton affinity in the gas phase (14). Some of the
fragment ions may be formed by high temperature reactions with the
volatile buffer (usually ammonium acetate) required for thermospray
ionization (10). Consistent with the fragmentation scheme general
to desorption techniques, $(M+NH_4-176)^+$ ions are important in the
cation spectra, as well as the class characteristic ion of mass
194, ammoniated glucuronic acid. Less fragmentation is observed in
anion spectra. The extent of fragmentation of glucuronides in
thermospray is dependent in part on the buffer concentration and on
the temperature.

In a comparison of thermospray with FAB (14), ammonium
acetate was used in the FAB matrix as well as the thermospray
buffer. Its presence significantly alters the cation FAB spectra

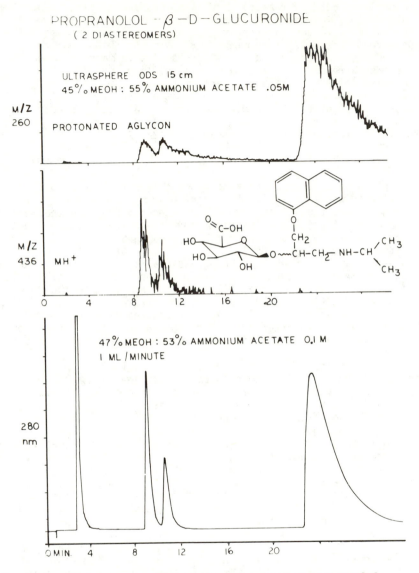

PROPRANOLOL - β—D—GLUCURONIDE
(2 DIASTEREOMERS)

M/Z 260

ULTRASPHERE ODS 15 cm
45% MEOH : 55% AMMONIUM ACETATE .05M

PROTONATED AGLYCON

M/Z 436   MH⁺

47% MEOH : 53% AMMONIUM ACETATE 0.1 M
1 ML / MINUTE

280 nm

0 MIN.   4   8   12   16   20

Figure 5. High pressure liquid chromatograms of propranolol
glucuronides recorded by thermospray mass spectrometry and
ultraviolet spectroscopy (47). Upper panel: selected ion
profile of (MH-176)⁺. Middle panel: selected ion profile of
(MH)⁺. Lower panel: profile of absorption at 280 nm.

of glucuronic acid conjugates of non-basic aglycons, providing $(M+NH4)^+$ ions with improved signal to noise ratios and sensitivities compared to $(M+H)^+$ ions generated in glycerol or thioglycerol without ammonium acetate (54).

FAB Analysis of Glucuronides of Carbinolamines and Oximes.

Numerous studies have been conducted to investigate the metabolic oxygenation of C-N systems, due to the widespread occurrence of this type of stucture in pharmaceutical agents and pesticides, and also because of the potential toxicological and pharmacological properties which N-oxygenated compounds posses. Oxidized metabolites include hydroxylamines, oximes and carbinolamines. All of these can be conjugated with glucuronic acid, which may result in detoxification. However, some of these N-O-glucuronides are very reactive, with the glucuronic acid moiety acting as a good leaving group. In order to examine the pharmacology of these glucuronides, representative C-N containing metabolites have been obtained and conjugated with glucuronic acid; the stability, chromatographic and mass spectral chacteristics of these compounds have been studied. A study of the O-glucuronides of a set of hydroxy acetylarylamines has been published (55). The characterization by fast atom bombardment of carbinolamine glucuronides derived from the herbicide diphenamid (56) the antitumor agent hexamethylmelamine (57) and the cholinesterase reactivator 2-pralidoxime (58) is discussed here, along with the characterization of glucuronides of phenylacetone oxime (a metabolite of amphetamine) (59,60) acetophenone oxime (61,62) and the antiviral agent enviroxime (63).

The first two carbinolamines named were synthesized using sequential catalysis by cytochrome P-450 oxygenase and glucuronyl transferase immobilized from rabbit liver microsomes onto sepharose beads (64,65) as shown in scheme 3. 2-Pralidoxime chloride was allowed to decompose at pH 7.4 in the presence of immobilized UDP-glucuronyl transferase and the cofactor uridine diphosphoglucuronic acid to obtain the conjugated carbinolimine shown in the scheme. Conjugation of the three oximes shown in scheme 4 was catalysed by immobilized transferase enzymes (65). Conjugates were purified by extraction and chromatography.

The electron impact spectrum of the volatile per(trimethyl) silylated derivative of enviroxime glucuronide is shown in Figure 6. The molecular ion is visible, providing a molecular weight which can be corrected for trimethylsiyl groups in the case of an unknown sample by analyzing a second portion of the sample derivatized with d_9-trimethylsilyl groups (1). The intense peak at m/z 375 is characteristic of trimethylsilylated glucuronides as a class (1). The loss of 481 mass units is characteristic of trimethylsilyated glucuronic acid linked to a hydroxyl or other oxygen-containing functional group. The analogous fragment lost from an amine linked glucuronide would have a different mass. This argues for conjugation of the hydroxyl groups of enviroxime and not the amino group. The presence of $(M-481)^+$ ions and the absence of $(M-392)^+$ ions is sustained in spectra of the other two oxime glucuronides as well, and may therefore be indicative of oxime conjugation. This electron impact analysis does require

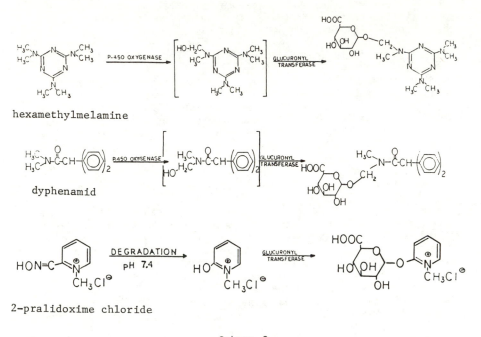

hexamethylmelamine

dyphenamid

2-pralidoxime chloride

Scheme 3

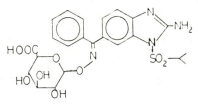

enviroxime glucuronide

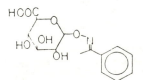

acetophenone oxime
glucuronide

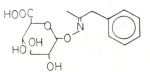

amphetamine oxime
glucuronide

Scheme 4

derivatization of the glucuronide. However, this approach is also compatible with gas chromatography and combined gas chromatography mass spectrometry.

   Positive and negative ion spectra of underivatized enviroxime glucuronide obtained by fast atom bombardment mass spectrometry are shown in Figures 7 and 8. These were run using a glycerol matrix. Typical of many FAB spectra, the cation spectrum reveals $(M+H)^+$ $(M+Na)^+$ at m/z 557 and $(M+K)^+$ at m/z 573, all even electron molecular ion species. Even electron fragment ions are observed notably $(M+H-176)^+$ at m/z 359 and $(M+H-192)^+$ at m/z 343. The anion spectrum contains an $(M-H)^-$ peak at m/z 533 unaccompanied by a natriated satallite. This and other peaks in the anion spectrum appear to correspond to the same ions seen in the cation spectrum, however two mass units lighter. The $(M-H-176)^-$ or $(m-177)^-$ ion reflects a class characteristic, fragmentation. Ions of mass 193 presumeably comprise anions of the glucuronic acid moiety. Anions were recorded with greater sensitivity than cations in this case.

   Figure 9 contains the FAB cation spectrum (glycerol) of the glucuronic acid conjugate of the hydrolysis product of 2-pralidoxime. In addition to being a carbinolimine, this metabolite also has quaternary ammonium center. This permanent charge results in facile detection of molecular cations. Anions are much harder to record. Interestingly no fragment ions are formed under the conditions used to record Figure 9.

   Spectra of positive and negative ions formed from the glucuronide of hydroxydiphenamid (a carbinolamide), by fast atom bombardment with glycerol matrices, are shown in Figures 10 and 11. Again the anion spectrum can be recorded with less sample. As in the case of enviroxime glucuronide, anions comprising the glucuronic acid moiety are detected (mass 193), and the loss of the glucuronic acid, $(M+H-192)^+$ is found in the cation spectrum. Both of these are class characteristic ions. Cleavage resulting in retention of both oxygen atoms by the glucuronic acid moiety (mass 192,193 or 194) combined with the absence of mass 177 or $(M+H-176)^+$ ions may be characteristic of carbinolamine conjugation. Enzymatic oxygenation in the diphenylmethine moiety can be ruled out by observation of the peak at m/z 167 in the anion specrum.

   The last FAB spectrum, Figure 12, is that of hydroxyhexamethylmelamine glucuronide measured with a thioglycerol matrix. The protonated molecular ion is observed, as well as the class characteristic ion $(M+H-194)^+$ of mass 209. When this analysis was run from a glycerol matrix, $(M-H)^-$ ions were desorbed in greater abundance than $(M+H)^+$ ions.

   In addition to presenting new observations on the analysis of some novel glucuronides conjugated to oxime and carbinolamine functional groups, this discussion is intended to illustrate the potential and limitations of fast atom bombardment, the most widely used of the new desorption mass spectrometry techniques. Molecular weights may be obtained and confirmed by redundant species. Class characteristic fragmentation may support identification as a glucuronide. In some cases fragmentation within the aglycon may provide an indication of the site of metabolic oxygenation and conjugation. It appears that conjugation with different functional groups leads to different patterns of cleavage within the

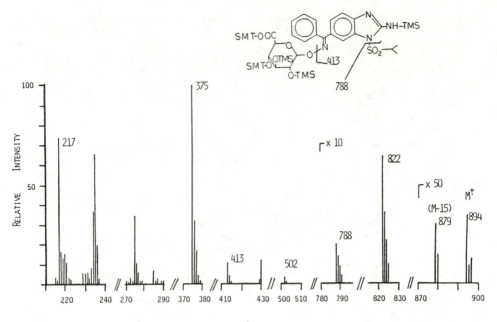

Figure 6. Electron impact mass spectrum of trimethylsilylated enviroxime glucuronide.

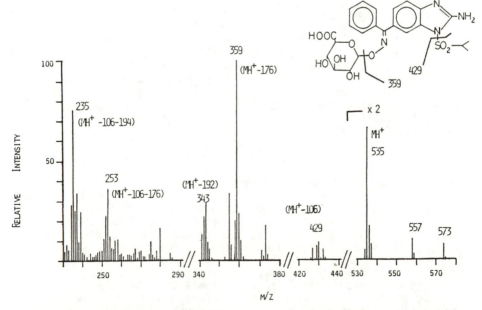

Figure 7. Positive ion fast atom bombardment mass spectrum of enviroxime glucuronide.

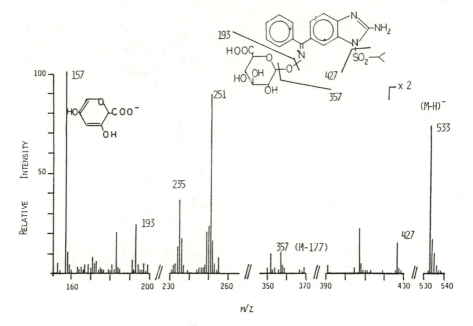

Figure 8. Negative ion fast atom bombardment mass spectrum of enviroxime glucuronide.

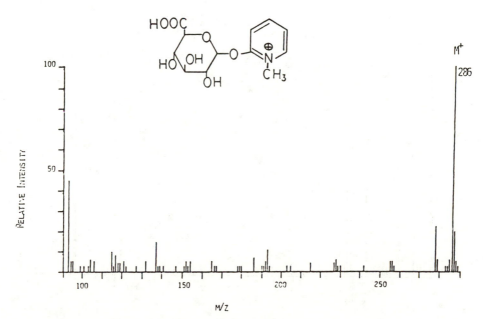

Figure 9. Fast atom bombardment mass spectrum of the glucuronide of the carbinolamine derived from pralidoxime chloride.

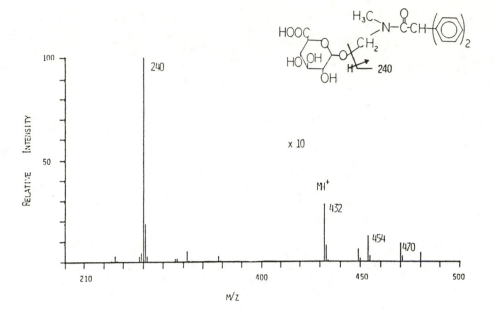

Figure 10. Positive ion fast atom bombardment mass spectrum of hydroxydiphenamide glucuronide.

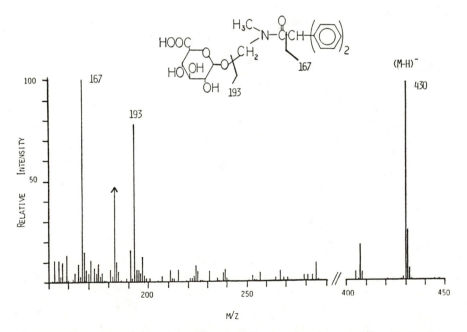

Figure 11. Negative ion fast atom bombardment mass spectrum of hydroxydiphenamide glucuronide.

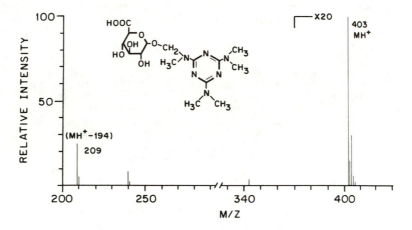

Figure 12. Fast atom bombardment mass spectrum of hydroxyhexamethylmelamine glucuronide.

glycosidic acetal group. When both can be obtained, anion and cation spectra provide complimentary and reinforcing information.

Summary

All of the more common conjugated xenobiotic metabolites are now known to be susceptible to mass spectral analysis using one or another of the desorption techniques. An investigator can expect to obtain molecular weight information and, in most cases, some structural information from fragmentation. Sensitivities currently range between 1 ng and 1 ug. Liquid chromatography mass spectrometry can be utilized as well as gas chromatography mass spectrometry.

Acknowledgments

We thank Donald Delong, Lilly Research Laboratories, for a sample of enviroxime, Patrick Callery, University of Maryland, School of Pharmacy, for a sample of the oxime of phenyl acetone; Robert J. Cotter, Dan Liberato, Gordon Hansen, Jeff Honovich, Ron Robbins, Carol Lisek, Mehrshid Alai and Deanne Dulik for assistance with mass spectra, especially those in Figures 1-5. This research was supported by USPHS grants NIH GM-21248 and GM07626.

Literature Cited

1. Fenselau, C.; Johnson, L.P. Drug Metab. Disp., 1980, 8, 274-283.
2. Paulson, G.; Bakke, J,; Giddings, J.; Simpson, M. Biomed. Mass Spectrom. 1978, 5, 128-132.
3. Paulson, G.; Simpson, M.; Giddings, J.; Bakke, J.; Stolzenberg, G.; Biomed. Mass Spectrom. 1978, 5, 413-417.
4. Damon, M.; Chavis, C.; Godard, P.; Michel, F.B.; Crastes de Paulet, A. Biochem. Biophys. Research Commun. 1983, 111, 518-524.
5. Maas, R.L.; Lawson, J.A.; Brash, A.R.; Oates, J.A. Adv. Prostag. Thrombox. and Leukotr. Res. 1983, 11, 229-234.
6. Giessman, U.; Rollgen, F.W. Int. J. Mass Spectrom. Ion Phys. 1981, 38, 267-279.
7. MacFarlane, R.D. Acc. Chem. Res. 1982, 15, 268-275.
8. Cotter, R.J. Anal. Chem. 1984, 56, 485A-504A.
9. Barber, M.; Bordoli, R.S.; Elliot, G.J.; Sedgwick, R.D.; Tyler, A.N. Anal. Chem. 1982, 54, 645A-657A.
10. Blakely, C.R.; Vestal, M.L. Anal. Chem. 1983, 55, 750-754.
11. Ens, W.; Standing, K.; Chait, B.T.; Field, F.H. Anal. Chem. 1981, 53, 1241-1244.
12. Wunsch, L.; Benninghoven, A.; Eicke, A.; Heinen, H.J.; Ritter, H.P.; Taylor, L.C.E.; Veith, J. Org. Mass Spectrom. 1984, 19, 176-182.
13. Balasunmugan, K.; Hercules, D.M.; Cotter, R.J.; Heller, D.; Benninghoven, A.; Sichterman, W.; Anders, V.; Keough, T.; MacFarlane, R.D.; McNeal, C.J. Anal. Chem. 1984, 56, 5759-5762.

14. Fenselau, C.; Liberato, D.J.; Yergey, J.A.; Cotter, R.J.; Yergey, A.L. Anal. Chem. 1984, 56, 5759-5762.
15. Fenselau, C., Yelle, L.; Stogniew, M.; Liberato, D.; Lehman, J.; Feng, P.; Colvin, M. Jr. Intern. J. Mass Spectrom. Ion Phys. 1983, 46, 411-414.
16. van Breemen, R.B.; Tabet, J.C.; Cotter, R.J. Biomed. Mass Spectrom. 1984, 11, 278-283.
17. Mumma, R.O.; Vastola, F.J. Org. Mass Spectrom. 1972, 6, 1373-1376.
18. Games, D.E.; Games, M.P.; Jackson, A.H.; Olavesen, A.H.; Rossiter, M.; Winterburn, P.J. Tetrahed Let. 1974, 2377-2380.
19. Schulten. H.R.; Lehmann, W.D. Anal. Chim. Acta. 1976, 87, 103-112.
20. Matcham, G.W.J.; Dodgson, K.S. Biochem. J. 1977, 167, 717-722.
21. Deutsch, J.; Gelboin, H.V. Biomed. Mass Spectrom. 1982, 9, 99-102.
22. Nelson, S.D.; Vaishnav, Y.; Kambara, H.; Baillie, T.A. Biomed. Mass Spectrom. 1981, 8, 244-251.
23. Gaskell, S.J.; Brownsey, B.G.; Brooks, P.W.; Green, B.N. Biomed. Mass Spectrom. 1983, 10, 215-219.
24. Ackermann, B.L.; Watson, J.T.; Newton, J.F.; Hook, J.B.; Braselton, W.E. Jr. Biomed. Mass Spectrom. 1984, 11, 502-511.
25. Jardine, I.; Scanlan, G.F.; Mattox, V.R.; Kumar, R. Biomed. Mass Spectrom. 1984, 11, 4-9.
26. Fenselau, C.; Cotter, R.J.; In "IUPAC Frontiers of Chemistry"; Laidler, K.J., Ed.; Pergammon: Oxford, 1982, p. 207-216.
27. Tunek, A.; Platt, K.L.; Przybylski, M.; Oesch, F. Chem. -Biol. Interactions 1980, 33, 1-17.
28. Pohl, L.R.; Branchflower, R.V.; Highet, R.J.; Martin, J.L.; Nunn, D.D.; Monks, T.J.; George, J.W.; Hinson, J.A. Drug Metab. Disp. 1981, 9, 334-339.
29. Przybylski, M.; Cysyk, R.L.; Shoemaker, D.; and Adamson, R.H. Biomed. Mass Spectrom. 1981, 8, 485-491.
30. Meerman, J.H.N.; Beland, F.A.; Ketterer, B.; Srai, S.K.S.; Bruins, A.P.; Mulder, G.J. Chem.-Biol. Interactions 1982, 39, 149-168.
31. Moss, E.J.; Judah, D.J.; Przybylski, M.; Neal, G.E. Biochem. J. 1983, 210, 227-233.
32. Gandich, K.; Przybylski, M. Biomed. Mass Spectrom. 1983, 10, 292-299.
33. Onkenhout, W.; Vermeulen, N.P.E.; Luijten, W.C.; deJong, H.J. Biomed. Mass Spectrom. 1983, 10, 614-619.
34. Przybylski, M.; Luderwald, I.; Kraas, E.; Voelter, W.; Nelson, S.D. Z. Naturforsch. Teil 1979, 34, 736-743.
35. Murphy, R.C.; Mathews, W.R.; Rokach, J.; Fenselau, C. Prostaglandins, 1982. 23, 201-206.
36. Larsen, G.L.; Ryhage, R. Xenobiotica 1982, 12, 855-860.
37. Ross, D.; Larsson, R.; Norbeck, K.; Ryhage, R.; Moldeus, P. Molecular Pharmacology 1985, 27, 277-286.
38. Pallante, S.L.; Lisek, C.A.; Dulik, D.M.; Fenselau, C. Drug Metab. Disp., 1986, in press.
39. Fairlamb, A.H.; Blackburn, P.; Ulrich, P.; Chait, B.T.; Cerami, A. Science, 1985, 227, 1485-1487.
40. Frear, D.S.; Swanson, H.R.; Mansager, E.R. Pest. Biochem. Physiol. 1985, 23, 56-65.

41. Bakke, J.E.; Bergman, A.L.; Larsen, G.L.   Science 1982, 217, 645-647.
42. Stogniew, M.; Fenselau, C.   Drug Metab. Disp. 1982, 10, 609-613.
43. Logan, C.J.; Cottee, F.H.; and Page, J.A.   Biochem. Pharmacol. 1984, 33, 2345-2346.
44. Hutson, D.H.; Logan, C.J.; Regan, P.D. Drug Metab. Disp. 1984, 12, 523-524.
45. Games, D.E.; Lewis, E.   Biomed. Mass Spectrom. 1980, 7, 433-436.
46. Kenyon, C.N.; Goodley, P.C.; Dixon, D.J.; Whitney, J.O.; Faull, K.F.; Barchas, J.D. American Laboratory, 1983, January, 38-49.
47. Liberato, D.J.; Fenselau, C.C.; Vestal, M.L.; Yergey, A.L. Anal. Chem. 1983, 55, 1741-1744.
48. Fenselau, C.; Larsen, B.S. In "Drug Metabolism"; Siest, G., Ed.; Pergamon Press: Oxford, 1985.
49. Thomson, B.A.; Iribarne, J.V.; Dziedzic, P.J.   Anal. Chem. 1982, 54, 2219-2224.
50. Whitehouse, C.M.; Dreyer, R.N.; Yamashita, M.; Fenn, J.B. Anal. Chem. 1985, 57, 675-679.
51. Iribarne, J.V.; Thomson, B.A. J. Chem. Phys. 1976, 64, 2287-2294.
52. Thomson, B.A.; Iribarne, J.V. J. Chem. Phys. 1979, 71, 4451-4463.
53. Fenselau, C. Anal. Chem. 1977, 49, 563A - 570A.
54. Fenselau, C. In "Mass Spectrometry in Health and Life Sciences"; Burlingame, A.L. and Castagnoli, N.J., Ed.; Elsevier: Amsterdam, 1985.
55. Feng, P.C.; Fenselau, C.; Colvin, M.; Hinson, J.   Drug Metab. Disp. 1981, 9, 521-524.
56. McMahon, R.E.; Sullivan, H.R. Biochem. Pharm.   1965, 14, 1085-1092.
57. Gescher, A,; Hickman, J.A.; Stevens, F.G.   Biochem. Pharmacol. 1979, 28, 3235-3238.
58. Brown, N.D.; Strickler, M.P.; Sleeman, H.; Doctor, B.P.   J. Chromatog. 1981, 212, 361-365.
59. Cho, A.K.; Wright, J. Life Science, 1978, 22, 363-371.
60. Hucker, H.B.; Michniewicz, B.M.; Rhodes, R.E. Biochem. Pharmacol. 1971, 20, 2123-2127.
61. Sternson, L.A.; Hes, J.   Pharmacology 1975, 13, 234-240.
62. Sternson, L.A.; Hincal, F.; Bannister, S.J.   J. Chromatog. 1979, 144, 191-200.
63. Wikel, J.H.; Paget, C.J.; Delong, D.C.   J. Med. Chem. 1980, 23, 368-372.
64. Lehman, J.P.; Ferrin, L.; Fenselau, C.; Yost, G.S. Drug Metab. Disp. 1981, 9, 15-18.
65. Yelle, L., M.A. Thesis, Johns Hopkins University, Baltimore, MD 1982.
66. Hamar-Hansen, C.; Fournel, S.; Magdalou, J.; Boutin, J.A.; Siest, G. submitted.
67. van Breemen, R.B.; Fenselau, C.; Dulik, D.M.   In "Biological Reactive Intermediates"; Snyder, R.; Witmer, C.M.; Jollow, D.J. and Kocsis, J.J. Eds.; Plenum Press: New York, 1986.
68. Dulik, D.M.; Hilton, J.; Fenselau, C.; Colvin, M.; Biochem. Pharm. in press.

RECEIVED October 30, 1985

Identification of Xenobiotic Conjugates
by Nuclear Magnetic Resonance Spectrometry

V. J. Feil

Metabolism and Radiation Research Laboratory, Agricultural Research Service,
U.S. Department of Agriculture, Fargo, ND 58105

A review of the literature that contains 120 references on the application of NMR spectrometry for the
identification of xenobiotic conjugates. The references (primarily 1980 to 1984) provide structures and
spectral summaries for sulfate, carbohydrate, amino
acid, and miscellaneous conjugates.

This review is primarily a compilation of examples where nuclear
magnetic resonance (NMR) has been used in some capacity for the
characterization of xenobiotic conjugates. Most references are
taken from the period 1980-1984; some earlier references are
included because they provide examples of metabolites that were not
available in more recent references. The following information,
when available, is included in the compilations: conjugate structure, source of the conjugate, NMR characteristics of the conjugate,
effect of the conjugating group on the spectrum of the xenobiotic,
NMR frequency, solvent, and literature reference.
        Related reviews have been published on the use of NMR in pesti-
cide analysis (1), metabolic studies (2,3), drug metabolism (4,5),
and medicinal chemistry (6,7). Since most of these reviews as well
as many texts contain introductions to NMR, none will be presented
here. Furthermore, an audio course on Fourier Transform NMR
Spectroscopy is also available (8). Since structural characterization is often based on empirical correlations, we have found compilations of proton spectra by Aldrich, Sadtler, and Varian (9-11)
and carbon spectra by Breitmaier et al. and Bruker (12,13) to be
useful. Becker (14) has presented extensive lists of references
that contain useful compilations of NMR data (pp 77 and 106).

Techniques

The use of NMR in the identification of xenobiotic conjugates is
quite limited. Many recent publications that include NMR in structural characterizations used high field instruments (200 MHz or
higher). Thus, the use of NMR in metabolite characterization may
increase with increased availability of high field instruments

because the potential for obtaining structural information from
high field spectra is so great.

No published report on the use of two-dimensional NMR (15),
2-D NMR, in xenobiotic conjugate identification was found, presumably because large samples are required. 2-D NMR should prove
useful in the assignment of peaks in the spectrum of the parent
xenobiotic such as in the identification of hernandulcin (16). The
accurate assignment of a spectrum is often a prerequisite to the
determination of the conjugate by NMR. 2-D NMR was used in conjunction with ^{13}C enrichment in a study of glycine metabolism in
tobacco suspension cells (17).

Sample requirements usually pose no problem in ^{1}H NMR, but may
be a problem with other nuclei. Even sub microgram samples are
feasible with high field instruments; however, sample and solvent
impurities may prevent good spectroscopy at that level. Micro
cells, exhaustive exchange with a deuterium solvent (D_2O or
CD_3OD), multiple pulse sequences , and presaturation pulses for
solvent peak suppression have all been reported in xenobiotic conjugate identification. High field instruments, unlike iron magnet
instruments, have not been available with micro cells; however, an
unsupported 1.8 mm tube placed inside a 5 mm tube has provided
spectra of adequate resolution with a 470 MHz spectrometer (18).

Nuclear Overhauser Enhancement

Nuclear Overhauser enhancement (NOE) was used in the assignment of
structures I-III. Irradiation of I at * produced a 27% increase of
the integrated intensity for the proton at + but had no effect on
the intensity of the remaining protons (19). Irradiation of the H
at * of II produced larger increases (22 and 41%) in intensities of
the two protons designated + than did irradiation of the CH_3 at *
(17 and 27%) (20). Irradiation of III at * caused a 12% increase in
the integrated intensity for the benzylic proton designated +, but
no increase in the intensity for the proton associated with the
hydroxyl group (21).

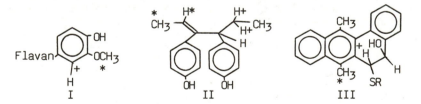

Shift Reagents

Shift reagents have not been used extensively in xenobiotic metabolism studies; however, they may find increased use for determination of isomer composition in studies on metabolic mechanisms
because metabolic rates are often different for geometric and
stereoisomers. Mizugaki et al. used ^{1}H NMR with Eu(fod)$_3$ to determine structures of cis- and trans-3-alkenoic acids (22). Moser et
al. used ^{1}H NMR and a chiral shift reagent for analysis of the

stereoisomers of Metolachlor (23). Wainer et al. used ^{1}H NMR and a
europium chiral shift reagent for analysis of d- and l-amphetamine
(24). Stec et al. used ^{31}P NMR and a praseodymium chiral shift
reagent for the determination of enantiomeric purity of ifosfamide
and two of its urinary metabolites (25).

Carbon-13 Enrichment

The use of compounds enriched with ^{13}C is common in biosynthetic
studies (26), but quite rare in xenobiotic metabolism studies.
Hernandez et al. (27) studied the enzymic and non-enzymic reaction
of glutathione with benzo(a)pyrene 4,5-oxide enriched with ^{13}C at
the 4 and 5 positions.   The oxide was prepared with an enrichment
of 70 atom % at position 4 and 97 atom % at position 5.   The isomeric compositions of the reaction mixtures could be determined
because of the different enrichment levels.   ^{13}C NMR spectra showed
two clusters containing five peaks each. Each cluster contained one
intense peak corresponding to either a hydroxy substituent or a glutathione substituent for the carbon at position 5 (97 atom %
enrichment). Each cluster also contained two doublets with
appropriate [13]C-^{13}C coupling constants corresponding to the doubly
labeled isomers at positions 4 and 5.
        Feil et al. (28) isolated a Propachlor metabolite from
germfree rats that had either structure IV or V (reaction with
butanol/HCl had yielded the butyl ether-butyl ester derivative of
IV). The proton decoupled ^{13}C NMR spectrum of the metabolite isolated from rats dosed with 2-chloro-N-isopropyl-acetanilide-2-[^{13}C]
showed in CD$_3$OD a broadened triplet at δ58.1 rather than the
expected large singlet. This result suggested isotope exchange was
occurring to afford a structure that contained only one deuterium on
the ^{13}C enriched carbon. A model compound, 4'-Methoxy-2-(methylsulfonyl)acetanilide, in CD$_3$OD initially yielded a singlet at δ59.2
for the comparable carbon, but a triplet at δ58.8 on standing overnight. Structure VI is one of several that can be drawn showing
either carbanion or sp^2 character. The observed chemical shift is
similar to those obtained by Matsuyama et al. (29) for some carbonyl
substituted sulfonium ylides.

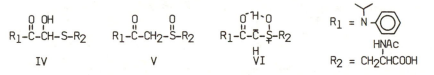

Russo et al. followed the metabolism of ^{13}C labeled hordenine
(p-hydroxy-N,N-dimethyl[^{13}C]phenethylamine) by root homogenates
with ^{13}C NMR. They observed a decrease in the signal due to the
dimethylamino group with a concomitant increase in a signal due
to the methylamino group, but observed no signal due to the metabolized methyl group (30).

Other Nuclei

Few applications of NMR with nuclei other than ^{1}H and ^{13}C have
been reported for xenobiotic metabolism studies. Some studies in

related fields are included here as they may serve as models for
xenobiotic metabolism studies.

No examples of the use of deuterium and tritium NMR in xenobiotic metabolism were found. Their use in biosynthetic studies
has been reviewed by Garson and Staunton (31). Sensitivity
problems exist with deuterium, but should not be a problem with
tritium since it is the most sensitive nucleus available (1.21 x
proton) and because of negligible tritium backgrounds. Tritium NMR
may be useful in the studies of xenobiotic-enzyme interactions as
shown by Scott et al. (32). Hazards due to the use of radioactivity should be minimal because 1 mCi of activity should provide
sufficient material for many experiments. However, isotope effects
may be a problem if the metabolic reaction directly involves the
tritium (or deuterium) atom because isotopes of hydrogen can greatly
affect enzymic reaction rates. Also, lability may be a problem as
Bakke and Feil have found with CD_3SO compounds, where exchange was
too rapid to permit metabolism studies (33).

Tsoupras et al. used ^{31}P NMR to detect the presence of phosphate groups in the 22-adenosinemonophosphoric ester of 2-deoxy-
ecdysone and 22-phospho-2-deoxyecdysone (34). Quantitative ^{31}P NMR
has been used in several studies. Wayne et al. used ^{31}P NMR for
quantitative analysis of organophosphorous pesticides (35). Zon et
al. used ^{31}P NMR to study chemical and microsomal oxidation of
cyclophosphamide (36), and to determine the half-life of a
cyclophosphamide analogue (37).

Mazzola et al. determined pesticide residues in foods by ^{19}F
NMR (38). Wyrwicz et al. made unsuccessful attempts at detecting
metabolites of methoxyflurane in circulating blood during 2 hours
of continuous anesthesia (39). The compound had previously been
reported to be highly metabolized by the liver. Spratt and Dorn
(40) have prepared a large number of p-fluorobenzoyl derivatives of
alcohols, phenols, carboxylic acids, amines and thiols. These compounds generally are formed in high yield and may be useful for
isolation and analysis of metabolites. Chemical shifts vary over
a 10 ppm range, but model compounds would likely be required, for
example, to distinguish between compounds containing SH and
OCH_3 groups from compounds containing OH and SCH_3 groups. This is
indicated by the following shifts relative to
1,2-difluoro-tetrachloroethane: PhS, -36.89; PhO, -37.23; PhN,
-40.38; $PhCh_2O$, -38.12.

Amino acid conjugates

Examples of the use of NMR in the identification of amino acid
conjugates are listed in Table I. Relatively few examples of the
use of ^{13}C NMR were found, presumably because of the large amounts
of sample required. Carbon NMR spectra obtained on an iron magnet
instrument (15-25 MHz) can provide useful information for identification of amino acid conjugates (27,28,42,45,48,50,63,64);
however, proton spectra obtained on these instruments (60-100 MHz)
are often of limited value because of inadequate spectral disper-
sion. Many of the examples in Table I are, therefore, of proton
spectra taken at 200 MHz or higher. Figure 1 lists approximate
proton and carbon shift values of glutathione-type conjugates
(proton and carbon shift values are generally found within 0.4 and

1.9 ppm respectively of these values). Mercapturate and cysteine conjugates show similar values for the cysteine absorbances. Because of a chiral center in cysteine, the protons on the β carbon are magnetically non equivalent; therefore, the protons of cysteine (also of the glutamyl group) yield an ABC coupling pattern. Chemical shift and coupling constant variations often render the three expected sets of doublet of doublets unrecognizable. Computer programs are available (often as part of FT NMR programs) for calculation of spectra from suspected shift values and coupling constants to determine the accuracy of assignments, and for conversion of spectra from one frequency to another. A useful hard copy source that may aid pattern recognition also exists (68).

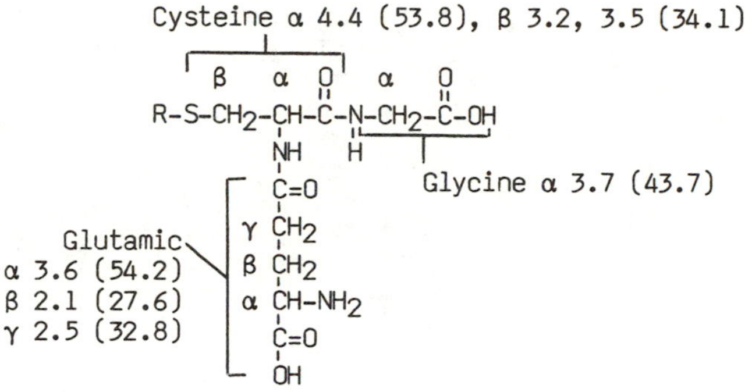

Cysteine α 4.4 (53.8), β 3.2, 3.5 (34.1)

Glycine α 3.7 (43.7)

Glutamic.
α 3.6 (54.2)
β 2.1 (27.6)
γ 2.5 (32.8)

Figure 1. Approximate chemical shifts of glutathione-type conjugates.

Table I. Examples of NMR usage in identification of amino acid conjugates. Cysteine, mercapturate (N-acetylcysteine), glutathione, cysteinylglycine, and glutamylcysteine are all conjugated through sulfur.

Compound	Ref.	Spectroscopic Summary
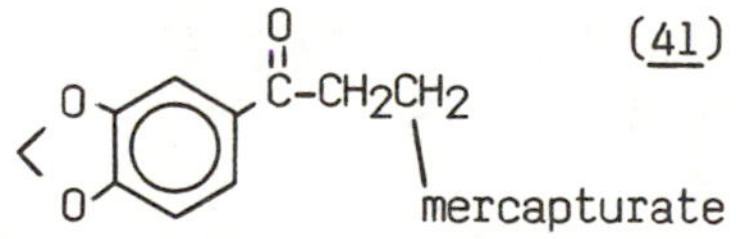 mercapturate — Rats and mice (bile and urine).	(41)	200 MHz D₂O — Cys α 4.4; Cys β 2.9–3.2--partially obscured by CH₂CH₂.
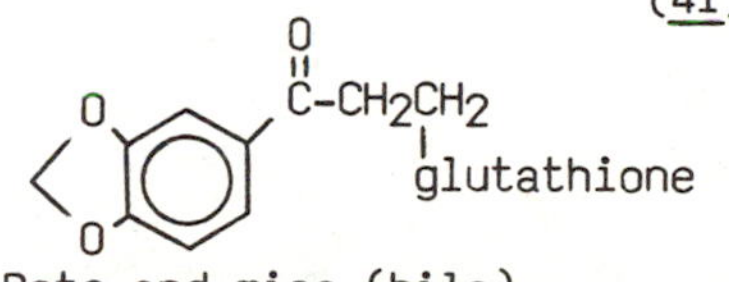 glutathione — Rats and mice (bile).	(41)	200 MHz D₂O — Cys α 4.5; Cys β 2.8–3.2--partially obscured by CH₂CH₂.
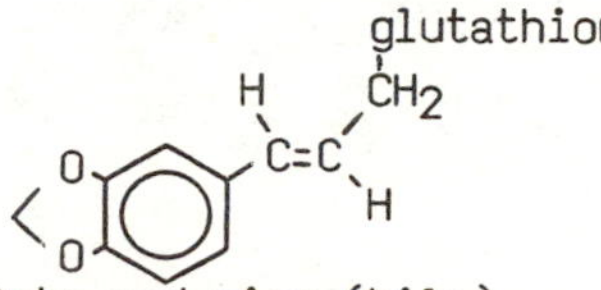 glutathione — Rats and mice (bile).	(41)	200 MHz D₂O — Cys α 4.5 m; Cys β 2.8–3.1 dq; Gly α 3.7 s; Glu α 3.75, β 2.1, γ 2.5

Continued on next page

Table I. Continued

Compound	Ref.	Spectoscopic Summary
CH-glutathione, CH, CH_2 — Rats and mice (bile).	(41)	200 MHz D_2O Cys α 4.3–4.5 Cys β 2.7–3.2
a —CH-glutathione, CH_2, b OH — Rat (liver).	(42)	Cys α 53.7 25 MHz Cys β 33.2 D_2O Gly α 42.8 Glu α 52.3, β 26.8, γ 31.9 a 53.7, b 65.3, Aryl C_1 140.1
a —CH-OH, CH_2, b glutathione — Rat (liver).	(42)	Cys α 53.7 25 MHz Cys β 33.9 D_2O Gly α 42.8 Glu α 52.0, β 26.6, γ 31.8 a 73.2, b 40.2, Aryl C_1 142.5
N-glutathione, H — Synthesis.	(43)	500 MHz Cys α 4.61 Cys β 2.80, 3.17 Gly α 3.75 Glu α 3.60, β 2.12, γ 2.54
N-glutathione-S-oxide, H — Synthesis.	(43)	500 MHz Cys α 4.88, 4.90 Cys β 3.62, 3.71 and 3.60, 3.69 Gly α 3.74, 3.79 and 3.72, 3.78 Glu α 3.51, 3.53 Glu β 2.08 Glu γ 2.50, 2.54 and 2.51, 2.55
O ‖ glutathione-C-glutathione Rat (bile and liver microsomes).	(44)	15.1 MHz D_2O Cys α 53.6, β 32.3 Gly α 44.2 Glu α 54.7, β 26.9, γ 32.1 C=O 191.8
HO— —CH_2-glutathione Rat (bile).	(45)	25 MHz D_2O Cys α 53.9 Cys β 33.4 Gly α 44.3 Glu α 54.9, β 27.1, γ 32.3 CH_2 36.7; t-Bu 30.6, 34.9
S ‖ CH_2=CHCH_2-N-C-mercapturate, H — Synth. Rats and mice (urine).	(46) (47)	360 MHz 90 MHz CD_3OD Cys α 4.65 m Cys β 3.97 dd, 3.59 dd Acetyl Me 2.0

Table I. Continued

Compound	Ref.	Spectroscopic Summary

(cyclohexyl-OH) mercapturate
Synthesis. Rat (urine).
	(48)	100 MHz ^1H not entirely 100 MHz
		  assignable 25 MHz
		^1H and ^{13}C spectra of rat DMSO
		  metabs. like synth. samples.
		Cyclohex CO 73.1, 73.5
		Cyclohex CS 52.8, 53.4

(cyclohexyl-OH) mercapturate
Rat (urine).
	(48)	Cyclohex CO 68.3, 68.9 100 MHz
		Cyclohex CS 51.4 25 MHz
		                    DMSO
		100 MHz ^1H not entirely
		  assignable.

$\overset{O}{\overset{\|}{HO-C}}-CH-CH_2CH_2-NH_2$ cysteine Human (urine).	(49)	360 MHz Cys α 3.95 dd D_2O Cys β 3.05-3.20 2dd GABA α 3.45 t GABA β 2.05 m, 2.16 m GABA γ 3.05-3.20 m
Palmitoyl-O-$\overset{c}{\overset{\|}{C}}H_2$ Palmitoyl-O-$\overset{\|}{C}$H b a CH_2-S-CH_2-$\overset{O}{\overset{\|}{C}H-C}$-OH Synthesis. Palmitoyl-$\overset{\|}{N}$H	(50)	20.1 MHz Cys α 51.8 d $CDCl_3$ Cys β 34.8, 35.09 Doublets due to diasteromers a 33.0, b 70.2, c 63.5 Palmitoyl resonances also assigned.
HO-(ring)-CH_2-cysteine Synthesis.	(51)	67.8 MHz Cys α 55.2 CH_3OD Cys β 34.0 CH_2 37.6
(ring)-N=N-(ring)-$\overset{H}{\overset{\|}{N}}$-$CH_2$ glutathione Synthesis. Rat (bile).	(52)	67.9 MHz Cys α 55.6 Acetone/D_2O Cys β 33.0 CD_3OD/D_2O Gly α 44.6 Glu α 54.6, β 27.8, γ 32.5 Aromatics also assigned.
$\overset{CH_3}{\overset{\|}{\underset{\overset{\|}{N}H}{C=O}}}$ ... glutathione OH Rat (bile).	(53)	360 MHz Cys α 4.3; 55.0 15.1 MHz Cys β 3.07, 3.20; 35.4 D_2O Gly α 3.42, 3.46; 44.3 Glu α 3.56; 54.3 Glu β 1.91; 27.1 Glu γ 2.25; 32.3 Acetyl Me 1.97; 23.6 Aromatic assignments also made.

(ring)-$\overset{a}{O}\overset{b}{CH_2}$-$\overset{}{CH}$-$\overset{c}{CH_2}$
      HO glutathione
Synthesis.
	(54)	360 MHz
		                     D_2O
		Cys α 4.56 TMSP
		Cys β 2.93, 3.11
		Gly α 3.97
		Glu α 3.82, β 2.16, γ 2.55
		a 4.05, 4.16; b 4.10; c 3.69, 3.78

Continued on next page

Table I. Continued

Compound	Ref.	Spectroscopic Summary
$CH_2-C\equiv N$ / CH-mercapturate (phenyl) Synthesis. Rat (urine). H and CH_3 analogues also reported.	(55)	90 MHz CDCl$_3$ Cys α 4.66–4.88 m Cys β 2.84–2.92 m Acetyl Me 1.94 s, 2.03 s Spectrum on methyl ester.
$HO-CH_2CH$-mercapturate / H Synthesis. Rat (urine). Me and Ph analogues also reported.	(55)	90 MHz CDCl$_3$ Cys α 4.72–4.90 m Cys β 3.04 Acetyl Me 2.04 Spectrum on methyl ester.
OH, glutathione, H–N, CH$_3$, CH$_3$, N$^+$CH$_3$ structure Rat (bile).	(56)	250 MHz D$_2$O Cys α 3.91 m Cys β 3.20 m, 3.48 m Gly α 3.56 d, 3.44 d Glu α 3.64 m Glu β 1.85 m, 1.77 m Glu γ 2.20 m Chem. shifts very sensitive to temperature and concentration.
H OH O / $C-OCH_3$ / H cysteine structure Synthesis.	(57)	360 MHz CDCl$_3$ Cys α 4.61 m Cys β 2.79 m, 3.04 m Gly α 4.08 Spectrum on Me ester, NCCF$_3$ deriv.
Rat liver microsomes. HO, H H, glutathione, O, OCH$_3$ structure	(58)	360 MHz D$_2$O Cys α 4.50 t Cys β 2.79 dd, 3.09 dd Gly α 3.91 s Glu α 3.68 Glu β 2.06 dt Glu γ 2.44 t
glutathione N, N–CH$_2$–CH–CH$_2$–OCH$_3$ / OH / NH$_2$ structure Synthesis- mixture of 4 and 5 isomers.	(59)	360 MHz D$_2$O Cys α 4.52 Cys β 3.02, 3.22 Gly α 3.74 Glu α 3.56, β 2.12, γ 2.56 4 and 5 imidazole 6.96, 7.15 N–CH$_2$ 3.76, 4.02
HO, H, glutathione, O, N–CH$_3$, H, HO structure Rat liver microsomes.	(60)	500 MHz D$_2$O DDS Cys α 4.8 dd Cys β 3.39 dd, 2.98 dd Gly α 3.74–3.84 Glu α 3.74–3.84, 2.16, 2.57 Conjugation site assigned on basis of NMR comparison with Morphine.

Table I. Continued.

Compound	Ref.	Spectroscopic Summary
Chicken (excreta).	(61)	360 MHz Acetone α 4.4 m β 1.7 m γ 1.7–1.9 m δ 3.4 q Acetyl Me 1.9 s Ester Me 3.64 s
Rat (urine).	(33)	470 MHz CD₃OD Cys α 4.845 dd, Cys β 3.718 dd, 3.362 dd Acetyl Me 1.95 s Spectrum on methyl ester.
Rat (urine).	(62)	90 MHz CDCl₃ Cys α 4.69–4.96 m Cys β 2.86–2.97 m Acetyl Me 1.92 s, 2.04 s Ester Me 3.71 s, 3.76 s a and b 3.8–4.2 m
Rat (urine).	(62)	90 MHz CDCl₃ Cys α 4.67–5.08 m Cys β 3.02 d Acetyl Me 2.04 s Ester Me 3.76 s, 3.77 s a 4.67–5.08 m, b 2.77–2.93 m
Rat (urine).	(62)	90 MHz CDCl₃ Cys α 4.76–4.96 m Cys β 3.05 d Acetyl Me 2.04 s Ester Me 3.72 s O=C–CH₂–S 3.90s, 3.91 s
Synthesis.	(63)	25.1 MHz D₂O Cys α 55.67 Cys β 35.08 Gly α 44.24 Glu α 56.39, β 28.68, γ 33.82 a 54.20, b 67.25 Also data on other stereoisomer.
Synthesis.	(63)	25.1 MHz D₂O Cys α 55.67 Cys β 34.98 Gly α 44.24 Glu α 56.39, β 28.68, γ 33.83 a 74.79, b 42.10 Also data on other stereoisomer.

Continued on next page

Table I. Continued

Compound	Ref.	Spectroscopic Summary
mercapturate(Me ester) Synthesis.	(64)	25.1 MHz Cys α 52.4 (4.83) (100 MHz) Cys β 33.4 (2.87) CDCl$_3$ Acetyl Me 22.8 (2.03) TMS Ester Me 51.9 (3.69) a 53.3 (3.87), b 65.8 (3.87) Also data on other stereoisomer.
(Me ester) mercapturate Synthesis.	(64)	25.1 MHz Cys α 51.9 (4.83) (100 MHz) Cys β 34.5 (2.98) CDCl$_3$ Acetyl Me 22.7 (2.01) TMS Ester Me 52.2 (3.75) a 72.5 (4.83), b 41.9 (2.58) Also data on other stereoisomer.
glutathione OH Mixture of 4 and 5 isomers Little skate liver - glutathione transferase.	(27)	25.1 MHz ^{13}C enriched compounds D$_2$O C-4 70%, C-5 97% dioxane C-4-OH 71.05, C-5-S 49.82 C-5-OH 71.19, C-4-S 48.67
 Rat (excreta).	(28)	C 22.6 MHz CD$_3$OD ^{13}C enriched cpd at *, 58.1
from cysteine Rabbit guinea-pig (urine).	(65)	H 90MHz H-5 4.4d and 4.8d C 22.6 MHz (J=15.6 Hz) DMSO H-6 3.1d and 3.8 (J=15.6 Hz) C-2 196.4 C-5 38.4 C-4 98.4 C-6 48.8
glutathione Rat (bile).	(66)	H 90MHz D$_2$O Conjugate site determined from aromatic patern.
Man (uremic sera).	(67)	H 90MHz CD$_3$OD Glycine α 4.13

Sulfates

Few examples of the use of NMR in the identification of xenobiotic
sulfates have been reported. Probably the most useful approach for
the identification of sulfates is to first determine the NMR spectrum of the sulfate, then remove the sulfate (sulfatase or acid) and
redetermine the NMR spectrum. The carbon to which the sulfate is
attached in aliphatic compounds is shifted downfield approximately 6
ppm while adjacent carbons are shifted upfield slightly when compared to the corresponding hydroxy compound. Likewise, protons on
the carbon to which the sulfate is attached are found approximately
0.5 ppm further downfield. The situation with aromatic sulfates is
somewhat more complex. Substituent effects for monosubstituted benzenes have been compiled (69-72). Benzene is given a value of
128.5, and additive values are given for C-1, ortho, meta, and para
positions (positive values downfield, negative values upfield).
Although the determinations were for monosubstituted benzenes, useful approximate spectra can be calculated for more highly substituted aromatics by the principle of additivity. Ragan (73) obtained
the following parameters in D_2O using external TMS as a reference:
C-1, +22.7; ortho, -6.8; meta +1.6; para, -119. Feil and Paulson
(74) obtained the following parameters as determined from 2,6-[^{2}H]-
and 2,4,6-[^{2}H]-phenyl sulfate in alkaline D_2O using internal sodium
3-trimethylsilyl propionate as a reference: C-1, +25.5; ortho, -4.1;
meta, +3.9; para, +0.5. Parameters for the substituents -OH and
-O⁻ are: C-1, +26.9; ortho, -12.7; meta, +1.4; para, -7.3 and C-1,
+39.6; ortho, -8.2; meta, +1.9; para, -13.6 (69-72). Thus, the
sulfate group has little effect on the shift of C-1 when compared to
the corresponding phenol in neutral solution, but shifts C-1
approximately 14 ppm upfield when compared to the phenol run in
alkaline solution. The use of ^{33}S in sulfate metabolite identification has not been reported and is unlikely to be useful
because of its low sensitivity (approximately one tenth that of
^{13}C). Table II lists some sulfates of xenobiotics and of natural
products.

Table II. Examples of NMR usage in identification of sulfate
ester conjugates; values in parentheses are for the corresponding OH
compound.

Compound	Ref.	NMR Summary	
	(75)		25 MHz
		24 40.3 (40.0)	CD₃OD
		25 72.4 (74.1)	
		26 75.2 (70.3)	
		27 23.9 (23.7)	
Toad (gallbladder).			

Continued on next page

Table II. Continued

Compound	Ref.	NMR Summary	
SO$_3^-$ ... O$_3$SO$^-$ ... SO$_3^-$ OSO$_3^-$ (positions 2,3,4,6,7) Okinawan sponge.	(76)	1 40.1 (40.4) 2 75.5 (69.3) 3 75.5 (69.9) 4 25.1 (25.3) 5 45.4 (45.8) 6 78.8 (70.4) 7 39.2 (41.6)	25 MHz CD$_3$OD
=CH$_2$ $^-$O$_3$SO (positions 4,3,2) Synthesis.	(77)	Spectrum similar to that of vitamin D$_3$. 3α 4.7 (3.95)	270 MHz CD$_3$OD
HO 3 2 4 1 + $^-$O$_3$SO—⬡—CH$_2$CH$_2$NH$_3$ 5 6 Synth. Ref. to animal metab.	(78)	2 6.73 5 7.07 6 6.63 1st order splitting pattern.	250 MHz DMSO-d$_6$
$^-$O$_3$SO 3 2 4 1 + HO—⬡—OCH$_2$CH$_2$NH$_3$ 5 6 Synth. Ref. to animal metab.	(78)	2 7.06 5 6.75 6 6.82 1st order splitting pattern.	250 MHz DMSO-d$_6$ TMS
3 2 OH CH$_3$ 4 1 | $^-$O$_3$SO—⬡—OCH$_2$CHCH$_2$N—CH H CH$_3$ 5 6 Rat, dog, man (urine).	(79)	2,6 7.0 3,5 7.3 Addition of NaOD caused upfield shift of all isopropyl protons.	80 MHz D$_2$O NaOD
6 OH $^-$O$_3$SO⬡ ⟨OH⟩ 4 2 NH OH ⟨⟩ Man (urine).	(80)	2 6.85 (6.57) 4 6.85 (6.43) 6 6.9 (6.57)	250 MHz D$_2$O
3 2 4 ⬡ 1 OSO$_3^-$ 5 6 Synthesis.	(73)	1 151.2 4 126.6 2,6 121.7 3,5 130.1 ^{1}H NMR and IR also reported.	25 MHz D$_2$O

Table II.   Continued

Compound	Ref.	NMR Summary
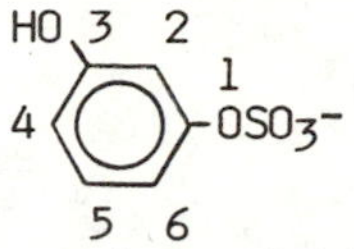 HO 3 2 / 1 / 4 ⟨○⟩-OSO₃⁻ / 5 6 Synthesis. Di- and tri-sulfate also reported.	(73)	25 MHz D₂O 1 152.9 4 101.4 2,6 101.2 3,5 158.4 ¹H NMR and IR also reported.
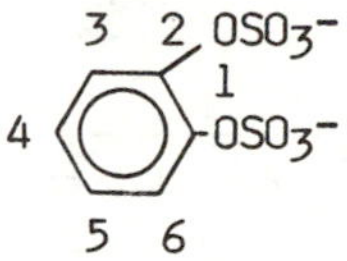 3 2 OSO₃⁻ / 1 / 4 ⟨○⟩-OSO₃⁻ / 5 6 Synthesis.	(73)	25 MHz D₂O 1,2 143.4 3,6 123.5 4,5 127.4 ¹H NMR and IR also reported.
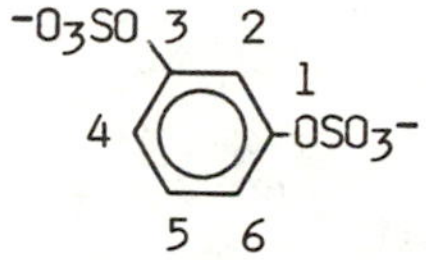 ⁻O₃SO 3 2 / 1 / 4 ⟨○⟩-OSO₃⁻ / 5 6 Synthesis.	(73)	25 MHz D₂O 1,3 151.8 4,6 119.5 2 115.3 5 130.7 ¹H NMR and IR also reported.
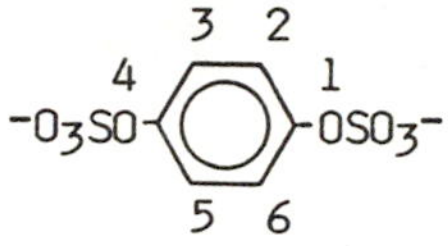 3 2 / 4 ⟨○⟩ 1 / ⁻O₃SO-⟨○⟩-OSO₃⁻ / 5 6 Synthesis.	(73)	25 MHz D₂O 1,4 149.0 2,3,5,6 123.0 ¹H NMR and IR also reported.
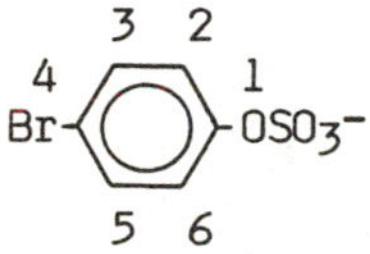 3 2 / 4 ⟨○⟩ 1 / Br-⟨○⟩-OSO₃⁻ / 5 6 Synthesis.	(74)	22.6 MHz D₂O/NaOD 1 153.2 2, 6 126.2 3.5 135.4 4 121.7
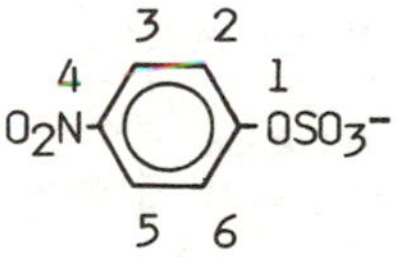 3 2 / 4 ⟨○⟩ 1 / O₂N-⟨○⟩-OSO₃⁻ / 5 6 Synthesis.	(74)	22.6 MHz D₂O/NaOD 1 159.2 2,6 124.3 3,5 128.4 4 147.5
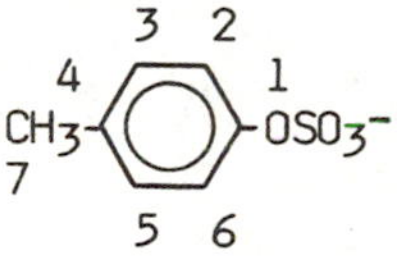 3 2 / 4 ⟨○⟩ 1 / CH₃-⟨○⟩-OSO₃⁻ / 7 / 5 6 Synthesis.	(74)	22.6 MHz D₂O/NaOD 1 151.7 2,6 124.0 3,5 132.9 7 22.7 4 139.1
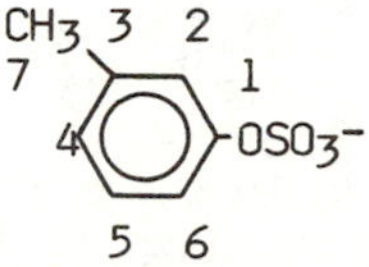 CH₃ 3 2 / 7 ⟨○⟩ 1 / 4 ⟨○⟩-OSO₃⁻ / 5 6 Synthesis.	(74)	22.6 MHz D₂0/NaOD 1 153.9 2 124.7 5 132.3 3 143.2 6 121.0 4 129.6 7 23.1

Continued on next page

Table II. Continued

Compound	Ref.	NMR Summary		
CH$_3$O (positions 3 2 4 1 7 5 6) OSO$_3^-$	(74) Synthesis.	1 146.2 2,6 124.0 3,5 116.1 4 158.0	7 57.0	22.6 MHz D$_2$O/NaOD
Cl (positions 3 2 4 1 5 6) OSO$_3^-$	(74) Synthesis.	1 152.6 2,6 125.8 3,5 133.4 4 133.6		22.6 MHz D$_2$O/NaOD
Cl 3 2 4 1 5 6 OSO$_3^-$	(74) Synthesis.	1 154.5 2 124.7 3 136.9	4 129.1 5 133.6 6 122.9	22.6 MHz D$_2$O/NaOD
OSO$_3^-$	(81) Rat (urine).	Protons of carbons 2 and 3 shifted downfield 0.28 and 0.15 ppm respectively.		80 MHz Acetone/D$_2$O

Carbohydrate Conjugates

Proton NMR has been used for the determination of configuration about the anomeric carbon and the conjugation site of carbohydrates with xenobiotics. Most xenobiotic conjugates appear to have the β configuration; however, in many cases this has been established only by enzymatic hydrolysis. Enzyme hydrolysis may give erroneous results because of impure enzyme or lack of specificity, as demonstrated by the hydrolysis of an α-glucoside by either α- or β-glucosidase (82). The determination of the conjugation site and the establishment of configuration are essential to the determination of metabolic mechanisms, biological activity and catabolism of xeno-biotic conjugates. Table III lists examples in which NMR has been used in a variety of ways. For thorough analysis, the NMR of the conjugate should be obtained, the conjugating group should be removed and identified (i.e. glucose oxidase, derivatization, GC, GC/MS), and the NMR of the unconjugated metabolite should be deter-mined. Comparison of spectra of the conjugated and unconjugated metabolite may yield information on the site of conjugation. For example, glucuronidation of a phenol shifts the ortho protons down-field approximately 0.15 ppm (83), and shifts the proton of a secon-dary alcohol downfield 0.1 - 0.3 ppm (84). Reviews on the use of ^{1}H NMR in the identification of carbohydrates indicate that caution should be exercized in assignment of configuration (85-87). The establishment of the anomeric configuration by NMR has usually been done by determining the coupling constant between the anomeric pro-ton and the proton on C-2 of the carbohydrate moiety. Chemical shift may also be indicative of configuration as protons in the α configuration usually absorb farther downfield than β protons, (e.g. the shifts for the α and β anomeric protons for p-nitrophenyl gluco-side are δ 5.9 and 5.2 respectively); however, chemical shifts can only provide supporting evidence in configurational assignments because of substituent and solvent effects. The magnitude of the coupling constant between the protons of C-1 and C-2, $J_{1,2}$, has most frequently been used to establish anomeric configuration (88a). Vicinal protons in a glucose or glucuronic acid conjugate in the β configuration (trans-diaxial) generally have coupling constants of 7 to 12 Hz while vicinal protons in the α configuration (axial-equatorial) have coupling constants of 2 to 4 Hz. Deviations from the normal configurations may be generated by intramolecular hydrogen bonding, interaction between the acid group of a glucuronic acid and an amino group, and by the anomeric effect (highly electronegative substituents on the anomeric carbon of a carbohydrate in the β configuration prefer an axial position) (89-90). These effects can be suppressed by using hydroxylic solvents (D_2O and CD_3OD) and by methylation of glucuronides. Derivatization (acetyl, methyl, and trimethylsilyl) of the conjugates to suppress intramolecular interactions in solvents such as chloroform and acetone may support the assignments. Derivatization may also be necessary because of interference of the HOD peak (ca δ4.8) with the anomeric proton resonance.

Carbon NMR has rarely been used in the identification of xenobiotic conjugates; therefore, appropriate model compounds must often be taken from the carbohydrate or natural product literature. Glucose and glucuronic acid are the most common carbohydrate conjugates. The α anomers of glucose, glucuronic acid, and their glycone derivatives absorb at higher field than the β anomers; however, the differences are not so great that assignments can be confidently made without additional information. This is shown by the following compounds: glucose α 92.8, β 96.7; methyl glucoside α 99.6, β 103.4; p-nitrophenyl glucoside α 100.4, β 102.7 (12). Single bond $^{13}C-^{1}H$ coupling constants at the anomeric carbon vary with configuration, being approximately 160 Hz for β anomers and 170 Hz for α anomers (88b); however, to the author's knowledge, this technique has not been applied to xenobiotic conjugate characterization.

The application of both carbon and proton NMR to the establishment of anomeric configurations in xenobiotic conjugates has been quite limited, and verification by synthesis has been even more limited. The NMR method appears to be in general agreement with synthesis as no clearly contradictory reports were found. Unfortunately, the usual method of synthesis, the Koenigs-Knorr reaction, may give mixtures (91), and it is conceivable that the incorrect isomer may have been isolated in some cases. A possible discrepency exists with the glucuronide of fluoresein. The reported proton NMR absorbance for the anomeric carbon at δ5.09 ($J_{1,2}$=2.6 Hz) suggests an α configuration while synthesis from the α bromo compound and hydrolysis by β-glucuronidase suggest a β configuration (98).

Table III. Examples of NMR usage in identification of carbohydrate conjugates.

Compound (Ref.)	NMR Summary
β-galactose (92) β-glucose −OCH$_2$ β-cellobiose ⊨O HO OH Hydrocortisone, prednisolone, dexamethasone, fludrocortisone. Also peracetyl derivatives. Synthesis.	200 MHz Anomeric H δ (J$_{1,2}$ in Hz): DMSO Glucosides 4.18 ± 0.01 (7.6) peracetyl 4.22 ± 0.02 (7.6 ± 0.1) Galactosides 4.17 ± 0.02 (7.6 ± 0.3) peracetyl 4.21 ± 0.04 (7.5 ± 0.5) Cellobioside 4.15 (7.6) peracetyl 4.28 (7.6) Other assignments also made.

CH$_3$ (93)
     O-β-glucuronide
CH$_3$ −N   =O
    N
        H (OH)
Man and rat.
	20 MHz
	Anomeric C 103.04 DMSO
	Other glucuronide assignments
	also made. NMR of norantipyrine
	tautomers aided in assignment of
	conjugation site.

(94)
     O-glucuronide
   OCH$_3$
N
H   Man (urine).
Metab. of 5,6-dihydroxyindole.
	360 MHz
	No anomeric NMR data. DMSO
	NMR used to establish MeO D$_2$O
	and conjugate sites.

F F (95) O O β-glucoside β-glucoside Mixture - fungal metabolites.	500 MHz NMR on mixture of Acetone isomers as acetyl derivatives. Conjugation sites determined by NMR. Anomeric H 5.57 (7.7) 5.65 (7.7)

(84)
H
  N−CH$_3$
  ∙ıılH
     O-β-glucuronide
 	
Man (urine).
Metabolite of oxaprotaline.
	100 MHz
	Anomeric H 360 MHz
	4.95 (J=8 Hz) 400 MHz
	Other assignments CD$_3$OD
	also made.

Table III. Continued

Compound	Ref.	NMR Summary
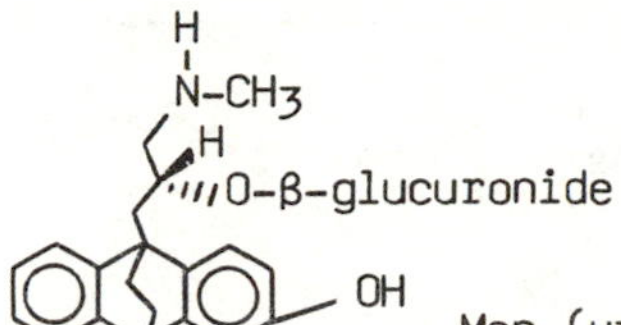 H N-CH$_3$ H ,,,,O-β-glucuronide OH Man (urine). Metabolite of oxaprotaline.	(84)	100 MHz Anomeric H 360 MHz 4.45 (J=8 Hz) 400 MHz Location of hydroxyl CD$_3$OD group based on aromatic shifts.
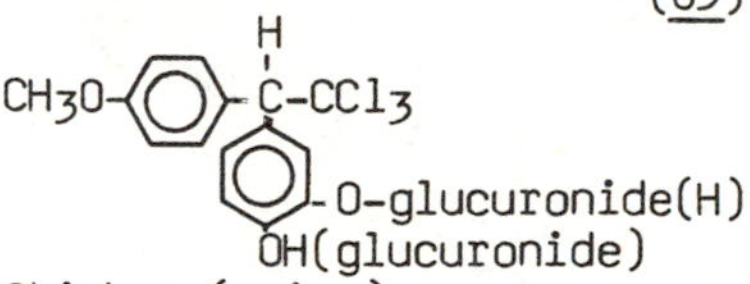 CH$_3$O-⟨⟩-C-CCl$_3$ O-glucuronide(H) OH(glucuronide) Chicken (urine).	(83)	90 MHz No anomeric data. NMR suggested a 4-methoxyphenyl group. Conju- gation sites determined by com- parison to synthetic cpds after following sequence: OH to O- benzyl; β-glucuronidase; CH$_2$N$_2$.
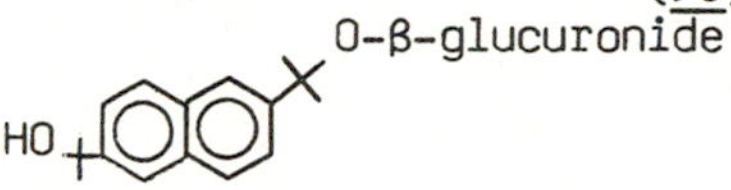 O-β-glucuronide HO Also conjugates of more highly oxidized compounds. Rat (urine).	(96)	60 MHz β-Glucuronidase, NMR CDCl$_3$ on unconjugated cpds to determine degree and site of oxidation. No anomeric NMR data.
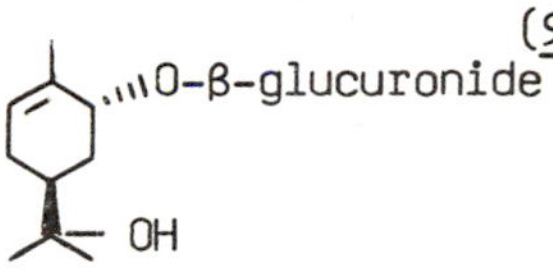 ,,,,O-β-glucuronide OH Rat (urine).	(97)	25 MHz Anomeric C 101.7 CD$_3$OD Anomeric H J$_{1,2}$ 7.5 Hz ^{13}C and ^{1}H NMR used to determine conjugation site. ^{1}H NMR used to establish original configuration of aglycone.
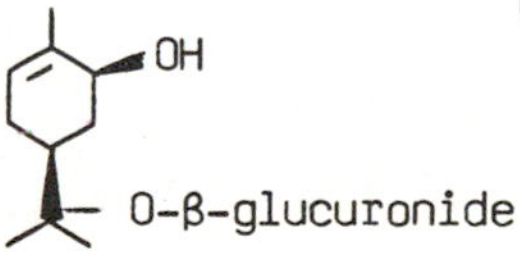 OH O-β-glucuronide Rat (urine).	(97)	25 MHz Anomeric C 98.1 CD$_3$OD Anomeric H J$_{1,2}$ 7.5 Hz ^{13}C and ^{1}H NMR used to determine conjugation site. ^{1}H NMR used to establish original configuration of aglycone.
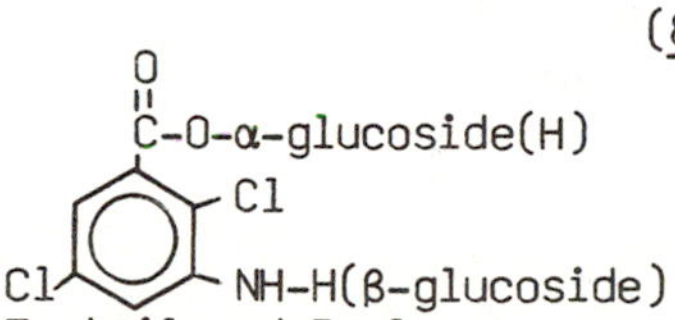 O ‖ C-O-α-glucoside(H) Cl Cl—⟨⟩—NH-H(β-glucoside) Foxtail and Barley.	(82)	90 MHz Spectra on acetyl derivs. CDCl$_3$ Acyl anomeric H 6.34 TMS (J$_{1,2}$=3.7 Hz) N- anomeric H 5.95 (J=8.0 Hz) Synthetic acyl β-glucoside, anomeric H 5.97 (J=8.0 Hz)

Continued on next page

Table III. Continued

Compound	Ref.	NMR Summary
HO, O-glucuronide Synthesis - ref. to rabbit and human metabolites.	(98)	100 MHz CD$_3$OD Hydrolyzed by β-glucuronidase but anomeric proton at 5.09 ($J_{1,2}$=2.6 Hz) suggests α.
β-glucoside, O=C-O Xanthium and spinach.	(99)	270 MHz Acetone Anomeric H (acetyl deriv) 5.52 ($J_{1,2}$ 8.5 Hz)
OH, O=C-OCH$_3$ β-glucoside Tomato (shoots).	(100)	100 MHz CDCl$_3$ Anomeric H 4.54 (7.0 Hz) for per-O-acetyl methyl ester. H-4 shifted downfield 0.2 ppm from OH cpd.
CH$_3$, Cl-C-, CH$_2$CO, CH$_3$ β-glucuronide Man (urine).	(101)	80 MHz CD$_3$OD Anomeric H 5.57 ($J_{1,2}$=7.3 Hz)
Pr, Pr, N-S-, CO-β-glucuronide Man (urine).	(102)	Anomeric C 95.4
glucuronide, S, CH$_2$COH, N, Cl Rats, dogs, and monkeys (urine).	(103)	80 MHz CD$_3$OD No anomeric NMR data. Spectrum similar to parent cpd except that protons on benzene ring of benzimidazole were shifted. Acyl glucuronide also reported - no anomeric NMR data.
6-O-malonyl-β-glucoside Cl, N-C-NCH$_3$, O H, Cl Spinach leaves.	(104)	180 MHz CDCl$_3$ Anomeric H 4.87 (J=7.7 Hz) Anomeric C 104.3 Malonyl CH$_2$, V 41.2, H 3.38 Spectra on peracetyl deriv. Other assignments also made.

Table III. Continued

Compound	Ref.	NMR Summary
β-glucoside-N—N (with structure: O, H, CH₃S, N, N) Tomato.	(105)	90 MHz DMSO Anomeric H 4.92 (J=7.2Hz) A mixture of two isomeric tris-t-butyldimethylsilyl derivatives yielded doublets at 4.34 and 4.50 (J=7.0 Hz).
(pyridine structure) S-β-glucoside (glucuronide) +N, O Synthesis. Rabbit, monkey, rat, and dog (urine). Synthesis.	(106) (107)	Anomeric C 83.1 for 25 MHz glucoside DMSO Anomeric C 82.6 for glucuronide Site of conjugation established by comparison to CH₃S, CH₃SO, and CH₃SO₂ cpds of pyr β pyr-N oxide. - - - - Anomeric C 106.7 for glucoside.
Cl, HC, CH β-glucuronide-C≡C-COH CH₂ CH₃ Rabbit (urine).	(108)	100 MHz D₂O Anomeric H 4.7 (J=7.7 Hz)
CH₃, N, O, N-S-( )-NH, CH₃, N, H, O, β-glucuronide Swine (liver and muscle).	(109)	360 MHz DMSO Anomeric H 4.40 (J₁,₂=9.0 Hz) NMR indicated unchanged pyrimidine ring and shifted p-aminophenyl protons.

Miscellaneous Conjugates

Table IV lists structures of some uncommon conjugates to which NMR has been applied in characterizations. Both the xenobiotics and the conjugates are of diverse structural types. Ecdysone-type compounds, although not xenobiotics in the strict definition of the term, are included because they represent conjugation with a phosphate group.

Table IV. Examples of NMR usage in identification of miscellaneous conjugates.

Compound	Ref.	NMR Summary
Insect eggs.	(110)	100 MHz CD$_3$OD — Protons on C-2 and C-3 shifted downfield 0.52 and 0.22 ppm from respective ecdysone values.
Adenosine-O, OH ... Insect eggs.	(34)	400 MHz CD$_3$OD — ^{31}P δ1.0 C-22 β78.5 (J$_{POC}$=4 Hz) - shifted downfield 4.6 ppm from ecdysone. ^{13}C aided ribose identification. ^{1}H aided adenine identification. Phosphate conj. at C-22 also reported.
Synthesis. (Hamster embryo).	(111)	270 MHz CD$_3$OD — Hydrocarbon and guanosine attachment sites based on ^{1}H NMR interpretation.
Mice (liver).	(112)	270 MHz DMSO — Saffrole and nucleotide attachment sites based on ^{1}H NMR interpretation.
-CNHCH=CH$_2$ and -CNHCH$_2$CH=CH$_2$ conjs. also reported. Radish and mustard.	(113)	360 MHz Pyridine — a 3.49, b 3.40, x 6.28. Conjs. of 2,4-D also reported, but NMR data only on p-hydroxy-styrylmustard of picloram.

Table IV. Continued

Compound	Ref.	NMR Summary
 CHCH₂OH / CH / (phenyl) / O / HCCH₂OH / αHC——N(H)—(3-chlorophenyl Cl) / (phenyl) OCH₃ / OH Synthetic lignin	(114) (115)	22.6 MHz α 59.5 (ref 114) for Dioxane/D₂O 3-chloroaniline α 56.9 (ref 115) for 4-chloro- aniline and 3,4-dichloro- aniline
CH_3 H / αC=Cβ / 3 O—C CH₃ / CH₃ O / O CH₃ / OCNHCH₃ / O Carrots.	(116)	80 MHz α–CH₃ 1.87 m CDCl₃ β–CH₃ 1.95 dq (1.5, 7 Hz) β–H 6.1 qq (1, 7 Hz) H–3 6.0 s
⁺N(CH₃)₃ / CH₂ γ / O / C–O—CH β / CH₂ α / COO⁻ Dog.	(117)	60 MHz N(CH₃)₃ 3.12 s D₂O α 2.56 d (6Hz) β 5.68 m γ 3.82 d (9Hz)
O——(phenyl)—Cl / (CH₂)₁₀ / O / CO– (steroid) Rat (liver).	(118)	H 250 MHz CDCl₃ Aromatic pattern similar to that of 4-benzyloxy-benzoic acid. Spectrum also similar to that of cholesterol.
Cl / (phenyl) / CH₃ / CH–CH / CH₃ C–O– (steroid) / O Mammals.	(119)	Proton NMR aided in the characterization of metabolite. One optical isomer of Fenvalerate was metabolized more rapidly than others.
H / (phenyl)—O—(phenyl)—C–N–(CH₂)₃ / O NH / (CH₂)₄ / NH / (phenyl)—O—(phenyl)—C–N–(CH₂)₃ / O H Earthworms. Cypermethrin metabolism.	(120)	Proton NMR similar to spermine and aromatic portion of Cypermethrin.

Concluding Remarks

The use of NMR in the characterization of xenobiotic conjugates
has been quite limited. At one point in time this was due to
sample requirements, but the introduction of Fourier transform
spectrometers largely eliminated this problem. Nevertheless the
application of NMR has been much less than that of mass spectrometry. The period of time covered by this review shows a small
increase in NMR usage, primarily in the high field area (200-500
MHz). Several publications dramatically demonstrate the usefulness
of high field NMR. For example, the identification of the glutathione conjugate of morphine by Correia et al. (60) would have been
nearly impossible without high field data. Equally impressive are
the determinations by Cerniglia et al. (95) of the conjugation sites
and anomeric configurations of a mixture of glucosides of 1-
fluoronaphthalene. The assignments were made without separation of
the mixture of these metabolites. The increased availability of
high field instruments should generate an increase in the application of NMR to the characterization of xenobiotic conjugates because
of the large amount of structural information available in these
spectra. NMR of other nuclei, especially carbon-13 and tritium, in
conjunction with isotope labeling, should find increased application
as experiments are designed to determine metabolic pathways.

Literature Cited

1. Wilson, N.K. In "Pesticide Analysis"; Das, K.G., Ed.; Marcel
    Dekker, Inc.: New York, 1981; p. 263.
2. Garlick, P.B.; Radda, G.K. Tech. Life Sci., (Sect.): Biochem.
    1979, B216, 1.
3. Scott, A.I.; Baxter, R.L. Ann. Rev. Biophys. Bioeng., 1981,
    10, 151-74.
4. Calder, I.C. In "Progress in Drug Metabolism"; Bridges, J.W.;
    Chasseaud, L.F., Ed.; Wiley and Sons: New York, 1979; Vol. 3,
    p. 303.
5. Hawkins, D.R. In "Isotopes: Essential Chemistry and
    Applications"; Elvidge, J.A.; Jones, J.R., Eds.; The Chemical
    Society Burlington House: London, 1980; p. 270.
6. Rackham, D.M. In "Isotopes: Essential Chemistry and
    Applications"; Elvidge, J.A.; Jones, J.R., Eds.; The Chemical
    Society Burlington House: London, 1980; p. 97.
7. Halliday, D.; Lockhart, I.M. In "Progress in Medicinal
    Chemistry"; Ellis, G.P.; West, G.B., Eds.; North-Holland
    Publishing Co.: New York, 1978; p. 28.
8. Becker, E.D. Fourier Transform NMR Spectroscopy; The American
    Chemical Society: ACS Audio Course, 1983.
9. Pouchert, C.J.; Campbell, J.R. "Aldrich Library of NMR
    Spectra"; Aldrich Chem. Co.: Milwaukee, Wisc., 1975; Vols.
    I-XI.
10. "The Sadtler Collection of High Resolution Spectra".
    Sadtler Research Laboratories: Philadelphia.
11. "Varian Associates. High Resolution NMR Spectra Catalog"; 1962;
    Vol. 1; 1963; Vol. 2.
12. Brietmaier, E.; Haas, G.; Voelter, W. "Atlas of Carbon-13 NMR
    Data"; Heyden: London, Vols. 1-2.

13. Formacek, V.; Desnover, L.; Kellerhals, H.P.; Keller, T.; Clerc, J.T. "^{13}C Data Bank"; Bruker Physik, 1976; Vol. 1.
14. Becker, E.D. "High Resolution NMR Second Edition"; Academic Press: New York, 1980.
15. Bax, A.D. "Two-Dimensional Nuclear Magnetic Resonance in Liquids"; D. Reidel Publishing Company: Dordrecht, Holland.
16. Compadre, C.M.; Pezzuto, J.M.; Kinghorn, A.D.; Kamath, S.K. Science 1985, 227, 417-19.
17. Ashworth, D.J.; Mettler, I.J. Biochemistry 1984, 23, 2252-57.
18. "Purdue University Biochemical Magnetic Resonance Laboratory" supported by NIH grant RRO1077 from the Biotechnology Resources Program of the Division of Research Resources".
19. Tschesche, R.; Braun, T.M.; v.Sassen, W. Phytochemistry 1980, 19, 1825-29.
20. Airy, S.C.; Sinsheimer, J.E. Steroids 1982, 38, 593-95.
21. Beland, F.A.; Harvey, R.G. J. Amer. Chem. Soc. 1976, 98, 4963-70.
22. Mizugaki, M.; Hoshino, T.; Ito, Y.; Sakamoto, T.; Shiraishi, T.; Yamanaka, H. Chem. Pharm. Bull. 1980, 28, 2347-50.
23. Moser, H.; Rihs, G.; Sauter, H. Z. Naturforsch., B 1982, 37B, 451-62.
24. Wainer, I.W.; Schneider, L.C.; Weber, J.D. J. Assoc. Off. Anal. Chem. 1981, 64, 848-50.
25. Misiura, K.; Okruszek, A.; Pankiewicz, K.; Stec, W.J.; Czownicki, Z.; Utracka, B. J. Med. Chem. 1983, 26, 674-79.
26. Leete, E.; Porwoll, J. Aldrichimica Acta 1985, 18, 13-19.
27. Hernandez, O.; Walker, M.; Cox, R.H.; Foureman, G.L.; Smith, B.R.; Bend, J.R. Biochem. Biophys. Res. Commun. 1980, 96, 1494-1502.
28. Feil, V.J.; Bakke, J.E.; Larsen, G.L.; Gustafsson, B.E. Biomed. Mass Spectrom. 1981, 8, 1-4.
29. Matsuyama, H.; Minato, H.; Kobayashi, M. Bull. Chem. Soc. Jap. 1977, 50, 3393-96.
30. Russo, C.A.; Burton, G.; Gros, E.G. Phytochemistry 1983, 22, 71-3.
31. Garson, M.J.; Staunton, J. Chem. Soc. Rev. 1979, 8, 539-61.
32. Evans, J.N.S.; Fagerness, P.E.; Mackenzie, N.E.; Scott, A.I. J. Am. Chem. Soc. 1984, 106, 5738-40.
33. Bakke, J.E.; Feil, V.J. unpublished studies.
34. Tsoupras, G.; Hetru, C.; Luu, B.; Constantin, E.; Lagueux, M.; Hoffmann, J. Tetrahedron 1983, 39, 1789-96.
35. Wayne, R.S.; Stockton, G.W.; Wilson, L.A. Pestic. Chem.: Hum. Welfare Environ. Proc. Int. 1983, 4, 381-4.
36. Brandt, J.A.; Ludeman, S.M.; Zon, G.; Todhunter, J.A.; Egan, W.; Dickerson, R. J. Med. Chem. 1981, 24, 1404-08.
37. Tsui, F-P.; Robey, F.A.; Engle, T.W.; Ludeman, S.M.; Zon, G. J. Med. Chem. 1982, 25, 1106-10.
38. Mazzola, E.P.; Borsetti, A.P.; Page, S.W.; Bristol, D.W. J. Agric. Food Chem. 1984, 32, 1102-03.
39. Wyrwicz, A.M.; Schofield, J.C.; Burt, C.T. In "Noninvasive Probes of Tissue Metabolism"; Wiley and Sons: New York, 1982; pp. 149-71.
40. Spratt, M.P.; Dorn, H.C. Anal. Chem. 1984, 56, 2038-43.
41. Fennell, T.R.; Miller, J.A.; Miller, E.C. Cancer Res. 1984, 44, 3231-40.

42. Watabe, T.; Ozawa, N.; Hiratsuka, A. Biochem. Pharmacol. 1983, 32, 777-85.
43. Mulder, G.J.; Unruh, L.E.; Evans, F.E.; Ketterer, B.; Kadlubar, F.F. Chem.-Biol. Interact. 1982, 39, 111-27.
44. Pohl, L.R.; Branchflower, R.V.; Highet, R.J.; Martin, J.L.; Nunn, D.S.; Monks, T.J.; George, J.W.; Hinson, J.A. Drug Metab. Dispos. 1981, 9, 334-39.
45. Tajima, K.; Yamamoto, K.; Mizutani, T. Chem. Pharm. Bull. 1983, 10, 3671-77.
46. Ioannou, Y.M.; Burka, L.T.; Matthews, H.B. Toxicol. Appl. Pharmacol. 1984, 75, 173-81.
47. Mennicke, W.H.; Görler, K.; Krumbiegel, G. Xenobiotica 1983,
48. van Bladeren, P.J.; Breimer, D.D.; Seghers, C.J.R.; Vermeulen, N.P.E.; van der Gen, A.; Cauvet, J. Drug Metab. Dispos. 1981, 9, 207-11.
49. Kamerling, J.P.; Wadman, S.K.; Duran, M.; de Bree, P.K.; Vliegenthart, J.F.G.; Przyrembel, H.; Bremer, H.J. J. Chromatogr. 1983, 277, 41-51.
50. Jung, G.; Carrera, C.; Brückner, H.; Bessler, W.G. Liebigs Ann. Chem. 1983, 1608-22.
51. Nakagawa, Y.; Hiraga, K.; Suga, T. Biochem. Pharmacol. 1983, 32, 1417-21.
52. Ketterer, B.; Srai, S.K.S.; Waynforth, B.; Tullis, D.L.; Evans, F.E.; Kadlubar, F.F. Chem.-Biol. Interact. 1982, 38, 287-302.
53. Hinson, J.A.; Monks, T.J.; Hong, M.; Highet, R.J.; Pohl, L.R. Drug Metab. Dispos. 1982, 10, 47-50.
54. Van den Eeckhout, E.; Sinsheimer, J.E.; Baeyens, W.; De Keukeleire, D.; De Bruyn, A.; De Moerloose, P. Anal. Letters 1983, 16, 785-98.
55. van Bladeren, P.J.; Delbressine, L.P.C.; Hoogeterp, J.J.; Beaumont, A.H.G.M.; Breimer, D.D.; Seutter-Berlage, F.; van der Gen, A. Drug Metab. Dispos. 1981, 9, 246-49.
56. Maftouh, M.; Monsarrat, B.; Rao, R.C.; Meunier, B.; Paoletti, C. Drug Metab. Dispos. 1984, 12, 111-19.
57. Rackham, D.M.; Morgan, S.E.; Paschal, J.W.; Elzey, T.K. Spectroscopy Letters 1981, 14, 589-95.
58. Moss, E.J.; Judah, D.J.; Przybylski, M.; Neal, G.E. Biochem. J. 1983, 210, 227-33.
59. Varghese, A.J. Biochem. Biophys. Res. Commun. 1983, 112,1013-20.
60. Correia, M.A.; Krowech, G.; Caldera-Munoz, P.; Yee, S.L.; Straub, K.; Castagnoli, Jr., N., Chem.-Biol. Interact. 1984, 51, 13-24.
61. Huckle, K.R.; Stoydin, G.; Hutson, D.H.; Millburn, P. Drug Metab. Dispos. 1982, 10, 523-28.
62. Seutter-Berlage, F.; Delbressine, L.P.C.; Smeets, F.L.M.; Ketelaars, H.C.J. Xenobiotica 1978, 8, 413-18.
63. Yagen, B.; Hernandez, O.; Bend, J.R.; Cox, R.H. Bioorganic Chem. 1981, 10, 299-310.
64. Yagen, B.; Hernandez, O.; Bend, J.R.; Cox, R.H. Chem.-Biol. Interact. 1981, 34, 57-67.
65. Görler, K.; Krumbiegel, G.; Mennicke, W.H.; Giehl, H.-U. Xenobiotica 1982, 12, 535-42.
66. Shoemaker, D.D.; Cysyk,R.L.; Padmanabhan, S.; Bhat, H.B.; Malspeis, L. Drug Metab. Dispos. 1982, 10, 35-9.

67. Lichtenwalner, D.M.; Suh, B.; Lichtenwalner, M.R.   J. Clin.
    Invest. 1983, 71, 1289-96.
68. Wiberg, K.B.; Nist, B.J.   "The Interpretation of NMR Spectra";
    W.A. Benjamin, Inc.: New York, 1962.
69. Silverstein, R.M.; Bassler, G.C.; Morrill, T.C.   "Spectrometric
    Identification of Organic Compounds", Wiley and Sons: New York,
    1981; Fourth Edition.
70. Levy, G.C.; Lichter, R.L.; Nelson, G.L.   "Carbon-13 Nuclear
    Magnetic Resonance Spectroscopy"; Wiley and Sons: New York,
    1980; Second Edition.
71. Memory, J.D.; Wilson, N.K.   "NMR of Aromatic Compounds"; Wiley
    and Sons: New York, 1982.
72. Nelson, G.L.; Williams, E.A.   Progress Phys. Org. Chem. 1976,
    12, 229.
73. Ragan, M.A.   Can. J. Chem. 1978, 56, 2681-85.
74. Feil, V.J.; Paulson, G.D.   Unpublished studies.
75. Kuramoto, T.; Kihira, K.; Matsumoto, N.; Hoshita, T.   Chem.
    Pharm. Bull. 1981, 29, 1136-39.
76. Fusetani, N.; Matsunaga, S.; Konosu, S.   Tetrahedron Lett.
    1981, 22, 1985-88.
77. Reeve, L.E.; DeLuca, H.F.; Schnoes, H.K.   J. Biol. Chem. 1981,
    256, 823-26.
78. Osikowska, B.A.; Idle, J.R.; Swinbourne, F.J.; Sever, P.S.
    Biochem. Pharmacol. 1982, 31, 2279-84.
79. Hoffmann, K-J.; Arfwidsson, A.; Borg, K.O.   Drug Metab. Dispos.
    1982, 10, 173-79.
80. Macgregor, T.R.; Nastasi, L.; Farina, P.R.; Keirns, J.J.   Drug
    Metab. Dispos. 1983, 11, 568-73.
81. Hanzlik, R.P.; Bhatia, P.   Xenobiotica 1981, 11, 779-83.
82. Frear, D.S.; Swanson, H.R.; Mansager, E.R.; Wien, R.G.
    J. Agric. Food Chem. 1978, 26, 1347-51.
83. Davison, K.L.; Lamoureux, C.H.; Feil, V.J.   J. Agric. Food
    Chem. 1984, 32, 900-08.
84. Dieterle, W.; Faigle, J.W.; Kriemler, H.-P.; Winkler, T.
    Xenobiotica 1984, 14, 311-19.
85. Kotowycz, G.; Lemieux, R.U.   Chem. Rev. 1973, 73, 669.
86. Angyal, S.J.   In "The Carbohydrates", Pigman, W.; Horton, D.,
    Eds.; Academic Press: New York, NY, 1972; 2nd Edition, Vol. IA,
    p. 200.
87. Bundle, D.R.; Lemieux, R.U.   In "Methods in Carbohydrate
    Chemistry, Vol. VII; Academic Press: New York, NY, 1976.
88. Hall, L.D.   In "The Carbohydrates"; Pigman, W.; Horton, D.;
    Wander, J.D., Eds.; Academic Press: New York, 1980; 2nd
    Edition, Vol. IB, (a) p. 1306, (b) p. 1318.
89. Angyal, S.J.   In "The Carbohydrates", Pigman, W.; Horton, D.,
    Eds.; Academic Press: New York, NY, 1972; 2nd Edition, Vol. IA,
    p. 204.
90. "Anomeric Effect Origin and Consequences"; Szarek, W.A.;
    Horton, D., Eds.; ACS Symposium Series 87, American Chemical
    Society: Washington, DC, 1979.
91. Igarashi, K. In "Advances in Carbohydrate Chemistry and
    Biochemistry"; Tipson, R.S.; Horton, D., Eds.; Academic Press:
    New York, 1977; 243-277.
92. Friend, D.R.; Chang, G.W.   J. Med. Chem. 1985, 28, 51-7.
93. Böttcher, J.; Bässmann, H.; Schüppel, R.; Lehmann, W.D.
    Naunyn-Schmiedeberg's Arch. Pharmacol. 1982, 321, 226-33.

94. Matous, B.; Budesinska, A.; Budesinsky, M.; Duchon, J.; Pavel, S. Neoplasma 1981, 28, 271-74.
95. Cerniglia, C.E.; Miller, D.W.; Yang, S.K.; Freeman, J.P. Appl. Environ. Microbiol. 1984, 48, 294-300.
96. Kojima, S.; Honda, T.; Nakagawa, M.; Kiyozumi, M.; Takadate, A. Drug Metab. Disp. 1982, 10, 429-33.
97. Ventura, P.; Shiavi, M.; Serafini, S. Xenobiotica 1983, 13, 139-46.
98. Matsunaga, I.; Nagataki, S.; Tamura, Z. Chem. Pharm. Bull. 1984, 32, 2832-35.
99. Boyer, G.L.; Zeevaart, J.A.D. Plant Physiol. 1982, 70, 227-31.
100. Milborrow, B.V.; Vaughan, G.T. Aust. J. Plant Physiol. 1982, 9, 361-72.
101. Hasegawa, J.; Smith, P.C.; Benet, L.Z. Drug Metab. Dispos. 1982, 10, 469-73.
102. Eggers, N.J.; Doust, K. J. Pharm. Pharmacol. 1981, 33, 123-24.
103. Janssen, F.W.; Kirman, S.K.; Fenselau, C.; Stogniew, M.; Hofmann, B.R.; Young, E.M.; Ruelius, H.W. Drug Metab. Dispos. 1982, 10, 599-604.
104. Suzuki, T.; Casida, J.E. J. Agric. Food Chem. 1981, 29, 1027-33.
105. Frear, D.S.; Mansager, E.R.; Swanson, H.R.; Tanaka, F.S. Pestic. Biochem. Physiol. 1983, 19, 270-81.
106. Jeffcoat, A.R.; Gibson, W.B.; Rodriguez, P.A.; Turan, T.S.; Hughes, P.F.; Twine, M.E. Toxicol. Appl. Pharmacol. 1980, 56, 141-54.
107. Warder, D.E. Carbohydr. Res. 1981, 94, 115-18.
108. Abolin, C.R.; Tozer, T.N.; Craig, J.C.; Gruenke, L.D. Science 1980, 209, 703-04.
109. Giera, D.D.; Abdulla, R.F.; Occolowitz, J.L.; Dorman, D.E.; Mertz, J.L.; Sieck, R.F. J. Agric. Food Chem. 1982, 30, 260-63.
110. Isaac, R.E.; Desmond, H.P.; Rees, H.H. Biochem. J. 1984, 217, 239-43.
111. Cary, P.D.; Turner, C.H.; Cooper, C.S.; Ribeiro, O.; Grover, P.L.; Sims, P. Carcinogenesis 1980, 1, 505-12.
112. Phillips, D.H.; Miller, J.A.; Miller, E.C.; Adams, B. Cancer Res. 1981, 41, 2664-71.
113. Chkanikov, D.I.; Pavlova, N.N.; Makeev, A.M.; Nazarova, T.A. Fiziol. Rast. 1984, 31, 321-27.
114. Still, G.G.; Balba, H.M.; Mansager, E.R. J. Agric. Food Chem. 1981, 29, 739-46.
115. Sandermann, H. Jr.; Scheel, D.; v.d. Trenck, T. J. Appl. Polym. Sci.: Appl. Polymer Symp. 1983, 37, 407-20.
116. Sonobe, H.; Kamps, L.R.; Mazzola, E.P.; Roach, J.A.G. J. Agric. Food Chem. 1981, 29, 1125-29.
117. Quistad, G.B.; Staiger, L.E.; Schooley, D.A. J. Agric. Food Chem. 1978, 26, 76-80.
118. Fears, R.; Baggaley, K.H.; Walker, P.; Hindley, R.M. Xenobiotica 1982, 12, 427-33.
119. Miyamoto, J.; Kaneko, H.; Okuno, Y. 189th ACS National Meeting, Division of Pesticide Chemistry.
120. Leahey, J.P., Curl, E.A.; Edwards, P.J. 189th ACS National Meeting, Division of Pesticide Chemistry.

RECEIVED December 9, 1985

NOVEL XENOBIOTIC CONJUGATES

9

Lipophilic Xenobiotic Conjugates

Gary B. Quistad[1] and David H. Hutson[2]

[1]Biochemistry Department, Zoecon Corporation, Palo Alto, CA 94304
[2]Department of Biochemistry, Shell Research Ltd., Sittingbourne,
 Kent ME9 8AG, England

Although foreign compounds generally are rendered
more hydrophilic by the vast majority of metabolic
conjugations, occasionally xenobiotics enter pathways of lipid biosynthesis and thereby are converted to more lipophilic derivatives. Specifically, the xenobiotic participates in normal routes
of fatty acid metabolism. Foreign acids may mimic
natural fatty acids (directly or after chain-
extension to homologous acids), thereby incorporat-
ing into glycerides or forming esters of cholesterol. Xenobiotics themselves may be esterified
with natural fatty acids. This report summarizes
the occurrence of this recently discovered class of
nonpolar conjugates in animals and plants.

Historically, metabolic conjugations have been regarded as an
organism's attempt to render an intruding chemical innocuous by
increasing its polarity, thus facilitating excretion. While this
maxim is still generally correct, important exceptions have been
described in the last decade. There are now several examples of
polar metabolites which are toxicologically significant. For
example, within certain chemical classes, glucuronidation, sulfation, and glutathione conjugation have each been implicated in the
formation of reactive metabolites which initiate carcinogenesis.
The generalization that conjugates are more hydrophilic than their
precursors must also be examined in the light of the discovery of
an increasing number of xenobiotics forming derivatives of natural
lipids. These xenobiotic lipids, formed from both acids, amines,
and alcohols, are certainly conjugates by virtue of their
mechanisms of formation, but their physical properties distinguish
them from the classical (hydrophilic) conjugates.

This review emphasizes recent examples of novel lipophilic
conjugates. Previously reviewed nonpolar conjugates (1-3) are
considered only briefly and although certain relatively simple
conjugations (e.g. methylation, formylation, and acetylation) also

produce more lipophilic metabolites, these reactions have been
known for many years and are not reviewed here.

Chain Extension of Acids

The first example of chain extension of a xenobiotic involved
furfural which was reported in 1887 to add a 2-carbon unit thus
producing 3-(2-furyl)acrylic acid which was isolated in rabbit
urine as its glycine conjugate (4). Subsequent workers were not
able to confirm this observation (see 3 for leading references).
    The hypotensive drug, 5-(4-chlorobutyl)picolinic acid, is
converted by rats into four metabolites which each contain the net
addition of one C_2 unit (5, Figure 1). The isolation of the β-keto
acid and α,β-unsaturated acid metabolites implies that alkyl-
substituted picolinic acids are alternative substrates in the
enzymatic chain elongation of natural fatty acids. Similarly,
horses excrete benzoic acid in urine as chain-extended metabolites
(6, Figure 1). These β-keto and β-hydroxy acids also could be
viewed as intermediates in a pathway toward the fully saturated
extended-chain metabolite. Thus, while β-hydroxy, β-keto, and α,β-
unsaturated intermediates are usually not abundant species in
natural fatty acid chain elongation (i.e. the fully saturated
product predominates), these intermediary metabolites may be
quantitatively important for xenobiotic acids.
    Because of its structural similarity to acetate, it is not
surprising that cyclopropanecarboxylic acid (from the miticide
cycloprate) is incorporated readily into normal fatty acid pathways. Six homologous ω-cyclopropyl fatty acids were identified
from treatment of rats, dogs, and a cow with cycloprate (7-10).
Generally, these ω-cyclopropyl fatty acids were not free, but
rather were esterified to produce modified hybrid lipids (vide
infra).
    There are few examples of chain elongation of xenobiotic acids
by plants. The herbicides 2,4-DB [4-(2,4-dichlorophenoxy)butanoic
acid] and 2,4-D (2,4-dichlorophenoxyacetic acid) are both converted
by alfalfa to 6-(2,4-dichlorophenoxy)hexanoic acid (11,12).
Fluoroacetate is converted by plants (13) and rats (14) into long-
chain ω-fluoro fatty acids. In plant seeds 18-fluoorooctadecenoic
acid was the main constituent from fluoroacetate. Unsaturated
long-chain ω-cyclopropyl fatty acids also have been isolated from
apples and oranges treated with cycloprate (15, Figure 1).

Glycerides

Several xenobiotic acids are known to form hybrid triacylglycerols
(i.e. a xenobiotic acid esterified with a natural diacylglycerol)
and adequate reviews of these conjugates have appeared elsewhere
(1-3). Such hybrid triacylglycerols have been reported for 3-
phenoxybenzoic acid (rats; 16) and acidic metabolites of cycloprate
(rat, cow, dog; 8-10), methoprene (chicken; 17), dodecylcyclohexane
(rat; 18), and ethyl 4-benzyloxybenzoate (rat; 19). After
discovering the formation of hybrid triacylglycerols from 4-
benzyloxybenzoate, Fears et al. (19) used an in vitro system
containing rat liver slices to probe the generality of this
pathway. When fifteen aromatic acids were screened, four acids

were incorporated into fat more efficiently than was 4-
benzyloxybenzoate while most of the remaining acids (including
nicotinic and acetylsalicylic) were incorporated poorly or not at
all. Pharmacologically important acids such as the antiinflamma-
tory drugs fenoprofen and ibuprofen formed novel CoA thioesters and
were incorporated into glycerolipids by substituting for natural
fatty acids.

Reports of the formation of hybrid diacylglycerols (i.e. glycerol esterified with a xenobiotic acid plus one natural fatty acid)
are rare. When chickens were dosed with the insect growth regulator methoprene, 4% of the radiolabel in egg yolks was identified
as a mixture of two hybrid diacylglycerols (17). When rats were
dosed at 1 mg/kg with the pyrethroid insecticide fluvalinate in
corn oil (20), 2% of the radiolabel in the fecal extract (0-1 day)
was identified as hybrid diacylglycerols resulting from esterification of both oleoylglycerol (Figure 2) and linoleoylglycerol by an
anilino acid metabolite. Hybrid 1,2-diacylglycerols were more
abundant than the corresponding 1,3-adducts. Interestingly, oleic
and linoleic acids are the primary fatty acid constituents of the
lipids in corn oil (the dose vehicle). A reduction of lipid in the
dosing medium resulted in the disappearance of these diglycerides,
demonstrating the effect of diet on their formation.

The anilino acid from fluvalinate forms the only reported
conjugate between a xenobiotic and glycerol (20; Figure 2). This
monoacylglycerol represented 2% of the radiolabel in the 0-1 day
fecal extract from rats dosed with fluvalinate in corn oil.
Curiously, although actively sought, hybrid triacylglycerols could
not be identified in the above studies with fluvalinate in rats.

Cholesterol Esters

This class of lipid conjugates is the most nonpolar yet identified,
a characteristic which is often useful in pursuing the identification of unknown metabolites. The first cholesterol ester of a
xenobiotic was reported in 1976 for a saturated methoprene metabolite which contributed 15% of the total ^{14}C-residue in the liver of
a chicken given a single oral dose of methoprene at 64 mg/kg (17).
The ω-cyclopropyl fatty acids derived from cycloprate also form
esters of cholesterol. Three such esters contributed 5% of the
total residual radiocarbon in rat carcasses four days after a
single oral dose of cycloprate at 21 mg/kg (8).

A cholesterol ester forms in the liver of rats given an oral
dose (250 mg/kg for seven days) of the hypolipidemic drug 1-(4-
carboxyphenoxy)-10-(4-chlorophenoxy)decane (21). This xenobiotic
cholesterol ester represented about 11% of the total lipid in the
liver and was neither further metabolized nor transported by lipoproteins. An additional 1% of the total liver lipids consisted of
hybrid triacylglycerols containing this xenobiotic. The authors of
this work suggest that the hypocholesterolemic activity of the drug
in rats results from hepatic accumulation of the xenobiotic cholesterol ester which appears to promote the hydrolysis of natural
cholesterol esters and thereby facilitate clearance of low-density
lipoproteins.

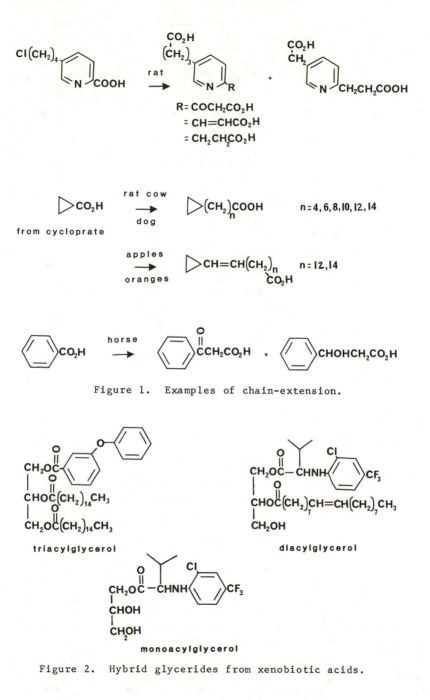

Figure 1.   Examples of chain-extension.

triacylglycerol

diacylglycerol

monoacylglycerol

Figure 2.   Hybrid glycerides from xenobiotic acids.

The antineoplastic drug chlorambucil is converted into a cholesterol ester by plasma from humans and dogs (22). While chlorambucil itself is not converted into the cholesterol ester, when it is administered as its prednisolone ester (prednimustine), such a conversion does occur by transesterification involving lecithin: cholesterol acyltransferase (Figure 3). It is suggested that the cholesterol ester may offer a slow-release mechanism for chlorambucil, but the toxicological significance of this lipid conjugate is still uncertain.

The most recent cholesterol ester of a xenobiotic is derived from an acidic metabolite of the pyrethroid insecticide fenvalerate and is discussed in detail elsewhere in this book (23). Importantly, this cholesterol ester also is produced via transesterification from a single isomer of fenvalerate and has important toxicological consequences.

Fatty Acid Derivatives

Certain xenobiotic alcohols and amines react with natural fatty acids to form nonpolar conjugates. Many of the early investigations in this area were conducted by Edith Leighty and her coworkers who found fatty acid derivatives of the following: DDOH, a DDT metabolite (24); several tetrahydrocannabinols (25,26), also cf. 27); cyclohexylamine, from cyclamate (28); codeine (29); trichloroethanol, from trichloroethylene (30); and pentachlorophenol (31). Since the work of Leighty is detailed elsewhere in this volume (32), further discussion here is unnecessary. However, as an epilogue to Leighty's discovery of a palmitic acid conjugate of pentachlorophenol produced by rat liver microsomes in vitro (31), the same conjugate has been demonstrated recently in human fat (33). Since pentachlorophenol has been widely used in industry and agriculture since the 1930's and was found previously in human fat, Ansari et al. (33) suspected that the formation of fatty acid derivatives of pentachlorophenol might occur also. Indeed, analysis of human fat obtained randomly at autopsy revealed a 240 ppb residue of the palmitate conjugate. The origin of this nonpolar conjugate is unknown. It could be formed in humans, but the authors could not exclude consumption of the intact conjugate in pentachlorophenol-contaminated food.

The list of xenobiotics acylated by fatty acids in animals continues to expand. Seven fatty acid esters of etofenamate, a nonsteroidal anti-inflammatory drug, were identified in dog feces (34). The esters utilized oleic, palmitic, linoleic, stearic, palmitoleic, myristic, and lauric acids, but collectively these lipophilic conjugates represented <0.1% of the applied dose. The hydroxyethyl groups of dipyridamol (a coronary vasodilator) and mopidamol (a cytostatic agent) are esterified with oleic and palmitic acids (35). These lipophilic derivatives of both drugs are excreted in feces (about 4% of the applied dose for rats and humans).

Four acyl hydrazine metabolites were isolated from the feces of sheep dosed at 50 mg/kg with the anthelmintic, p-toluoyl chloride phenylhydrazone (36, Figure 4). Collectively, these amide derivatives (acylation by stearic, palmitic, myristic, and lauric

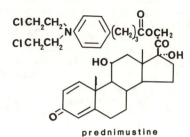

prednimustine

cholesterol

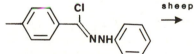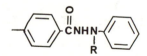

Figure 3.   Formation of a cholesterol ester from prednimustine
(prednisolone ester of chlorambucil) by
transesterification.

human →

p-toluoyl chloride
phenylhydrazone

sheep →

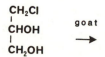

R for acyl moiety of stearate,
palmitate, myristate, laurate

CH₂Cl
|
CHOH   goat
| →
CH₂OH

stearate

and palmitate esters

Figure 4.   Fatty acid derivatives of xenobiotics in animals.

acids) represented 27% of the applied dose. Thus, not only
alcohols, but also N-H groups are acylated by natural fatty acids.

As with xenobiotic acylglycerols (vide supra), a number of
fatty acid derivatives of xenobiotics have been identified in
feces. These neutral lipophilic compounds probably would not be
expected to be secreted with the bile and so their presence in
feces requires explanation. It is possible that they are formed by
the action of enterobacteria and thus may not be found in tissues.

An intriguing story has arisen recently concerning fatty acid
esters of 3-chloro-1,2-propanediol. Diesters of this compound have
been isolated in small (<1% total neutral lipid), but consistent
levels from goat's milk (37). Such residues are absent from the
milk of cows and sheep. The origin of these lipophilic compounds
is quite uncertain. 3-Chloro-1,2-propanediol would be an unusual
natural product, but its formation on treatment of glycerol with
HCl is known. Since chlorine-based sanitizers are used widely in
dairy operations, this halodiol may be produced therefrom. It is
even possible that 3-chloro-1,2-propanediol may occur in dietary
sources. A mixture of C_{16} and C_{18} fatty acid esters of this
chloropropanediol has been isolated from adulterated Spanish
cooking oils (38).

There are only a few reported examples of lipophilic conjugates produced by plants. Two polyethoxylated detergents are
esterified by corn enzyme preparations with palmitic and linoleic
acids accounting for >85% of the fatty acid component (39, Figure
5). An experimental acaricide, Ro 12-0470, is converted by apples
into esters of C_{16}, C_{18}, C_{20}, and C_{22} fatty acids (40). These
lipophilic esters represented up to 22% of the applied dose just
one day after treatment of fruit and foliage. A unique angelic
acid conjugate was identified in carrots treated with carbofuran
(41). This conjugate increases with time and has a half-life
significantly longer than that of other carbofuran residues (42).
The angelic acid conjugate appears to occur only in carrots (not
found in potato or radish). Since angelic acid is abundant in the
essential oil of carrots, the identification of this lipophilic
residue is a good example of how knowledge of an organism's natural
constituents can lead to discovery of novel conjugates.

The Significance of Xenobiotic Lipid Conjugates

Consideration of the biochemical and toxicological significance of
these conjugates (3,43) has been largely speculative and the subject cannot be treated in depth at this time. Clearly, the ability
of several classes of xenobiotics to participate in the processes
of lipid biochemistry has been demonstrated. In view of the high
rate and quantity of turnover of natural lipids, it would seem
unlikely that competitive effects of xenobiotics will cause many
problems for lipid metabolism. Lipid conjugation is a mechanism
for retention of xenobiotic acids and alcohols (32). However, where
measured, the rates of removal of xenobiotic acylglycerols (8,43)
and fatty acids (8) are similar to those of the natural analogs.

The cholesterol esters appear to be more persistent. The only
unequivocal example of a causative relationship between a tissue
lesion (granulomata) and a xenobiotic lipid is described later in

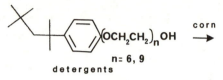

corn → palmitate and linoleate esters

n= 6, 9
detergents

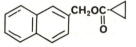

miticide

apples →

R for
esters of

C_{16} C_{18} C_{20} C_{22}
saturated fatty acids

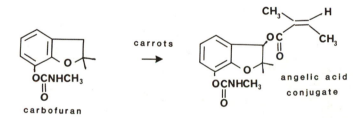

carrots →

carbofuran

angelic acid
conjugate

Figure 5.   Esters derived from xenobiotics and natural acids in plants.

this volume (23). Hypolipidemic action is reported to be associated with the formation of xenobiotic lipids and a direct therapeutic relationship even has been postulated for a cholesterol ester (21). However, generally the relationship may well be indirect, a xenobiotic acyl-CoA ester being the common feature and possibly the causative agent in hypolipidemic action. Potentially, more important targets may be cellular control mechanisms (perhaps diacylglycerols) and membranes (perhaps phospholipids), but there is as yet no evidence for the deleterious involvement of xenobiotic lipids at these levels. Currently, possibly the most important aspect of these novel conjugation reactions is their potential value as a source of explanations for the biochemical toxicologist and for exploitation by pharmacologists.

Literature Cited

1. Hutson, D. H. In "Progress in Pesticide Biochemistry"; Hutson, D. H.; Roberts, T. R., Eds.; Wiley: Chichester, 1982; Vol. 2, pp. 171-184,
2. Schooley, D. A.; Quistad, G. B. In "IUPAC Pesticide Chemistry: Human Welfare and the Environment"; Miyamoto, J.; Kearney, P. C., Eds.; Pergamon: Oxford, 1983; Vol. 3, pp. 301-306.
3. Caldwell, J.; Marsh, M. V. Biochem. Pharmacol. 1983, 32, 1667-1672.
4. Jaffe, M.; Cohn, R. Ber. Dt. Chem. Ges. 1887, 20, 2311.
5. Miyazaki, H.; Takayama, H.; Minatogawa, Y.; Miyano, K. Biomed. Mass Spectrom. 1976, 3, 140-145.
6. Marsh, M. V.; Caldwell, J.; Hutt, A. J.; Smith, R. L.; Horner, M. W.; Houghton, E.; Moss, M. S. Biochem. Pharmacol. 1982, 31, 3225-3230.
7. Schooley, D. A.; Quistad, G. B.; Staiger, L. E. Science 1978, 199, 544-545.
8. Quistad, G. B.; Staiger, L. E.; Schooley, D. A. J. Agric. Food Chem. 1978, 26, 60-66.
9. Quistad, G. G.; Staiger, L. E.; Schooley, D. A. J. Agric. Food Chem. 1978, 26, 71-75.
10. Quistad, G. B.; Staiger, L. E.; Schooley, D. A. J. Agric. Food Chem. 1978, 26, 76-80.
11. Linscott, D. L.; Hagin, R. D. Weed Sci. 1970, 18, 197-198.
12. Linscott, D. L.; Hagin, R. D.; Dawson, J. E. J. Agric. Food Chem. 1968, 16, 844-848.
13. Peters, L. A.; Hall, R. J. Biochem. Pharmacol. 1959, 2, 25-36.
14. Gal, E. M.; Drewes, P. A.; Taylor, N. F. Arch. Biochem. Biophys. 1961, 93, 1-14.
15. Quistad, G. B.; Staiger, L. E.; Schooley, D. A. J. Agric. Food Chem. 1978, 26, 66-70.
16. Crayford, J. V.; Hutson, D. H. Xenobiotica 1980, 10, 349-354.
17. Quistad, G. B.; Staiger, L. E.; Schooley, D. A. J. Agric. Food Chem. 1976, 24, 644-648.
18. Tulliez, J. E.; Bories, G. F. Lipids 1979, 14, 292-297.
19. Fears, R.; Baggaley, K. H.; Alexander, R.; Morgan, B.; Hindley, R. M. J. Lipid Res. 1978, 19, 3-11.

20. Quistad, G. B.; Staiger, L. E.; Jamieson, G. C.; Schooley, D. A. J. Agric. Food Chem. 1983, 31, 589-596.
21. Fears, R.; Baggaley, K. H.; Walker, P.; Hindley, R. M. Xenobiotica 1982, 12, 427-433.
22. Gunnarsson, P. O.; Johansson, S.-A.; Svensson, L. Xenobiotica 1984, 14, 569-574.
23. Miyamoto, J.; Kaneko, H.; Okuno, Y. Chapter 13, this volume.
24. Leighty, E. G.; Fentiman, A. F.; Thompson, R. M. Toxicol. 1980, 15, 77-82.
25. Leighty, E. G.; Fentiman, A. F.; Foltz, R. L. Res. Commun. Chem. Pathol. Pharmacol. 1976, 14, 13-28.
26. Leighty, E. G. Res. Commun. Chem. Pathol. Pharmacol. 1979, 23, 483-492.
27. Yisak, W.; Agurell, S.; Lindgren, J. E.; Widman, M. J. Pharm. Pharmacol. 1978, 30, 462-463.
28. Leighty, E. G.; Fentiman, A. F. Food Chem. Toxicol. 1983, 21, 251-254.
29. Leighty, E. G.; Fentiman, A. F. J. Pharm. Pharmacol. 1983, 35, 260-261.
30. Leighty, E. G.; Fentiman, A. F. Res. Commun. Chem. Pathol. Pharmacol. 1981, 32, 569-572.
31. Leighty, E. G.; Fentiman, A. F. Bull. Environ. Contam. Toxicol. 1982, 28, 329-333.
32. Chang, M.J.W.; Leighty, E. G.; Haggerty, G.; Fentiman, A. F. In "Advances in Xenobiotic Conjugation Chemistry"; Paulson, G. D.; Caldwell, J.; Hutson, D. H.; Menn, J. J., Eds.; ACS SYMPOSIUM SERIES; this volume; American Chemical Society: Washington, D.C., 1985.
33. Ansari, G.A.S.; Britt, S. G.; Reynolds, E. S. Bull. Environ. Contam. Toxicol. 1985, 34, 661-667.
34. Dell, H.-D.; Fiedler, J.; Kamp, R.; Gau, W.; Kurz, J.; Weber, B.; Wuensche, C. Drug Metab. Dispos. 1982, 10, 55-60.
35. Schmid, J.; Prox, A.; Baeur, E.; Koss, F. W. Abstracts 7th Eur. Workshop on Drug Metab., Zurich, 1980, Abstract 434.
36. Jaglan, P. S. J. Agric. Food Chem. 1980, 28, 682-684.
37. Cerbulis, J.; Parks, O. W.; Liu, R. H.; Piotrowski, E. G.; Farrell, H. M. J. Agric. Food Chem. 1984, 32, 474-476.
38. Gardner, A. M.; Yurawecz, M. P.; Cunningham, W. C.; Diachenko, G. W.; Mazzola, E. P.; Brumley, W. C. Bull. Environ. Contam. Toxicol. 1983, 31, 625-630.
39. Frear, D. S.; Swanson, H. R.; Stolzenberg, G. E. 174th National Meeting of the American Chemical Society, Chicago, IL, August 1977, PEST. 16.
40. Pryde, A.; Hänni, R. P. J. Agric. Food Chem. 1983, 31, 564-567.
41. Sonobe, H.; Kamps, L. R.; Mazzola, E. P.; Roach, J.A.G. J. Agric. Food Chem. 1981, 29, 1125-1129.
42. Sonobe, H.; Carver, R. A.; Krause, R. T.; Kamps, L. R. J. Agric. Food Chem. 1983, 31, 96-104.
43. Hutson, D. H.; Dodds, P. F.; Logan, C. J. Biochem. Soc. Trans. 1985, 13, in press.

RECEIVED August 19, 1985

10

Acylation of Xenobiotic Alcohols: A Metabolic Pathway for Drug Retention

M. J. W. Chang, E. G. Leighty[1], G. C. Haggerty, and A. F. Fentiman, Jr.

Battelle Columbus Laboratories, Columbus, OH 43201

A rat liver Coenzyme A-fortified microsomal system has been characterized and utilized in our laboratory for mechanistic studies of long-term retention of xenobiotics such as metabolites of drugs of abuse and chlorine-containing pesticides. Δ^9-Tetrahydrocannabinol (Δ^9-THC), Δ^8-THC, codeine, phencyclidine, trichloroethylene, pentachloro-phenol, and 2,2-bis($\underline{p}$-chlorophenyl)-1,1,1-trichloroethane (DDT) have been shown to conjugate to fatty acids either directly or through their hydroxylated metabolites using this CoA-fortified microsomal system. Conjugation was also found to occur at the nitrogen atom of the cyclamate metabolites. The structure of each conjugate was eluci-dated using a gas chromatography/mass spectrometry (GC/MS) method. The fatty acid conjugation of the hydroxylated metabolites of both THC and DDT has been demonstrated in vivo in Sprague-Dawley rats. The esterase system involved in cholesterol esterification is suggested to be the responsible enzyme system. The effects of both 11-palmitoyloxy-Δ^9-THC (11-palm-Δ^9-THC) and palmitoyl-codeine on behavioral and biochemical measures in rat were investigated. Both conjugates induced catalepsy and analgesia in the rat. The effects were similar to those seen with their non-conjugated counterparts, but the conjugates were less active.

Residual metabolites retained in the animal body long after the initial exposure have been documented for a variety of drugs and xenobiotics (1-3). Many metabolic pathways have yet to be identified.

The objective of this report is to summarize the work initiated by the late Dr. Edith G. Leighty and colleagues at the Battelle Columbus Laboratories, in which the investigation of a novel meta-bolic pathway for drug retention was conducted.

[1] Deceased

Historical Perspective

Indirect Evidence for the Existence of a Long-Lasting Metabolite.
In the past, studies of the distribution and metabolism of cannabinoids had been hampered by the lack of pure radiolabeled compounds.
In 1973, shortly after the radiolabeled cannabinoids became available, Dr. E. G. Leighty reported the recognition of some unidentified
metabolites that were retained in rat liver and spleen up to 15 days
after an acute intravenous or multiple subacute intraperitoneal
injections of either Δ^9-THC or Δ^8-THC (4). The metabolites could
be extracted by methanol and had a greater lipophilicity than
the parent THC. These long-retained metabolites were also found
in bone marrow (4) and represented an increasing percentage of
the total cannabinoids found in feces, but not urine (5).

Direct Demonstration of the Long-Lasting Metabolites as Fatty
Acid Conjugates. The lipophilicity of the unidentified metabolites
suggested that they may have corresponded to fatty acid-conjugated
metabolites (6). In order to produce quantities of the retained
metabolites sufficient for structural identification, large quanti-
ties of 11-hydroxy-Δ^8-THC (11-OH-Δ^8-THC) or 11-hydroxy-Δ^9-THC (11-
OH-Δ^9-THC) were incubated in a conventional in vitro rat liver
microsomal NADPH-generating system (6). A nonpolar metabolite
fraction was isolated from the in vitro system by thin-layer chromatography (TLC) and high-pressure liquid chromatography (HPLC).
Structural identification of the in vitro metabolites was achieved
by mass spectrometry (MS), proton magnetic resonance (PMR) spectrometry, and infrared (IR) spectrometry. It was found that the
metabolites were primarily conjugates of palmitic and stearic
acids with lesser amounts of C_{18}-unsaturated fatty acid conjugates
(presumably with oleic and linoleic acids). Cannabinoid metabolites
showing the same TLC mobilities and mass spectra were isolated
from rat liver after chronic intraperitoneal injections with [14]C-
labeled Δ^9-THC (6). In 1978, Yisak and co-workers also reported
the isolation and characterization of fatty acid conjugates of
4"-hydroxy-, 5"-hydroxy- and 7-hydroxycannabinol as in vivo metabolites of cannabinol isolated from rat feces (7). Gas chromatographic data indicated palmitate esters to be lesser components
compared with the oleates with a ratio of 1:5 in the rat excreta.
In 1980, another in vivo conjugation of fatty acids to a hydroxylated toxin was identified. Palmitic, stearic, oleic, and linoleic
fatty acid conjugates of a 2,2-bis(p-chlorophenyl)-ethanol metabolite (DDOH) were isolated from the livers and spleens of DDT-
treated male and female rats (8).

The Biochemical Mechanism of Formation

The In Vitro Esterase System. In earlier work 11-OH-Δ^9-THC and
11-OH-Δ^8-THC were incubated in an in vitro rat liver microsomal
NADPH-generating system (6). Despite the low yield, the presence
of the fatty acid-conjugated cannabinoids in the incubation mixture
led our laboratory to assume that this commonly used microsomal
mixed function oxidase system was involved in the conjugating
process. Soon, however, it was confirmed that an energy-dependent

microsomal esterase system, similar to that used to esterify long-
chain fatty acids to cholesterol, was involved (9-11). The system
showed a cofactor requirement for CoA and ATP with an optimum
pH range of 6.8-7.2 (9). The optimally developed incubation mixture
contained the following components per ml of 0.1 M sodium phosphate
buffer (pH 7.0): 1 µmole CoA, 5 µmoles ATP, 5 µmoles $MgCl_2 \cdot 6H_2O$,
microsomes containing 1 mg protein, and 0.5 µmole radiolabeled
hydroxylated metabolite. Fatty acids or fatty acyl CoA were not
required (9-11).

The Mechanistic Action of the Esterase System. There are three
known pathways for the biosynthesis of cholesterol esters: (a)
acyl-CoA: cholesterol O-acyltransferase (ACAT), the main intracellular cholesterol ester synthesis system (12), (b) lecithin:
cholesterol O-acyltransferase (LCAT), the principle cholesterol
esterifing enzyme in plasma (13), and cholesterol esterase which
can both synthesize and hydrolyze the cholesterol esters (14).
Although palmitoyl-CoA could substitute for CoA in our in vitro
system, considerably less 11-palmitoyl-Δ^9-THC was produced. This
observation was in agreement with the report of Swell et al. (10)
in which it was suggested that the mechanism for cholesterol conjugation with a fatty acid did not appear mainly to proceed directly
from the fatty acyl CoA to the fatty acid-conjugated cholesterol.
However, this was in contrast to the observations of Goodman and
colleagues (11) who reported that the cofactors ATP and CoA could
be completely replaced by the addition of preformed fatty acyl
CoA. Their results supported the hypothesis proposed a decade
earlier by Mukherjee et al. in 1958 (15) that cholesterol esterification proceeded by a simple condensation between cholesterol and
fatty acyl CoA.
     The 11-palm-Δ^9-THC can be hydrolyzed to 11-OH-Δ^9-THC by cho-
lesterol esterase and triacylglycerol lipase but not by phospholipase
A, acetylesterase or phosphotransacetylase (16). An attempt to
modify the retention of fatty acid-conjugated DDT metabolites
was carried out by injecting the DDT-treated rats with sodium
salt of various bile acids, heparin or lecithin of which all were
known to affect the esterification or ester hydrolysis by the
cholesterol esterase system. The results indicated a significant
decrease in the retention of the conjugated DDT metabolites in
the rat liver and spleen (17).
     The CoA-fortified microsomal enzyme system was inhibited
by SKF 525-A (β-diethylaminoethyl-diphenylpropyl acetate) and
by 2,4-dichloro-6-phenylphenoxyethylamine·HBr (DPEA) in a dose-
dependent fashion (18). Both are known inhibitors of the classical
hepatic microsomal mixed-function oxidase system. This inhibition
has been suggested to involve a nonspecific binding to liver microsomal proteins and/or phospholipids (18,19). Binding to phospholipids would support the hypothesis proposed by Swell and co-workers
(10) that fatty acyl CoA derivatives were not necessarily the
only immediate source for the fatty acid moieties of the conjugates;
that phospholipids may act as the fatty acids reservoir in the
CoA-fortified enzyme system. However, it is also known that cholesterol esterification is very sensitive to the fluidity of the
microsomal membrane (12). The reduced conjugation produced when

a palmitoyl-CoA was substituted for CoA and ATP, and the inhibition
elicited by SKF 525-A and DPEA do not necessarily rule out the
involvement of the ACAT enzyme system in the fatty acid conjugation
to xenobiotic alcohols.

The possibility of nonenzymatic esterification of free fatty
acid by the hydroxylated metabolite was ruled out by two experiments:
(a) hydroxylated metabolites were incubated with washed microsomes
with no added cofactors, and (b) hydroxylated metabolites were
incubated with different concentrations of fatty acid in buffer
without microsomes. In both cases, no detectable fatty acid conjugated metabolites were found by TLC analyses (6).

Other Fatty Acid Conjugates Produced by the In Vitro System.
The following (hydroxylated) compounds have been successfully
conjugated to fatty acids with the in vitro CoA-fortified rat
liver microsomal preparations: N-(1-phenylcyclohexyl)-4-hydroxy-
piperidine and N-(1-phenyl-4-hydroxycyclohexyl)piperidine (20),
2,2-bis(p-chlorophenyl)ethanol (DDOH) (8), codeine (21), trichloroethanol (22), pentachlorophenol (23), 11-OH-Δ^8-THC and 11-OH-Δ^9-THC
(6), cannabinol (CBN), 11-OH-CBN, cannabidiol, 8β-OH-Δ^9-THC, 8β,11-
diOH-Δ^9-THC, 8α-OH-Δ^9-THC, and 8α,11-diOH-Δ^9-THC (24). Two metabolites of cyclamate, cyclohexylamine and N-cyclohexylhydroxylamine
were also shown to conjugate with palmitic acid in vitro at both
nitrogen and oxygen atoms with a preference for the nitrogen in
the latter metabolite (25). The fatty acid conjugates of the
above mentioned cannabinoid compounds (24) were identified and
distinguished from other metabolites of cannabinoids by their
migratory behavior in two different TLC systems.

The results of these in vitro studies demonstrate that conjugation of fatty acids can occur not only with a hydroxyl functional
group but also with an amine group. In addition, the isolation
of the in vivo fatty acid-conjugated metabolites of DDT, cannabinol
and Δ^9-THC and Δ^8-THC further supports that the in vitro CoA-fortified liver microsomal system could be used as an assay method
to predict whether certain compounds, or their metabolites, are
retained in vivo in lipid-containing tissues as fatty acid conjugates.

Studies on the Biological Effects of Some Persistent Fatty Acid
Conjugates

Effects of 11-palm-Δ^9-THC on the Hepatic Microsomal Drug Metabolizing Enzyme System. A comparison of the effects of 11-palm-Δ^9-THC
with Δ^9-THC and 11-OH-Δ^9-THC on the hepatic microsomal drug metabolizing enzyme system was performed and reported in 1979 (26).
With a treatment of three intraperitoneal injections of 50 mg
cannabinoid per kilogram body weight per day, it was found that
11-palm-Δ^9-THC did not affect the metabolism of either Type I
or Type II substrates but appeared to affect benzpyrene hydroxylase
activity and that cytochrome P-448 was apparently increased two-fold
by treatment with the conjugate (Table 1).

Table I.   Effects of 11-palm-Δ^9-THC on the rat hepatic microsomal
enzyme system

Treatment	Amino-pyrine	Ethyl-morphine	Aniline Hydroxylase	Benzpyrene Hydroxylase
Vehicle (10% Tween 80 in Saline)	100 ± 3%	100 ± 11%	100 ± 12%	100 ± 3%
11-OH-Δ^9-THC (50 mg/kg x 3, ip)	25 ± 7%	70 ± 9%	40 ± 13%	197 ± 32%
11-palm-Δ^9-THC (50 mg/kg x 3, ip)	95 ± 16%	103 ± 10%	97 ± 10%	203 ± 23%

Effects of 11-palm-Δ^9-THC on Behavioral and Neurochemical Parameters.
Studies were conducted to evaluate the effects of 11-palm-Δ^9-THC
on behavioral and neurochemical parameters in male Sprague-Dawley
rats.   Preliminary results indicated that 11-palm-Δ^9-THC produced
catalepsy and analgesia in intravenously injected animals.   These
effects were similar to those seen with 11-OH-Δ^9-THC but were
less pronounced.   Neither 11-OH-Δ^9-THC nor 11-palm-Δ^9-THC exerted
an effect on biogenic amine levels or catecholamine synaptosomal
uptake by this route of administration.   When 11-palm-Δ^9-THC was
given intracisternally, behavioral changes were observed soon
after injection and suggested that the fatty acid conjugate is
psychoactive (27).   Since it has been shown that 11-palm-Δ^9-THC
is hydrolyzed in vivo to 11-OH-Δ^9-THC (28), it is also possible
that the behavioral effects seen with the fatty acid conjugate
are the result of hydrolytic conversion to 11-OH-Δ^9-THC, the primary
psychoactive metabolite of marihuana (29).

Effects of Palmitoylcodeine on Behavioral and Neurochemical Parameters.   The fatty acid conjugate of codeine, namely palmitoylcodeine,
was also found to induce behavioral changes in the rat.   Following
intravenous injection of palmitoylcodeine, animals were observed
to exhibit catalepsy, analgesia and hypoactivity, responses similar
to those seen with codeine but to a lesser extent.   Neither codeine
nor palmitoylcodeine affected biogenic amine levels.   While palmitoylcodeine had no effect on catecholamine synaptosomal uptake,
codeine treatment was found to decrease striatal dopamine uptake.
The behavioral results suggest that the conjugate palmitoylcodeine
is also psychoactive in the rat (30).   Studies have yet to be
conducted to determine if palmitoylcodeine is hydrolyzed to codeine
either in vitro or in vivo.

Physiological Significance of Conjugation with Fatty Acids

Our previous studies indicated that the same metabolic pathway(s) involved in the microsomal esterification of cholesterol was also involved in the fatty acid conjugation of many hydroxylated metabolites and compounds. We have also demonstrated that these fatty acid conjugates, specifically 11-palm-Δ^9-THC, 11-palm-Δ^8-THC, and palmitoyl-DDOH, are preferentially retained over their metabolites in the tissues of rats, primarily in the spleen, liver, and fat. While TLC analysis (30) indicated that some fatty acid conjugates of codeine were retained in these same rat tissues, the retention of the radiolabeled phencyclidine (PCP) was confined mainly to the lung. Unfortunately there was not sufficient tissue available for structure identification of the retained codeine metabolites and PCP metabolites. Interestingly, a search for the fatty acid conjugated Δ^9-THC in dog tissues indicated that it was not present and if the same mechanism for drug retention does exist in the dog, it must be negligible (30).

The concern that chronic and long-term use of a drug could increase the concentration of retained fatty acid-conjugated metabolites to a level sufficient to exert adverse biological effects will exist until scientists prove or disprove the existence of the same retention pathway in humans. To address the concern that these conjugates and other long lasting derivatives of drugs or various pollutants such as DDT can be biologically active in various species, more research is warranted.

Postscript: A fatty acid-conjugated pesticide palmitoyl pentachlorophenol was isolated from human fat and reported in the literature after this manuscript was presented at the 189th National ACS Meeting (31).

Acknowledgments

The authors would like to express their gratitude to Drs. P. J. Kurtz and R. Deskin for their helpful suggestions in the designing of neurochemical and neurobehavioral studies. This work was supported in part by a grant from the National Institute on Drug Abuse, No. DA00793-07.

Literature Cited

1. Misra, A.L.; Pontani, R.B.; Bartolomeo, J. Res. Comm. Chem. Pathol. Pharmacol. 1979, 24, 431-45.
2. Matthews, H.B.; Birnbaum, L.S. In "Human and Environmental Risks of Chlorinated Dioxins and Related Compounds"; Tucker, R.E.; Young, A.L.; Gray, A.P., Eds.; ENVIRONMENTAL SCIENCE RESEARCH Vol. 26, Plenum Press: New York, 1983; pp. 463-75.
3. Youngs, W.G.; Gutenmann, W.H.; Lisk, D.J. Environ. Sci. Technol. 1972, 6, 451-52.
4. Leighty, E.G. Biochem. Pharmacol. 1973, 22, 1613-21.
5. Klausner, H.A.; Dingell, J.V. Life Sci. 1971, 10/I, 49-59.

6. Leighty, E.G.; Fentiman, A.F. Jr.; Foltz, R.L. Res. Comm.
    Chem. Pathol. Pharmacol. 1976, 14, 13-28.
7. Yisak, W.; Agurell, S.; Lindgren, J.E.; Widman, L.M.;
    J. Pharm. Pharmac. 1978, 30, 462-63.
8. Leighty, E.G.; Fentiman, A.F. Jr.; Thompson, R.M. Toxicology
    1980, 15, 77-82.
9. Leighty, E.G. Res. Comm. Chem. Pathol. Pharmacol. 1979,
    23, 483-92.
10. Swell, L.; Law, M.D.; Treadwell, C.R. Arch Biochem. Biphys.
    1964, 104, 128-38.
11. Goodman, D.S.; Deykin, D.; Shiratori, T. J. Biol. Chem. 1964,
    239, 1335-45.
12. Suckling, K.E. Biochem. Soc. Trans. 1983, 11, 651-53.
13. Assmann, G.; Jabs, H.U. Biochem. Soc. Trans. 1985, 13, 19-20.
14. Treadwell, C.R.; Vahouny, G.V. In "Handbook of Physiology";
    Code, C.F., Ed., American Physiological Society: Washington,
    D.C., 1968; 1407-38.
15. Mukherjee, S.; Kunitake, G.; Alfin-Slater, R.B. J. Biol.
    Chem. 1958, 230, 91-96.
16. Leighty, E.G. Res. Comm. Chem. Pathol. Pharmacol. 1979, 24,
    393-96.
17. Leighty, E.G. Res Comm. Chem. Pathol. Pharmacol. 1981, 31,
    69-75.
18. Leighty, E.G. Biochem. Pharmacol. 1980, 29, 1071-73,
19. Gillette, J.R. Adv. Pharmacol. 1966, 4, 219-61.
20. Leighty, E.G.; Fentiman, A.F. Jr. Res. Comm. Substance Abuse
    1980, 1, 139-48.
21. Leighty, E.G.; Fentiman, A.F. Jr. J. Pharm. Pharmacol. 1983,
    35, 260-1.
22. Leighty, E.G.; Fentiman, A.F. Jr. Res. Comm. Chem. Pathol.
    Pharmacol. 1981, 32, 569-72.
23. Leighty, E.G.; Fentiman, A.F. Jr. Bull Environm. Contam.
    Toxicol. 1982, 28, 329-33.
24. Leighty, E.G. Res. Comm. Substance Abuse 1980, 1, 169-75.
25. Leighty, E.G.; Fentiman, A.F. Jr. Chem. Toxicol. 1983, 21,
    251-54.
26. Leighty, E.G. Res. Comm. Chem. Pathol. Pharmacol. 1979, 25,
    525-35.
27. Haggerty, G.C.; Deskin, R.; Kurtz, P.J.; Fentiman, A.F. Jr.;
    Leighty, E.G. Presented at the 24th Annual Meeting of the
    Society of Toxicology, 1985. (Manuscript submitted.)
28. Leighty, E.G. Res. Comm. Substance Abuse 1980, 1, 49-63.
29. Christensen, H.D.; Freudenthal, R.I.; Gidley, J.T.; Rosenfeld,
    R.; Boegli, G.; Testino, L.; Brine, D.R.; Pitt, C.G.; Wall,
    M.E. Science 1971, 172, 165-67.
30. Chang, M.J.W.; Leighty, E.G.; Haggerty, G.C.; Fentiman, Jr.,
    A.F. unpublished data.
31. Ansari, G.A.S.; Britt, S.G.; Reynolds, E.S. Bull. Environ.
    Contam. Toxicol. 1985, 34, 661-67.

RECEIVED July 29, 1985

Novel Polar Xenobiotic Conjugates

Gary B. Quistad

Biochemistry Department, Zoecon Corporation, Palo Alto, CA 94304

This paper summarizes the occurrence of unusual xenobiotic conjugates which are substantially ionized in
animals and plants. Such conjugates generally are
deemed "unusual" either because they occur rarely or
because their existences seldom are documented, often
by virtue of recent discoveries.
    Although the unique features of polar xenobiotic
conjugates often defy a facile assignment of the
compounds to general categories, certain classes of
conjugates are discussed. For animals, the discussion focuses on unusual conjugates of urea, carnitine, amino acids, phospholipids, bile acids, and
multiple conjugates of foreign compounds. For
plants, conjugates containing unusual sugars, amino
acids, and malonic acid are featured.

There are numerous reviews of conjugation biochemistry available
(1-8). These reviews contain miscellaneous unusual conjugates
interspersed with a primary discussion of the major conjugative
pathways. The bias of most reviews is on conjugations in animals
although several reviews of conjugates in plants are available also
(9-12). There are even reviews devoted to novel conjugation
reactions (13-15). This paper emphasizes unusual conjugates which
have not been reviewed adequately elsewhere and provides references
to other uncommon, but well-known conjugation pathways.

Animals

Amino Acids and Peptides. Xenobiotic conjugates have been reported
for about two-thirds of the common α-amino acids. In general,
glycine is the single most common amino acid utilized in metabolic
conjugation although certain animal groups exhibit preferential
usage of other amino acids (16). For example, ornithine conjugates
are common in reptiles and some birds (Galliformes and
Anseriformes, e.g. chickens, turkeys, ducks) while other birds
(Columbiformes, e.g. pigeons) use glycine (16). Huckle et al. (17)

0097–6156/86/0299–0221$06.00/0

have shown that the major metabolite of 3-phenoxybenzoic acid from chickens was an N-acetylornithine conjugate (8% of the applied dose). Methionine adducts have been postulated as the source of methylthio groups in metabolites of phenanthrene and several other xenobiotics (18 and references therein). Jaglan and Arnold (19) proposed that nucleophilic displacement of a chlorine in dichloran by methionine gave a sulfonium ion adduct (Figure 1) which represented up to 50% of the ^{14}C-residues in goat milk after dosage with [^{14}C]dichloran. This unusual structure was supported by mass spectral and acid hydrolysis data. Some Arthropods (e.g. scorpions, spiders) utilize arginine which can also be decarboxylated to an agmatine conjugate (4,16). There are reports of conjugation with less abundant amino acids such as histidine; benzoylhistidine is formed in the fruit-eating bat (20). Valine and threonine conjugates were found in rat feces following treatment with fluvalinate (21). Thus, with more sophisticated instrumentation it is likely that soon conjugates of all natural amino acids will be found although most will be present at trace levels.

Taurine and carnitine are two less common amino acids which are becoming more widely recognized as important in conjugative metabolism. Only a few years ago there were few examples of taurine conjugation, but now there are frequent reports involving this sulfonic acid (22). Since our reports of carnitine conjugation with cyclopropanecarboxylic acid in rats, dogs, mites, and a cow (23-26), several other acids have been shown to react with carnitine (Figure 2). Pivaloylcarnitine has been isolated as a major metabolite from the urine of humans dosed with a methyldopa prodrug (27). Interestingly, this carnitine conjugate occurs in humans, but not other species which excrete pivalic acid as a glucuronide. Valproylcarnitine has been isolated from the urine of children treated with valproic acid, a widely used anticonvulsant (28).

Excluding conjugates derived from glutathione adducts, few xenobiotic acids are reported to form peptide conjugates involving more than one amino acid, and these examples have been reviewed amply (2,16,29). Cats produce glycylglycine and glycyltaurine conjugates from both quinaldic and kynurenic acids. Rats excrete a dipeptide conjugate of DDA (from DDT) containing aspartic acid and serine. 3-Phenoxybenzoic acid is excreted as a glycylvaline dipeptide conjugate in Mallard ducks (29) and chickens (17). The antineoplastic analogs methotrexate and aminopterin form polyglutamates in rats, mice, and man (15). Recently four peptide conjugates were identified from the urine of calves dosed with phenothiazine (30).

Although numerous structural types of metabolites are derived from reactivity with glutathione (e.g. 31), several recent examples of novel mercapturates have been reported (Figure 3). 4-Cyano-N,N-dimethylaniline is the substrate for formation of two unusual mercapturates (32-34). After monodesmethylation, a stable thioether mercapturate is produced (presumably via the N-methylol) as a major urinary metabolite by rats and mice (15 and 33% of the applied dose, respectively). Rats also convert up to 10% of the applied 4-cyano-N,N-dimethylaniline to a mercapturate arising from the reaction of glutathione with an acetyl group (possibly via a glycolylaniline). The corresponding methylsulfonyl derivative from the mercapturate was recovered in urine also (33). Mercapturic

acid sulfoxides are produced by rats from propachlor (35,36) and N-
acetyl-S-(2,3,5,6-tetrachlorophenyl)cysteine (37). The first acyl-
linked mercapturate found in human urine was reported for clofibrate (38). A disulfide with cysteine was the major metabolite of
captopril in humans (39).

For many years the formation of premercapturic acids has been
recognized for the metabolism of aromatic compounds (1,3). These
hydroxydihydro derivatives of aromatic systems are acid labile and
revert to more stable aromatic mercapturates unless precautions are
taken during isolation. Work with phenanthrene (18) and naphthalene (40,41) has demonstrated that premercapturates are metabolized
further in a manner similar to mercapturates (i.e. to mercapto-
acetic, mercaptolactic, and methylthio compounds, each retaining
the hydroxydihydro character of one of the original aromatic rings;
Figure 4). In rats these secondary metabolites of premercapturates
are produced primarily by the intestinal microflora (41).

Amides and Urea Adducts. There are only a few examples of amide
formation from xenobiotic acids (Figure 5). In 1969 it was
reported that humans and rabbits convert metiazinic acid to an
amide (42). Subsequently, amide metabolites have been identified
for clofibrate (43), anthranilic acid (44), and an anilino acid
(21). It is not surprising that such amides are isolated considering that nicotinic acid is converted readily to nicotinamide
in healthy animals and numerous natural peptides contain amide
functionality at the C-terminus. It is still unclear whether
amides are produced enzymatically or merely chemically. If ammonia
is responsible for amide formation, it may also contribute to the
production of an unusual acetylated amino acid from n-butylglycidyl
ether (45, Figure 5). The acetylated amino acid represented 23% of
the urinary metabolites from rats and presumably arose by ammonolysis of the epoxide followed by oxidation and acetylation. Recently
it was shown in earthworms that spermine, a polyamine, forms an
amide with 3-phenoxybenzoic acid from cypermethrin (46).

Several urea adducts have been identified in metabolism
studies of xenobiotics (Figure 6). The addition of urea to isoproturon (47), formaldehyde (48), and oxazapam (49) is chemically
mediated and occurs when primary metabolites remain in urine for
prolonged periods. In the case of caroxazone (50), it is unclear
whether the urea adduct is a chemical artifact or is produced
enzymatically. The strongest evidence that urea adducts can be
genuine metabolites comes from work with rosaramicin (51). This
macrolide antibiotic contains an aldehyde group that reacts with
two equivalents of urea to form 20-bisureidorosaramicin which is
excreted in urine. The bisureido metabolite did not form when
rosaramicin was incubated with urine in vitro. One may expect urea
adducts from certain ketones also since 11-β-hydroxy-4-androstene-
3,17-dione is converted by humans to a ureasterone. The 3α-ureido
steroid forms nonenzymatically from the glucuronide in urine at pH
5 (52).

Glycosides. Since glucuronide formation is often the predominant
metabolic pathway in animals, this process has been reviewed
frequently (e.g. 3 and references therein). The last decade has

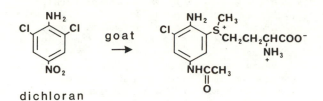

dichloran

Figure 1. Formation of a methionine conjugate.

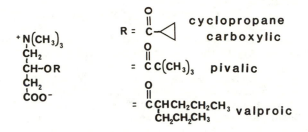

Figure 2. Carnitine conjugates.

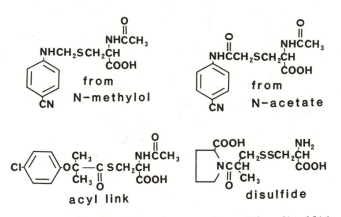

from
N-methylol

from
N-acetate

acyl link

disulfide

Figure 3. Unusual mercapturates and a disulfide.

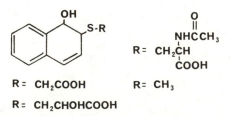

R = CH₂COOH   R = CH₃

R = CH₂CHOHCOOH

Figure 4. Premercapturate metabolites from naphthalene.

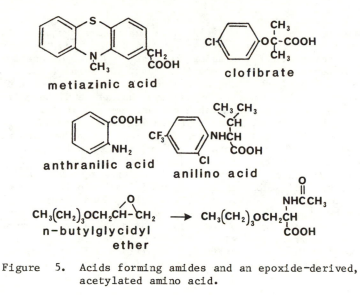

metiazinic acid

clofibrate

anthranilic acid

anilino acid

$CH_3(CH_2)_3OCH_2CH\text{-}CH_2$
n-butylglycidyl
ether

$\longrightarrow$

$CH_3(CH_2)_3OCH_2CH$
COOH
NHCCH$_3$

Figure 5. Acids forming amides and an epoxide-derived, acetylated amino acid.

caroxazone

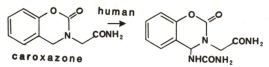

human $\longrightarrow$

CONH$_2$
NHCONH$_2$

oxazapam

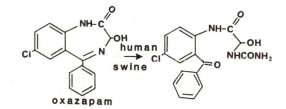

human
swine $\longrightarrow$

isoproturon

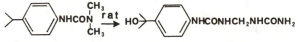

rat $\longrightarrow$

HO—〉—NHCONHCH$_2$NHCONH$_2$

$\overset{O}{\underset{\parallel}{HCH}}$ $\overset{rat}{\longrightarrow}$ $\overset{O}{\underset{\parallel}{NH_2CNHCH_2OH}}$ + $\overset{O}{\underset{\parallel}{HOCH_2NHCNHCH_2OH}}$

human   NHCONH$_2$
RCHO $\longrightarrow$ RCH
NHCONH$_2$
bisureido adduct

Figure 6. Urea adducts.

witnessed several examples of quaternary ammonium N-glucuronides
which have been well reviewed (13,53). Quaternary N-glucuronides
are known for the following drugs: cyproheptadine (13), amitripty-
lene (53), imipramine (53), chlorpromazine (53), tripelennamine
(13), cyclobenzaprine (13), ketotifen (54), clozapine (15), a
thiazoleacetic acid (55), and an analgesic with a pyridine ring
(13; Figure 7). It should be emphasized that these quaternized
glucuronides are not minor metabolites and are particularly
abundant in human urine, often representing more than 25% of the
applied dose (56). Quaternary N-glucuronides can be synthesized by
rabbit hepatic UDPGA transferase (57) and fast-atom bombardment
mass spectrometry is useful in structure identifications (57,53).

The formation of C-glucuronides (i.e. direct C-C linkage) has
been reviewed (13). Most of the reported C-glucuronides are derived from drugs which are relatively acidic. Thus, C-glucuronides
are produced from 1,3-dicarbonyl systems (sulfinpyrazone, phenylbutazone, feprazone) and an acetylenic compound (ethychlorvynol;
Figure 7). An aromatic C-glucuronide was identified from mice
after dosage with Δ^6-tetrahydrocannabinol (58). This aromatic C-
glucuronide is particularly interesting since the carbon-carbon
bond forms at a non-acidic site and an adjacent phenolic hydroxyl
is by-passed during glucuronidation.

An unusual carbamylglucuronide is excreted by humans given the
antiarrhythmic drug tocainide (59; Figure 8). This glucuronide has
acquired an extra carbonyl group of unknown origin prior to glucuronidation and is degraded further to a cyclic hydantoin derivative
(60). An unusual glucuronylglucuronide is excreted by dogs given
an intravenous dose of carazolol (61, Figure 8).

While glucuronides are by far the most prevalent glycoside
conjugates in higher animals, there are a few examples of conjugation with other sugars, most of which have been reviewed (13,3).
O-Glucosides are formed by steroids, bilirubin, and 4-nitrophenol.
N-Glucosides arise from amobarbitone, phenobarbitone, 5-(4-
pyridyl)-3-(4-pyrimidinyl)-1,2,4-triazole, and sulfamethazine
(62,63). The N-glucoside from sulfamethazine (Figure 9) is particularly interesting since it was a major product in swine tissue
(liver, muscle), but not in excreta. The genesis of this glucoside
is somewhat uncertain since sulfamethazine reacts with endogenous
components of the liver in vitro to generate the glucose adduct
(62). N-Ribosides have been reported for pyrazole, 2-hydroxynico-
tinic acid, and 6-mercaptopurine (64; Figure 9). Bilirubin is
reported to conjugate with xylose in several animal species (see 5)
and conjugates are known with N-acetylglucosamine for several
steroids (65).

In contrast to most animals, insects form glucosides, not
glucuronides. Ngah and Smith (66) have shown that six insect
species in five different orders convert 4-nitrophenol to a
glucoside-6-phosphate (Figure 10). This reaction seems to be
limited to phenols since similar conjugates were not found with
benzoic acid. Tenebrio larvae also are believed to convert 4-
nitrophenol to a glucoside-6-sulfate.

Sulfates and Phosphates. There are numerous reviews of sulfation,
a quantitatively important conjugation process (3,67-69). There are

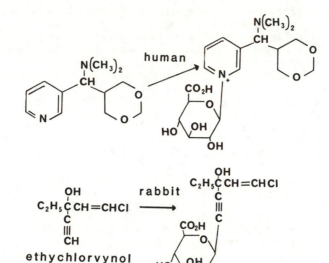

Figure 7.   Quaternary ammonium C- and N-glucuronides.

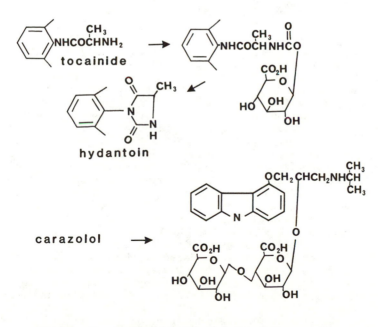

Figure 8.   Carbamyl- and glucuronylglucuronides.

natural products with the following types of sulfate bonding: $P-O-SO_3^-$, $C-O-SO_3^-$, $N-SO_3^-$, $N-O-SO_3^-$, and $S-SO_3^-$ (70). Of these possibilities, most xenobiotic sulfates reported to date are of the type $C-O-SO_3^-$, particularly with aryl substrates. There are several examples of $N-SO_3^-$ conjugates (i.e. sulfamates). The best known sulfamates are from aniline, 4-aminobenzoic acid, and 2-naphthylamine (see 1,4,6, and 7 for reviews). The stability of some sulfamates is questionable (3,4). Sulfamate formation is a major metabolic pathway for pentachloronitrobenzene in goats which excrete in urine up to 30% of the applied dose as the sulfamate of the corresponding aniline (Figure 11, 71). The first report of a sulfamate from a nonaromatic, alicyclic amine involves the drug tiaramide (72). While female mice excreted up to 28% of the dose of the tiaramide as the sulfamate, males excreted only 6%. Sulfation of an N-oxide is reported in rat liver for minoxidil (Figure 11) which is believed to react via a tautomer (73,74).

While formation of phosphate conjugates is well-known for insects (75,76), this pathway is documented rarely for other animals. Ngah and Smith (66) have shown that eight insect species in five different orders can produce phosphates from 4-nitrophenol. The mealworm, Tenebrio, also phosphorylates 8-quinolinol, 1-naphthol, and 4-methylumbelliferone. The few known examples of phosphates from xenobiotics in mammals have been reviewed (2) and include: phenol (cat, sheep), 2-naphthylamine (dog, human), and ethanol (rat). 2-Naphthylamine is converted into an unusual conjugate, bis(2-amino-1-naphthyl)phosphate.

Polar Lipids. Several xenobiotic acids have been shown to occur in tissue as modified phospholipids where a natural fatty acid is replaced by the foreign acid (Figure 12). When a beagle dog was dosed with cycloprate (24), 7% of the applied dose remained in the kidney after four days. Two-thirds of that radiolabel was associated with phospholipids and consisted largely of phosphoglycerides containing cyclopropanecarboxylic acid and its homologs. Phospholipases were used to characterize those unnatural phosphoglycerides. DeKant et al. (77) have suggested that trichloroethylene is converted to an amide from glyoxalic acid and ethanolamine via a phosphoglyceride (Figure 13). D,L-2-Fluoropalmitic acid accumulates in membrane lipids of cultured mammalian cells (78). The fluoropalmitate was incorporated without modification into phosphatidylcholine (1.6%) and sphingomyelin (0.6%), but not into other phospholipids or neutral lipids. Leblanc et al. (79) showed that L-cells incorporated tritiated ω-diazirinophenoxy derivatives of fatty acids into the major phospholipids (Figure 12). Irradiation with light allowed photoaffinity labeling of proteins which are normally acylated by fatty acids. Dodecylcyclohexane underwent extensive ω-oxidation in rats to afford cyclohexyldodecanoic acid which incorporated into both neutral and phospholipids in liver and fat (80).

Free radicals from certain halogenated xenobiotics incorporate into phospholipid. Trudell et al. (81,82) have shown that free radicals from carbon tetrachloride and halothane add to the double bonds of fatty acyl chains of phospholipids in the membrane surrounding cytochrome P-450. Oleic acid moieties were converted

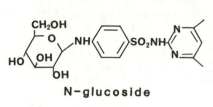
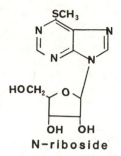

N-glucoside

from sulfamethazine

N-riboside

from 6-mercaptopurine

Figure   9.   N̲-Glycosides from sulfamethazine and 6-mercaptopurine.

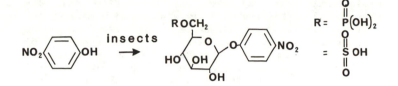

Figure 10.   Glucoside-6-phosphate and glucoside-6-sulfate from p-nitrophenol.

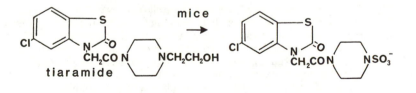

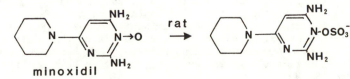

Figure 11.   Sulfamates from pentachloronitrobenzene and tiaramide, and a sulfate from minoxidil.

to a mixture of 9- and 10-(1-chloro-2,2,2-trifluoroethyl)stearates
for halothane while carbon tetrachloride gave a mixture of 9- and
10-(trichloromethyl)stearates. Link et al. (83) suggest that free
radical adducts may be responsible for cross-linkage of fatty acid
chains in phospholipids.

      Bile acid conjugation with a xenobiotic acid is a recently
discovered metabolic pathway (84). An anilino acid from the
pyrethroid insecticide fluvalinate acylates the 3-position of
natural bile acids in rats (85), chickens (86), and a cow (87), and
these conjugates represent 5-12% of the ^{14}C-residue in feces
(Figure 14). Eight different bile acids have been shown to
conjugate with this anilino acid in these animal species, but so
far no other xenobiotic acids have been reported to follow a
similar pathway.

Miscellaneous and Multiple Conjugates. Several aromatic amines are
converted to acidic amidates. Rabbits transform 18% of the applied
4-chloroacetanilide into an oxamic acid derivative (88). The oxa-
mate of sulfanilamide (Figure 15) is also a major metabolite from
rabbits (89). The oxamates are believed to arise from N-acetyl
groups which are oxidized to glyoxamates and subsequently to oxamates. Only one example of conjugation with succinic acid is known
and that involves the anorectic, 4-chloro-2-(ethylamino)propio-
phenone, which is oxidized to the desethyl metabolite prior to
conversion to the amide by succinic acid (90).

      α-Keto acids react with amino groups in xenobiotics to form
Schiff bases. Hydralazine (a hydrazine) and isoniazid (a
hydrazide) react with α-ketoglutaric and pyruvic acids (91,92).
Formation of these adducts may be purely chemical (i.e. not
enzymatic).

      Egger et al. (93) have reported that methylphenidate is
carbamylated by dogs and rats (Figure 16). Carbamylated methylphenidate was excreted in urine and represented 1% of the applied
dose in both rats and dogs. An additional 1% of the dose was
identified as the ester-hydrolyzed, carbamylated metabolite in rat
urine.

      The additions of several 1- and 2-carbon fragments to
xenobiotics are widely recognized as conjugation reactions. These
include formylation, methylation, acetylation, hydroxymethylation,
and glycolic acid formation. Since these conjugations result in
less polar or only slightly more polar (but not ionized) compounds,
they will not be considered further in this review. However, some
of the above occur with metabolites which exist as double
conjugates.

      Conjugation usually renders xenobiotics sufficiently polar for
facile excretion, but occasionally double conjugates are produced
(Table I). Two of the most unusual double conjugates arise from
dichloran (Figure 1) and 4-dimethylaminoazobenzene (Figure 17).
The sulfate-glutathione adduct represented 27% of the biliary
metabolites from 4-dimethylaminoazobenzene. Another novel double
conjugate involves viloxazine which is converted to a sulfate-
hippurate (98,99).

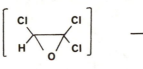

Figure 12.   Xenobiotic acids which incorporate into phospholipids.

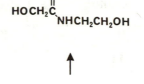

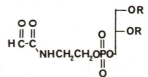

Figure 13.   Conversion of trichloroethylene into an amide _via_ a phospholipid derivative.

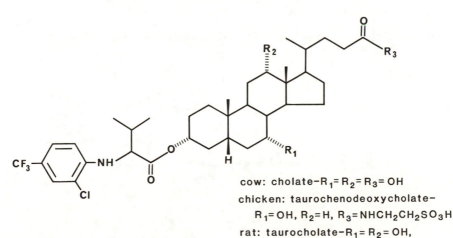

cow: cholate–$R_1 = R_2 = R_3 = OH$

chicken: taurochenodeoxycholate–
   $R_1 = OH$, $R_2 = H$, $R_3 = NHCH_2CH_2SO_3H$

rat: taurocholate–$R_1 = R_2 = OH$,
   $R_3 = NHCH_2CH_2SO_3H$

Figure 14.   Bile acid conjugates of an anilino acid.

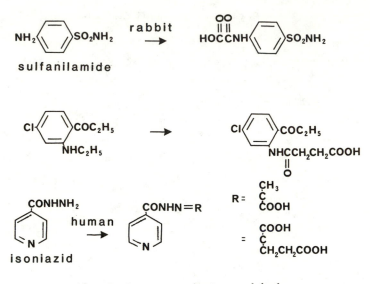

Figure 15. Oxalate, succinate, and hydrazones.

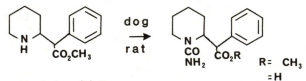

methylphenidate

Figure 16. Carbamylated metabolites from methylphenidate.

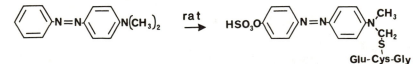

Figure 17. A sulfate-gluthathione adduct of 4-
dimethylaminoazobenzene.

Table I. Double Conjugates in Animals

Substrate	Conjugate	Ref.
diethylstilbestrol	sulfate - glucuronide	94
	diglucuronide	75
4-hydroxybenzoic acid	diglucuronide	6
biphenyl sulfate	sulfate-glucuronide	75
4-cyclohexylphenyl sulfate	sulfate-glucuronide	13
2-hydroxy-5-nitrophenyl sulfate	sulfate-glucuronide	1
phenoltetrabromophthalein	sulfate-glucuronide	1
monosulfate	sulfate-glutathione	
4-dimethylaminoazobenzene	O-sulfate-C-glutathione	95
1-bromopentane	O-sulfate-glutathione	95
2-naphthylamine	O-sulfate-N-formyl	4
dapsone	N-acetyl-N-sulfamate	14
o-toluidine	N-acetyl-N-glucuronide	13
phosgene	diglutathione	96
chloroform, bromotrichloromethane,	diglutathione	96,97
and carbon tetrachloride		
dichloran	acetyl-methionine	19
viloxazine	sulfate-hippurate	98,99

Plants

Amino Acids and Peptides. The formation of peptides from xenobiotic acids and single amino acids is a well-recognized metabolic
pathway, particularly for plant callus cultures (100). Although
numerous amino acids form such conjugates, aspartic acid is perhaps
the most common (101,102). Berlin et al. (103) reported an unusual
putrescine conjugate with 4-fluorocinnamic acid from L-4-fluoro-
phenylalanine (Figure 18). This putrescine conjugate is a major
metabolite in tobacco cell cultures, but it may be specific for
Nicotiana species which contain caffeoyl and feruloyl putrescine as
the main phenolic natural products. Two α-alanine conjugates have
been reported for the herbicide 3-amino-1,2,4-triazole (beans,
tomatoes; 11) and for the insecticide methidathion (tomatoes, soil;
104). The α-alanine conjugate from methidathion (Figure 18) represented 84% of the residue after seven days in tomatoes treated with
5-methoxy-2-oxo-3-(2H)-1,3,4-thiadiazole, a major metabolite of
methidathion. The mechanistic origin of these α-alanine conjugates
is unclear.
     Glutathione conjugates and related metabolites are common in
many agricultural crops (31). Almost the whole spectrum of
metabolic possibilities from glutathione-derived products occurs
with pentachloronitrobenzene which forms the following conjugates
(105,106): cysteine, dicysteine, diglutathione, malonylcysteine,
lactate, acetate, and methylthiol. Cysteine conjugates have also
been identified after treatment of sorghum with atrazine (107) and
tomatoes with methidathion (108,109). The cysteine conjugate of
atrazine is converted further by sorghum to an unusual lanthionine
conjugate (107, Figure 19). Lamoureux and Rusness (110) have shown
that malonylcysteine conjugates occur generally for numerous pesti-

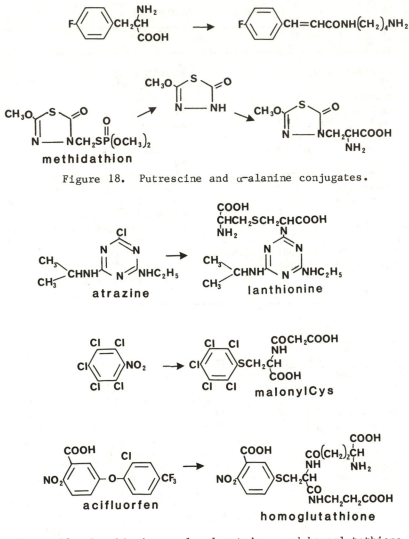

Figure 18. Putrescine and α-alanine conjugates.

Figure 19. Lanthionine, malonylcysteine, and homoglutathione
conjugates.

cides and plant species. Certain legumes such as soybeans use
homoglutathione (γ-Glu-Cys-β-Ala) instead of glutathione. Homoglutathione conjugates have been reported for acifluorfen (111,
Figure 19), metribuzin (112), and acetochlor (113).

Glycosides. There are several reviews of sugar conjugates from
pesticide application to plants (9,10,12,114). By far the most
commonly identified glycosides are β-D-glucosides (9). An exception to this generality is the formation of an α-D-glucoside from
chloramben by foxtails and barley (115). Other identified
glycosides include disaccharides, trisaccharides, polysaccharides,
glucosamines, and malonyl glucosides (Table II). Two unusual β-D-

Table II.   Unusual Glycosides from Xenobiotics in Plants

Sugar	Aglycone	Plant	Ref
disaccharide glucosylxylose & glucosylarabinose	D-p-fluorophenylalanine	tobacco	103
	3-(2',2'-dichlorovinyl) -2,2-dimethylcyclopropane- carboxylic acid	cotton	117
	3-phenoxybenzoic acid	vine	118
		cotton	119
	3-phenoxybenzyl alcohol	cotton	120
	2-(4-chlorophenyl)-3- methylbutyrate	cotton	121
gentiobiose	3-phenoxybenzoic acid	tomato, cotton, cabbage, kidney bean, cucumber	119
	2-(4-chlorophenyl)-3- methylbutyrate	tomato, cotton, cabbage, kidney bean, cucumber	121
	picloram	spurge	122
	diphenamid	tomato	10
cellobiose	3-phenoxybenzoic acid	tomato	119
sophorose	2-(4-chlorophenyl)-3- methylbutyrate	cotton,tomato	121
triglucoside	3-phenoxybenzoic acid	tomato	119
	2-(4-chlorophenyl)-3- methylbutyrate	tomato	121
	naphthylene acetic acid		9
polysaccharide	oxime from oxamyl	peanut	123
malonyl glucoside	3-phenoxybenzoic acid	cotton, tomato, cabbage, kidney bean	119
	2-(4-chlorophenyl)-3- methylbutyrate	cotton, kidney bean, cabbage cucumber	121
	metribuzin	tomato	124
	methazole metabolite	spinach	125
	flamprop	wheat	126
	p-nitrophenol	peanut	10
	2-chloro-4-(trifluoro- methyl)phenol	soybean	111
glucosamine	pyrazone		9
	3,4-dichloroaniline		9
glucosyl sulfate	phenmedipham metabolites	sugar beet	116

glucopyranos-2-yl sulfates have been proposed as metabolites of
phenmedipham in sugar beets (116), Figure 20).

Miscellaneous Conjugates. Several xenobiotic acids conjugate with
malonic acid alone. Since D-amino acids often are sequestered as
N-malonyl derivatives by plants, it is not surprising that D-p-
fluorophenylalanine is converted to such a malonate by tobacco cell
cultures (103). N-Malonylaniline is a metabolite of carboxin in a
peanut cell suspension (127). Phosphates have been suggested as
possible metabolites from two pyrimidines, ethirimol and
dimethirimol, in barley and cucumbers, respectively (11).

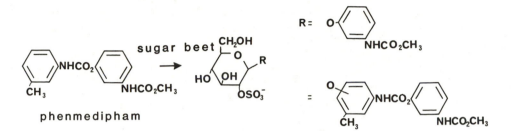

Figure 20. Glucosyl sulfate conjugates from phenmedipham.

Conclusions

Unusual conjugates generally occur when a xenobiotic is metabolized
by animals or plants which accentuate atypical metabolic pathways
adapted to the organism's own natural products. The xenobiotic
merely becomes an alternate substrate in a pathway which may be
important in the test organism, but unusual when compared to other
organisms. Thus, a knowledge of comparative biochemistry will
assist greatly in the prediction and identification of future novel
conjugates. Conversely, the identification of a novel conjugate
may afford information about the fundamental biochemistry of the
species under study. A further benefit in some instances may be an
indication of the mode of therapeutic, pesticidal, or toxic action
of the xenobiotic.

Literature Cited

1.   Parke, D. V. "The Biochemistry of Foreign Compounds";
     Pergamon: Oxford, 1968.
2.   Hirom, P. C.; Millburn, P. In "Foreign Compound Metabolism
     in Mammals"; Royal Society of Chemistry: London, 1981; Vol.
     6, pp. 111–132.
3.   Paulson, G. D. Residue Rev. 1980, 76, 31–72
4.   Caldwell, J. In "Metabolic Basis of Detoxication"; Jakoby,
     W. B.; Bend, J. R.; Caldwell, J., Eds.; Academic: New York,
     1982; pp. 271–290.
5.   Caldwell, J. In "Concepts in Drug Metabolism, Part A";
     Jenner, P.; Testa, B., Eds.; Marcel Dekker: New York, 1980;
     pp. 211–250.

6. Testa, B.; Jenner, P. "Drug Metabolism: Chemical and Biochemical Aspects"; Marcel Dekker: New York, 1976.

7. Dorough, H. W. In "Differential Toxicities of Insecticides and Halogenated Aromatics"; Matsumura, F., Ed.; Pergamon: Oxford, 1984; pp. 291-329.

8. Caldwell, J. Drug Metab. Rev. 1982, 13, 745-779.

9. Edwards, V. T.; McMinn, A. L.; Wright, A. N. In "Progress in Pesticide Biochemistry"; Hutson, D. H.; Roberts, T. R., Eds.; Wiley: Chichester, 1982; Vol. 2, pp. 71-125.

10. Frear, D. S. In "Bound and Conjugated Pesticide Residues"; Kaufman, D. D.; Still, G. G.; Paulson, G. D.; Bandal, S. K., Eds.; ACS SYMPOSIUM SERIES No. 29, American Chemical Society: Washington, D.C., 1976; Chap. 3.

11. Vonk, J. W. In "Progress Pesticide Biochemistry and Toxicology"; Hutson, D. H.; Roberts, T. R., Eds.; Wiley: Chichester, 1983; Vol. 3, pp. 111-162.

12. Edwards, V. T.; McMinn, A. L. In "Progress in Pesticide Biochemistry and Toxicology"; Hutson, D. H.; Roberts, T. R., Eds.; Wiley: Chichester, 1985; Vol. 4, Ch. 4.

13. Hawkins, D. R. In "Progress in Drug Metabolism"; Bridges, J. W.; Chasseaud, L. F., Eds.; Wiley: Chichester, 1981; Vol. 6, pp. 111-196.

14. Jenner, P.; Testa, B. Xenobiotica 1978, 8, 1-25.

15. Israili, Z. H.; Dayton, P. G.; Kiechel, J. R. Drug Metab. Dispos. 1977, 5, 411-415.

16. Huckle, K. R.; Millburn, P. In "Progress in Pesticide Biochemistry"; Hutson, D. H.; Roberts, T. R., Eds.; Wiley: Chichester, 1982; Vol. 2, pp. 127-169.

17. Huckle, K. R.; Stoydin, G.; Hutson, D. H.; Millburn, P. Drug Metab. Dispos. 1982, 10, 523-528.

18. Lertratanangkoon, K.; Horning, M. G.; Middleditch, B. S.; Tsang, W.-S.; Griffin, G. W. Drug Metab. Dispos. 1982, 10, 614-623.

19. Jaglan, P. S.; Arnold, T. S. J. Agric. Food Chem. 1982, 30, 1051-1056.

20. Collins, M. W.; French, M. R.; Hirom, P. C.; Idle, J. R.; Bassir, O.; Williams, R. T. Comp. Biochem. Physiol. 1977, 56C, 103.

21. Quistad, G. B.; Staiger, L. E.; Jamieson, G. C.; Schooley, D. A. J. Agric. Food Chem. 1983, 31, 589-596.

22. Emudianughe, T. S.; Caldwell, J; Smith, R. L. Xenobiotica 1983, 13, 133-138.

23. Quistad, G. B.; Staiger, L. E.; Schooley, D. A. J. Agric. Food Chem. 1978, 26, 60-66.

24. Quistad, G. B.; Staiger, L. E.; Schooley, D. A. J. Agric. food Chem. 1978, 26, 76-80.

25. Quistad, G. B.; Staiger, L. E.; Schooley, D. A.; Sparks, T. C.; Hammock, B. D. Pestic. Biochem. Physiol. 1979, 11, 159-165.

26. Quistad, G. B.; Staiger, L. E.; Schooley, D. A. J. Agric. Food Chem. 1978, 26, 71-75.

27. Vickers, S.; Duncan, C. A. H.; Smith, J. L.; Walker, R. W.; Flynn, H.; Arison, B. H. Fed. Proc., 1984, 43, 339; Xenobiotica, 1985, in press.

28.   Millington, D. S.; Bohan, T. P.; Roe, C. R.; Yergey, A. L.;
      Liberato, D. J. Clin. Chim. Acta 1985, 145, 69-76.
29.   Huckle, K. R.; Climie, I. J. G.; Hutson, D. H.; Millburn,
      P. Drug Metab. Dispos. 1981, 9, 147-149.
30.   Mitchell, S. C.; Waring, R. H. 189th National Meeting of
      the American Chemical Society, Miami, FL, May 1985, PEST 64;
      Xenobiotica, 1985, in press.
31.   Hutson, D. H. In "Bound and Conjugated Pesticide Residues";
      Kaufman, D. D.; Still, G. G.; Paulson, G. D.; Bandal, S. K.,
      Eds.; ACS SYMPOSIUM SERIES No. 29, American Chemical
      Society: Washington, D.C., 1976; pp. 103-131.
32.   Logan, C. J.; Cottee, F. H.; Page, J. A. Biochem.
      Pharmacol. 1984, 33, 2345-2346.
33.   Hutson, D. H.; Logan, C. J.; Regan, P. D. Drug Metab.
      Dispos. 1984, 12, 523-524.
34.   Hutson, D. H.; Lakeman, S. K.; Logan, C. J. Xenobiotica
      1984, 14, 925-934.
35.   Larsen, G. L.; Bakke, J. E. Xenobiotica 1981, 11, 473-480.
36.   Bakke, J. E.; Rafter, J.; Larsen, G. L.; Gustafsson, J. A.;
      Gustafsson, B. E. Drug Metab. Dispos. 1981, 9, 525-528.
37.   Renner, G.; Nguyen, P.-T. Xenobiotica 1984, 14, 693-704.
38.   Stogniew, M.; Fenselau, C. Drug Metab. Dispos. 1982, 10,
      609-613.
39.   Kripalani, K. J.; Dean, A. V.; Migdalof, B. H. Xenobiotica
      1983, 13, 701-705.
40.   Stillwell, W. G.; Horning, M. G.; Griffin, G. W.; Tsang, W.-
      S. Drug Metab. Dispos. 1982, 10, 624-631.
41.   Bakke, J.; Struble, C.; Gustafsson, J.-A.; Gustafsson, B.
      Proc. Natl. Acad. Sci. 1985, 82, 668-671.
42.   Populaire, P.; Terlain, M. N.; Pascal, S.; Lebreton, G.;
      Deconvelaere, B. Arzneim. Forsch. 1969, 19, 1214-1221.
43.   Siek, T. S.; Rieders, E. F. J. Anal. Toxicol. 1981, 5, 194.
44.   Naito, J.; Sasaki, E.; Ohta, Y.; Shinohara, R.; Ishiguro,
      I. Biochem. Pharmacol. 1984, 33, 3195-3200.
45.   Eadsforth, C. V.; Logan, C. J.; Page, J. A.; Regan, P. D.
      Drug Metab. Dispos. 1985, 13, 263-264.
46.   Leahey, J. P.; Curl, E. A.; Edwards, P. J. 189th National
      Meeting of the American Chemical Society, Miami, FL, May
      1985, PEST. 102.
47.   Muecke, W. In "Progress in Pesticide Biochemistry and
      Toxicology"; Hutson, D. H.; Roberts, T. R., Eds.; Wiley:
      Chichester, 1983; Vol. 3, pp. 279-366.
48.   Mashford, P. M.; Jones, A. R. Xenobiotica 1982, 12,
      119-124.
49.   Sisenwine, S. F.; Tio, C. O.; Shrader, S. R.; Ruelius, H.
      W. Arzneim. Forsch. 1972, 22, 682-687.
50.   Bernardi, L.; Coda, S.; Nicolella, V.; Vicario, G. P.;
      Gioca, B.; Minghetti, A.; Vigevani, A.; Arcamone, F.
      Arzneim. Forsch. 1979, 29, 1412-1416.
51.   Lin, C.; Puar, M. S.; Schuessler, D.; Prananik, B. N.;
      Symchowicz, S. Drug Metab. Dispos. 1984, 12, 51-56.
52.   Fukushima, D. K.; Noguchi, S.; Bradlow, H. L.; Zumoff, B.
      Nozuma, K.; Hellman, L.; Gallagher, T. F. J. Biol. Chem.
      1966, 241, 5336-5340.

53. Lehman, J. P.; Fenselau, C.; Depaulo, J. R. Drug Metab.
     Dispos. 1983, 11, 221–225.
54. Le Bigot, J. F.; Cresteil, T.; Kiechel, J. R.; Beaune, P.
     Drug Metab. Dispos. 1983, 11, 585–589.
55. Janssen, F. W.; Kirkman, S. K.; Fenselau, C.; Stogniew, M.;
     Hofmann, B. R.; Young, E. M.; Ruelius, H. W. Drug Metab.
     Dispos. 1982, 10, 599–604.   `
56. Fisher, L. J.; Thies, R. L.; Charkowski, D.; Donham, K. J.
     Drug Metab. Dispos. 1980, 8, 422–424.
57. Lehman, J. P.; Fenselau, C. Drug Metab. Dispos. 1982, 10,
     446–449.
58. Levy, S.; Yagen, B.; Mechoulam, R. Science 1978, 200,
     1391–1392.
59. Elvin, A. T.; Keenaghan, J. B.; Byrnes, E. W.; Tenthorey, P.
     A.; McMaster, P. D.; Takman, B. H.; Lalka, D.; Manion, C.
     V.; Baer, D. T.; Wolshin, E. M.; Meyer, M. B.; Ronfeld, R.
     A. J. Pharm. Sci. 1980, 69, 47.
60. Venkataramanan, R.; Axelson, J. E. Xenobiotica 1981, 11,
     259–265.
61. Senn, M.; Hindermayr, H.; Koch, K. Proc. 27th Ann. Conf.
     Mass Spectro. and Allied Topics, 1979, p. 649.
62. Giera, D. D.; Abdulla, R. F.; Occolowitz, J. L.; Dorman, D.
     E.; Mertz, J. L.; Sieck, R. F. J. Agric. Food Chem. 1982,
     30, 260–263.
63. Paulson, G. D.; Giddings, J. M.; Lamoureux, C. H.; Mansager,
     E. R.; Struble, C. B. Drug Metab. Dispos. 1981, 9, 142–146.
64. Zimm, S.; Grygiel, J. J.; Strong, J. M.; Monks, T. J.;
     Poplack, D. G. Biochem. Pharmacol. 1984, 33, 4089–4092.
65. Layne, D. S. In "Metabolic Conjugation and Metabolic
     Hydrolysis"; Fishman, W. H., Ed.; Academic: New York, 1970;
     pp. 21–52
66. Ngah, W. Z. W.; Smith, J. N. Xenobiotica 1983, 13, 383–389.
67. Mulder, G. J. In "Sulfate Metabolism and Sulfate
     Conjugation"; Mulder, G. J.; Caldwell J.; Van Kempen, G. M.
     J.; Vonk, R. J., Eds.; Taylor and Francis: London, 1982.
68. Mulder, G. J. In "Sulfation of Drugs and Related
     Compounds"; Mulder, G. J., Ed.; CRC Press: Boca Raton, 1981.
69. Paulson, G. D. In "Bound and Conjugated Pesticide
     Residues"; Kaufman, D. D.; Still, G. G.; Paulson, G. D.;
     Bandal, S. K., Eds.; ACS SYMPOSIUM SERIES No. 29; American
     Chemical Society: Washington, D.C., 1976; Ch. 6.
70. Dodgson, K. S.; Rose, F. A. In "Metabolic Conjugation and
     Metabolic Hydrolysis"; Fishman, W. H., Ed.; Academic: New
     York, 1970; Vol. I, pp. 239–325.
71. Aschbacher, P. W.; Feil, V. J. J. Agric. Food Chem. 1983,
     31, 1150–1158.
72. Iwasaki, K.; Shiraga, T.; Noda, K.; Tada, K.; Noguchi, H.
     Xenobiotica 1983, 13, 273–278.
73. Johnson, G. A.; Barsuhn, K. J.; McCall, J. M. Biochem.
     Pharmacol. 1982, 31, 2949–2954.
74. Johnson, G. A.; Barsuhn, K. J.; McCall, J. M. Drug Metab.
     Dispos. 1983, 11, 507–508.
75. Mulder, G. J. In "Metabolic Basis of Detoxication; Jakoby,
     W. B.; Bend, J. R.; Caldwell, J., Eds.; Academic: New York,
     1982; pp. 247–269.
76. Isaac, R. E.; Rees, H. H. Insect Biochem. 1985, 15, 65–72.

77. Dekant, W.; Metzler, M.; Henschler, D. Biochem. Pharmacol.
     1984, 33, 2021-2027.
78. Soltysiak, R. M.; Matsuura, F.; Bloomer, D.; Sweeley, C.
     C. Biochem. Biophys. Acta 1984, 792, 214-226.
79. Leblanc, P.; Capone, J.; Gerber, G. E. J. Biol. Chem. 1982,
     257, 14586-14589.
80. Tulliez, J. E.; Bories, G. F. Lipids 1979, 14, 292-297.
81. Trudell, J. R.; Bösterling, B.; Trevor, A. J. Proc. Natl.
     Acad. Sci. 1982, 79, 2678-2682.
82   Trudell, J. R.; Bösterling, B.; Trevor, A. J. Mol.
     Pharmacol. 1982, 21, 710-717.
83. Link, B.; Dürk, H.; Thiel, D.; Frank, H. Biochem. J. 1984,
     223, 577-586.
84. Quistad, G. B.; Staiger, L. E.; Schooley, D. A. Nature
     1982, 296, 462-464.
85. Quistad, G. B.; Staiger, L. E.; Jamieson, G. C.; Schooley,
     D. A. J. Agric. Food Chem. 1983, 31, 589-596.
86. Staiger, L. E.; Quistad, G. B.; Duddy, S. K.; Schooley, D.
     A. J. Agric. Food Chem. 1982, 30, 901-906.
87. Quistad, G. B.; Staiger, L. E.; Jamieson, G. C.; Schooley,
     D. A. J. Agric. Food Chem. 1982, 30, 895-901.
88. Kiese, M.; Lenk, W. Biochim. Biophys. Acta 1970, 222,
     549-551.
89. Fries, W.; Kiese, M.; Lenk, W. Xenobiotica 1971, 1,
     241-256.
90. Weinstock, J.; Parker, S. E.;. Lucyszyn, G. W.; Intoccia, A.
     P. Pharmacologist 1969, 11, 240.
91. Haegele, K. D.; McLean, A. J.; du Souich, P.; Barron, K.;
     Laquer, J.; McNay, J. L.; Carrier, O. Brit. J. Clin.
     Pharmacol. 1978, 5, 489.
92. Colvin, L. B. J. Pharm. Sci. 1969, 58, 1433.
93. Egger, H.; Bartlett, F.; Dreyfuss, R.; Karliner, J. Drug
     Metab. Dispos. 1981, 9, 415-423.
94. Hirom, P. C.; Millburn, P. In "Foreign Compound Metabolism
     in Mammals"; Royal Society of Chemistry: London, 1979; Vol.
     5, pp. 132-158.
95. Coles, B.; Srai, S. K. S.; Ketterer, B.; Waynforth, B.;
     Kadlubar, F. F. Chem. Biol. Interactions 1983, 43, 123-129.
96. Branchflower, R. V.; Nunn, D. S.; Highet, R. J.; Smith, J.
     H.; Hook, J. B.; Pohl, L. R. Toxicol. Appl. Pharmacol.
     1984, 72, 159-168.
97. Pohl, L. R.; Branchflower, R. V.; Highet, R. J.; Martin, J.
     L.; Nunn, D. S.; Monks, T. J.; George, J. W.; Hinson, J.
     A. Drug Metab. Dispos. 1981, 9, 334-339.
98. Case, D. E.; Illston, H.; Reeves, P. R.; Shuker, B.; Simons,
     P. Xenobiotica 1975, 5, 83-111.
99. Case, D. E.; Reeves, P. R. Xenobiotica 1975, 5, 113-129.
100. Buly, R. L.; Mumma, R. O. J. Agric. Food Chem. 1984, 32,
     571-577.
101. Mumma, R. O.; Hamilton, R. H. In "Bound and Conjugated
     Pesticide Residues"; Kaufmann, D. D.; Still, G. G.; Paulson,
     G. D.; Bandal, S. K., Eds.; ACS SYMPOSIUM SERIES No. 29;
     American Chemical Society: Washington, D.C., 1976; Ch. 5.

102. Mumma, R. O.; Davidonis, G. H. In "Progress in Pesticide
      Biochemistry and Toxicology"; Hutson, D. H.; Roberts, T. R.,
      Eds.; Wiley: Chichester, 1983; Vol. 3; Ch. 5.
103. Berlin, J.; Witte, L.; Hammer, J.; Kukoschke, K. G.; Zimmer,
      A.; Pape, D. Planta 1982, 155, 244-250.
104. Simoneaux, B. J.; Szolics, I. M.; Cassidy, J. E.; Marco, G.
      J. J. Toxicol.-Clin. Toxicol. 1982-1983, 19, 557-570.
105. Lamoureux, G. L.; Rusness, D. G. J. Agric. Food Chem. 1980,
      28, 1057-1070.
106. Rusness, D. G.; Lamoureux, G. L. J. Agric. Food Chem. 1980,
      28, 1070-1077.
107. Lamoureux, G. L.; Stafford, L. E.; Shimabukuro, R. H.;
      Zaylskie, R. G. J. Agric. Food Chem. 1973, 21, 1020-1030.
108. Simoneaux, B. J.; Martin, G.; Cassidy, J. E.; Ryskiewich, D.
      P. J. Agric. Food Chem. 1980, 28, 1221-1224.
109. Chopade, H. M.; Dauterman, W. C.; Simoneaux, B. J. Pestic.
      Sci. 1981, 12, 17-26.
110. Lamoureux, G. L.; Rusness, D. G. In "Pesticide Chemistry:
      Human Welfare and the Environment"; Miyamoto, J.; Kearney,
      P. C., Eds.; Pergamon: Oxford, 1983; Vol. 3, pp. 295-300.
111. Frear, D. S.; Swanson; H. R.; Mansager, E. R. Pestic.
      Biochem. Physiol. 1983, 20, 299-310.
112. Frear, D. S.; Swanson, H. R.; Mansager, E. R. Pestic.
      Biochem. Physiol. 1985, 23, 56-65.
113. Breaux, E. J. First International Symposium on Foreign
      Compound Metabolism; Oct. 30 - Nov. 4, 1983; West Palm
      Beach, FL; Program Abstracts; p. 20.
114. Quistad, G. B.; Menn, J. J. Residue Rev. 1983, 85, 173-197.
115. Frear, D. S.; Swanson, H. R.; Mansager, E. R.; Wien, R. G.
      J. Agric. Food Chem. 1978, 26, 1347-1351.
116. Celorio, J. I.; Hoyer, G. A.; Iwan, J.; Baltes, W. 187th
      National Meeting of the American Chemical Society; St.
      Louis, MO; April, 1984, PEST 11.
117. Wright, A. N.; Roberts, T. R.; Dutton, A. J.; Doig, M. V.
      Pestic. Biochem. Physiol. 1980, 13, 71-80.
118. More, J. E.; Roberts, T. R.; Wright, A. N. Pestic. Biochem.
      Physiol. 1978, 9, 268-280.
119. Mikami, N.; Wakabayashi, N.; Yamada, H.; Miyamoto, J.
      Pestic. Sci. 1984, 15, 531-542.
120. Roberts, T. R.; Wright, A. N. Pestic. Sci. 1981, 12,
      161-169.
121. Mikami, N.; Wakabayashi, N.; Yamada, H.; Miyamoto, J.
      Pestic. Sci. 1985, 16, 46-58.
122. Frear, D. S.; Swanson, H. R.; Mansager, E. R. 187th
      National Meeting of the American Chemical Society, St.
      Louis, MO.; April, 1984, PEST 12.
123. Harvey, J. Jr.; Han, J.C-Y.; Reiser, R. W. J. Agric. Food
      Chem. 1978, 26, 529-536.
124. Frear, D. S.; Mansager, E. R.; Swanson, H. R.; Tanaka, F.
      S. Pestic. Biochem. Physiol. 1983, 19, 270-281.
125. Suzuki, T.; Casida, J. E. J. Agric. Food Chem. 1981, 29,
      1027-1033.
126. Dutton, A. J.; Roberts, T. R.; Wright, A. N. Chemosphere
      1976, 195-200.
127. Larson, J. D.; Lamoureux, G. L. J. Agric. Food Chem. 1984,
      32, 177-182.

RECEIVED July 29, 1985

12

Hidden Xenobiotic Conjugates

H. Wyman Dorough[1] and John D. Webb

Graduate Center for Toxicology, University of Kentucky, Lexington, KY 40546

With few exceptions, diagrammatic presentations of
metabolic pathways for xenobiotics include a number of
structures that are enclosed within brackets. This is
the means by which the metabolism chemist designates a
proposed intermediate, or hidden metabolite, in that
particular pathway. Often, the proposed intermediate
is considered so transient in nature that its isolation
would be impractical, if not impossible, and its reactivity so great that its occurrence as a terminal
metabolite would be virtually inconceivable. When
secondary metabolic processes are involved, these transient metabolites are referred to as hidden conjugates.
These intermediate, transitory metabolites have their
own chemical and toxicological properties and it is
reasonable to assume that they may influence, at least
in part, the toxic action of the parent xenobiotic.

One of the most intriguing facets associated with the study of
xenobiotic metabolism is the occasional discovery of metabolites
whose pathways of formation defy logical explanation. One of the
least intriguing is the report by a colleague that a major metabolite was overlooked in one's own study. Each of these situations
may arise from what might be termed "hidden" metabolites. These
hidden metabolites are common to both primary and secondary biotransformation processes, but this paper deals solely with the
latter where the metabolite is referred to as a hidden conjugate.
    Perhaps the most common reason for the existence of hidden
conjugates, especially in the past, is the failure of the analytical
methods employed to detect total xenobiotic residues present in a
sample. This problem has been solved to a great extent by the use
of radiotracer techniques in xenobiotic investigations. Once its
existence is established, the analytical chemist can usually isolate
and identify the metabolite. This, however, does not mean that

[1]Current address: Department of Biological Sciences, Mississippi State University,
Mississippi State, MS 39762

0097–6156/86/0299–0242$06.75/0
© 1986 American Chemical Society

hidden conjugates did not occur in vivo during the formation of the
newly identified tissue residue or excretory product. One must
recognize that conjugates are not always resistant to further metabolism and that products recovered from the excreta do not represent
each step in a metabolic pathway. It is quite possible that the
endocon moiety of a conjugate is metabolized, sometimes extensively,
before being voided from the body. The intermediates leading to
the form excreted are hidden conjugates. The same is true when a
conjugate formed in vivo is subsequently deconjugated and the
exocon conjugated in a different manner before elimination from the
body occurs.

       Hidden conjugates may also result from the use of animal model
systems and incorrectly assuming that the model is valid for humans.
Species differ in their metabolic capabilities and extreme caution
must be exercised in extrapolating animal data to humans, or for
that matter, from any one species to another. Just because a particular conjugate is not formed in an animal model does not mean it is
not formed in humans. Too often, however, metabolism studies in
humans, especially involving nontherapeutic chemicals, are limited or
lacking completely and hidden conjugates (those formed specifically
in humans) may remain as such indefinitely.

       Finally, hidden conjugates may occur simply because of chemical
instability during sample storage and handling. Failure to consider
conjugate degradation in a sample (tissue extract, bile, urine, etc.)
and to take steps to prevent such degradation from occurring may
cause the analyst to miss its presence or to identify the degradation
products as being derived from in vivo biotransformation. It may
well be that some of the very bizarre metabolic pathways proposed in
the literature stem from attempts to explain from a biochemical
viewpoint the formation of certain degradation products.

       While there may be additional causes for the existence of
hidden conjugates, those mentioned above will be addressed in this
paper. What we do not know can hurt us, and the author is well
aware that the most important hidden conjugates possibly have yet
to be postulated.

Analytical Problems

When conjugates are present in a sample extract but are insensitive to the analytical methods being utilized, they are indeed
hidden from the analyst. Such was the case in one of the first
scientific projects in which the present author participated.
The study dealt with the stability and recovery of insecticides
from milk (1) and one of the compounds was the carbamate insecticide carbaryl. These studies were being conducted in 1959 just
after the introduction of this material which was touted insecticidally as the new DDT without the persistence problems.

       Interest in the stability and recovery of carbaryl from milk
was generated by reports, some of which were later published (2),
that carbaryl in the diet of dairy cows at levels in excess of 400
ppm for 2 weeks did not result in carbaryl residues in the milk.
These findings were based on a colorimetric method for 1-naphthol
because at the time hydrolysis was the assumed sole pathway for
carbaryl metabolism in mammals (Figure 1). 1-Naphthol was measured
in samples before and after alkaline treatment to hydrolyze carbaryl

to l-naphthol and the difference was taken as the concentration of
carbaryl per se. Our studies demonstrated that the lack of reported
residues was not a function of carbaryl instability and recovery,
and added support to the reports that carbaryl residues were not
present in the milk of cows fed the insecticide in the diet.

Several years later, the problem was again addressed using
carbaryl radiolabeled with carbon-14 and liquid scintillation counting as the method of residue detection (3). The results obtained
(Figure 2) were typical of those observed for hundreds of chemicals
after carbon-14 analysis by scintillation counting became commonly
used in laboratories around the world. Radioassay of the whole
milk sampled 6 hr after treatment (3 mg/kg, oral) showed that 950
ppb ^{14}C-carbaryl equivalents were present (Figure 2). The colorimetric method for carbaryl used in the previous studies (2) indicated that only 3% of the total radiocarbon was carbaryl or other metabolites which responded to the fluoborate chromogenic reagent. This
study and many others made it clear that the carbamate insecticides
are extensively metabolized by oxidative and conjugation mechanisms.
Numerous hidden metabolites would still be present if the analytical
methodologies were limited to detection of the parent molecule.

The use of radioisotopes in xenobiotic metabolism studies has
largely eliminated the problem of certain metabolites totally escap-
ing detection. A word of caution, however, is necessary if total
accountability is to be achieved. Most protocols call for the use
of a radioactive compound having the labeled atom at a non-labile
site on the molecule. The advantages of this approach is apparent,
but there are some disadvantages. For example, what if that labile
portion of the molecule is a reactive intermediate that reacts with
macromolecules in the cell to induce necrosis, carcinogenesis, or
some other toxic effect? One concern of the author in this regard
is the assumed fate of the carbamate moiety (Figure 1) after being
cleaved from the alcohol/phenol. If the carbamic acid is immediate
ly degraded in vivo to carbon dioxide and methylamine, then, there
is no apparent cause for alarm. Unfortunately, this has not been
adequately demonstrated using a variety of carbamate chemicals. In
fact, the opposite is indicated by studies using certain carbamates
radiolabeled in the carbonyl and N-methyl positions. There is
almost always radiocarbon in tissues which is derived solely from
the carbamate portion of the molecule. This is usually attributed
to incorporation of radioactive carbon dioxide into normal biochemical pathways to yield naturally occurring cell components. In some
cases the levels of such residues are quite high (4) and it may be
unwise to consider all the metabolites as innocuous natural cellular
chemicals.

Further Metabolism of Conjugates

Conjugates of xenobiotics eliminated in the urine and feces of
exposed animals normally provide the basis upon which secondary
metabolic pathways are proposed. In many cases, the mechanism of
formation of the conjugate is readily apparent. Such is the case
with phenols which are excreted as sulfate and glucuronide conjugates. The biochemical donors are known, the enzymes are well
characterized and the substrate specificities of these enzymes are
generally well established. Although this does not rule out the

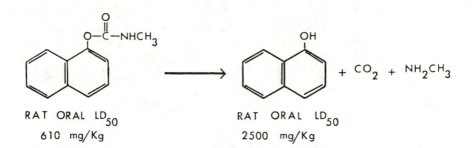

RAT ORAL LD$_{50}$
610 mg/Kg

CARBARYL

RAT ORAL LD$_{50}$
2500 mg/Kg

1-NAPHTHOL

Figure 1. Hydrolytic pathway of metabolism of the insecticide carbaryl. Once considered the only route of metabolism, radio-tracer studies later showed that oxidation, hydrolysis and conjugation reactions resulted in over a dozen metabolites being formed by some organisms.

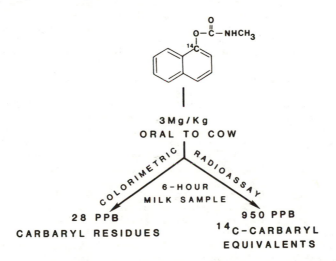

3 Mg/Kg
ORAL TO COW

COLORIMETRIC   RADIOASSAY

6-HOUR
MILK SAMPLE

28 PPB
CARBARYL RESIDUES

950 PPB
^{14}C-CARBARYL
EQUIVALENTS

Figure 2. Hidden residues in cow's milk as a result of the failure of colorimetric analysis to detect total ^{14}C-carbaryl equivalents present in sample.

possibility that sulfate and glucuronide conjugates in the excreta
were not formed by some unusual, and unknown, pathway, there is
clearly no need to suspect that such is the case. It is when a
conjugates does not fit within known pathways of biochemical synthesis that the metabolism chemist turns to hypothetical intermediates
to explain his/her findings. The soundness of the hypothetical
pathways vary, but in most instances it is purely creative chemistry
with a dash of logic thrown in to give some sense of legitimacy to
the proposed scheme. Nonetheless, it is often these hypothetical
intermediates, or unknown conjugates for purposes of this paper,
that are eventually isolated, identified, and consequently, removed
from the list of hidden conjugates.

The source of conjugates previously considered as hidden has
for the most part been the bile. A classical example is the pro-
gressive establishment of the pathway leading to the presence of 1-
naphthyl mercapturic acid in the urine of mammals treated with
naphthalene (5, 6). Initially, it was known only that 1-naphthyl
mercapturic acid was the major metabolite of naphthalene and that it
was formed from an acid-labile product voided in the urine. This
product was eventually identified as naphthyl premercapturic acid
(Figure 3) and there is some evidence that the N-acetyl cysteine
substituent may be at the number 2 position of the naphthalene ring
rather than the number 1 position as shown in Figures 3 and 4 (7).
Over the years, its formation from the glutathione conjugate of
naphthalene (isolated from the bile) has been well documented. As
demonstrated in Figure 4, the epoxide of napthalene must be generated by the cytochrome P-450 system prior to its reaction with
glutathione. While we have isolated the intact glutathione conjugate from rat bile and from S-15 liver enzymes to which glutathione
was added, only those products marked with an asterisk in Figure 4
were detected in the urine of intact rats. Almost 60% of the ^{14}C-
naphthalene dose was voided in the urine and almost 70% of this was
as the premercapturic acid. These in vitro and in vivo studies with
rats are used in our laboratory as positive controls for glutathione
conjugation when assessing similar conjugation reactions with other
substrates (8).

There are many other examples of documented glutathione conjugates of xenobiotics where the final proof was gained by analysis of
the bile. One which involves a compound recognized by almost every
one and where glutathione conjugation is of definite toxicological
importance is outlined in Figure 5. The compound is the widely used
analgesic acetaminophen (N-hydroxyacetanilide) which at high doses
has been shown to cause liver necrosis in man, rats and other mammals. Mitchell and coworkers (9) demonstrated that acetaminophen
was bioactivated to a reactive intermediate (Figure 5,A) by the
cytochrome P-450 system. This hypothetical intermediate (N-acetyl-
p-quinoneimine) was proposed as the hepatotoxin and it was further
proposed that at low doses the intermediate was detoxified by reaction with glutathione. While there are differences of opinion as to
the mechanism of formation of the reactive intermediate (10), its
identity as N-acetyl-p-quinoneimine has been confirmed (11). That
the intermediate was detoxified by glutathione conjugation was supported by the isolation of acetaminophen-glutathione (Figure 5,B)
from the bile of rats (12). These findings support the proposal (9)
that high doses of acetaminophen deplete hepatic glutathione and,

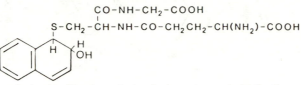

Naphthalene

| Glycine
CO-NH-CH$_2$-COOH
HS-CH$_2$-CH-NH-CO-CH$_2$-CH$_2$-CH(NH$_2$)-COOH
Cysteine | Glutamic Acid

Glutathione

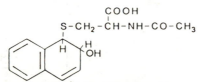

CO-NH-CH$_2$-COOH
S-CH$_2$-CH-NH-CO-CH$_2$CH$_2$-CH(NH$_2$)-COOH

S-(1,2-dihydro-2-hydroxy-1-naphthyl) glutathione

Naphthalene-GSH conjugate

COOH
S-CH$_2$-CH-NH-CO-CH$_3$

N-acetyl-S-(1,2,-dihydro-2-
hydroxy-1-naphthyl) cysteine

1-Naphthyl premercapturic acid

COOH
S-CH$_2$-CH-NH-CO-CH$_3$

N-acetyl-S-(1-naphthyl)
cysteine

1-naphthyl mercapturic acid

Figure 3. Glutathione conjugation of naphthalene in mammals is a
classic example of in vivo conjugate formation (Nap-GSH) that is
not evident by the nature of metabolites (mercapturic acids)
recovered from the urine.

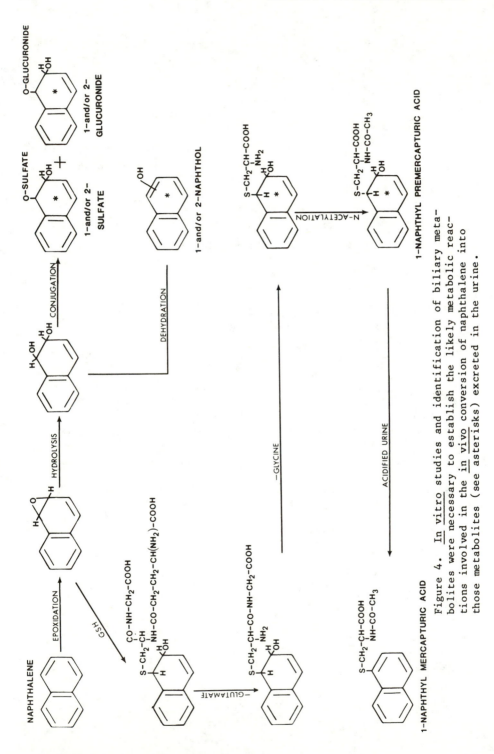

Figure 4. In vitro studies and identification of biliary metabolites were necessary to establish the likely metabolic reactions involved in the in vivo conversion of naphthalene into
those metabolites (see asterisks) excreted in the urine.

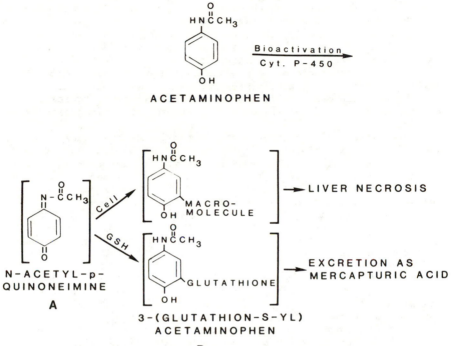

Figure 5. Acetaminophen overdose induces liver necrosis by formation of more hepatotoxin (metabolite A) than can be deactivated by glutathione conjugation (metabolite B).

consequently, the toxic metabolite is free to react with other
tissue molecules. While the conversion of acetaminophen-glutathione
to its mercapturic acid (the excreted form) still involves several
hidden conjugates, neither the glutathione conjugate (Figure 5,B)
nor its precursor (Figure 5,A) can any longer be considered as
hidden or hypothetical metabolites.

A situation similar to acetaminophen detoxification by glutathione also is proposed for chloroform (Figure 6). This compound is
both hepatotoxic and nephrotoxic to mammals. In the liver, the
toxicity is thought to result from the reaction of phosgene, formed
by cytochrome P-450, with other tissue components to induce the
damage. Phosgene is further metabolized in the liver to form diglu-
tathionyl dithiocarbonate (GSCOSG) which undergoes additional metabolism to form 2-oxothiazolidine-4-carboxylic acid (OTZ). This same
pathway has been proposed for chloroform-induced nephrotoxicity (13)
and the mechanism of conversion of chloroform to OTZ is presented in
Figure 6.

The purpose of the study was to determine if the kidney did
indeed metabolize chloroform in the same manner as did the liver.
It was thought unlikely that phosgene formed in the liver was stable
enough to be transported in the blood to the kidney and the authors
demonstrated that the metabolites GSCOSG and OTZ were nontoxic to
the kidney. The conclusion, therefore, was that phosgene is formed
in the kidney where it can directly act as a nephrotoxicant. Deac-
tivation of phosgene in kidney homogenates was accomplished by
glutathione conjugation followed by its further metabolism through a
series of proposed hidden conjugates to yield the cyclized product
OTZ (Figure 6).

In a study of the metabolism of the insecticide dimethylvinphos
in rats and dogs, a hidden glutathione conjugate was proposed which,
unlike the previous examples, leads to a urinary metabolite which
contains no portion of the glutathione molecule (14). That part of
the metabolic pathway is shown in Figure 7. According to the proposed scheme for formation of the urinary metabolite 1-(2,4-di-
chlorophenyl)ethanol, the parent compound is first hydrolyzed to
yield 2,4-dichlorophenacyl chloride. This metabolite was shown by
in vitro studies to react spontaneously with glutathione to form the
hidden conjugate S-(2,4-dichlorophenacyl)glutathione (Figure 7).
Subsequent loss of the glutathione moiety to form the ketone was
shown to be an enzyme-catalyzed, glutathione-dependent reaction.
Conversion of the ketone to the ethanol derivative [1-(2,4-dichloro-
phenyl)ethanol] was also an enzyme-catalyzed reaction, but in this
case the reaction was NADPH-dependent. In both cases, the required
enzymes were present in the rat liver cytosol. In rats, the 1-(2,4-
dichlorophenyl)ethanol was voided in the free form in the feces (19%
of the dose) and as the glucuronide (31%) in the urine. Very small
quantities of the metabolite (0.5%) were excreted in the feces by
dogs and only 6% in the urine was as the glucuronide. The major
urinary metabolite in dogs was desmethyl dimethylvinphos (44%),
whereas in rats this demethylated derivative of the insecticide
accounted for only 3% of the dose. Demethylation required both
glutathione S-methyl transferase in the cytosol and microsomal mono-
oxygenase (14).

An interesting hypothetical (hidden) metabolite of 4-nitro-
toluene (4 NT) has been proposed based on recent studies of its

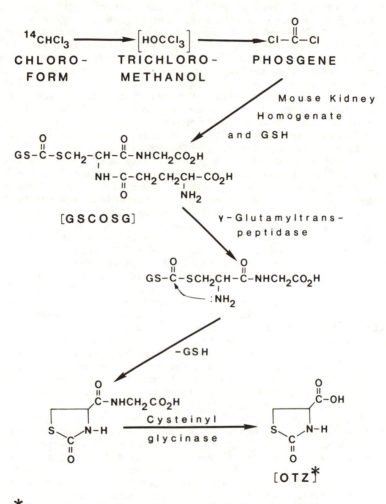

*Only metabolite detected in homogenate.

Figure 6. Liver and kidney toxicity caused by chloroform is thought to result from phosgene, a metabolite which at low levels may be rendered harmless by glutathione conjugation and further metabolism to yeild 2-oxothiazolidine-4-carboxylic acid (OTZ).

metabolism by rat hepatocytes (Figure 8). The major metabolite
formed by the hepatocytes was S-(4-nitrobenzyl)glutathione (68%)
which suggests that the methyl group of 4NT simply reacted directly
with a molecule of glutathione (15). The investigators proposed
that the glutathione conjugate was formed by the reaction of glutathione with 4-nitrobenzyl alcohol sulfate. It was found that the
conversion of 4-nitrobenzyl sulfate to S-(4-nitrobenzyl)glutathione
required glutathione-S-transferase. Because sulfoconjugation is so
common in mammals, these data suggest that unexpected glutathione-
type products may be among the unknown metabolites (insensitive to
enzyme or acid/base hydrolysis) often encountered with xenobiotics
that are metabolized in vivo to sulfate conjugates. More important,
the findings imply that sulfate conjugates may serve as reactive
intermediates that could result in various types of cellular damage.
        This has been specifically proposed by Hutson et al. (16) as a
result of their findings on the fate of 4-cyano-N,N-dimethylaniline
(CDA) in rats. That portion of the metabolic pathway pertinent to
the proposed electrophilic sulfate conjugate formation is depicted
in Figure 9. The postulated reactive intermediate was put forth as
an explaination for the presence of the mercapturic acid metabolite
(N-acetyl-S-[2-keto-2-(4-cyanoanilino)ethyl]cysteine. Rats treated
with a single oral dose (18.5 mg/kg) of CDA-^{14}C excreted 10% of the
administered radiocarbon in the urine as this metabolite (Figure
9,C). Further support of the involvement of the N-acetyl group was
the isolation of minor quantities of 4-cyano-N-methylsulphinyl-
acetyl-aniline and its methylsulphonyl analog. In the proposed
pathway, CDA undergoes demethylation to form 4-cyanoaniline (Figure
9,A) which is then acetylated to form N-acetyl-4-cyanoaniline
(Figure 9,B). Both of these metabolites were isolated and identified. The next 2 steps, however, are hidden metabolites, the first
being the N-hydroxymethyl derivative of metabolite B which, theo-
retically, undergoes sulfate conjugation to form an electrophilic
intermediate that reacts with glutathione. The mercapturic acid
metabolite (Figure 9,C) isolated from the urine leaves no doubt that
glutathione was involved in CDA metabolism.
        In the past, the glutathione/mercapturic acid pathway has been
generally considered a means for detoxification of reactive intermediates. However, there is ever increasing evidence that this
pathway may often generate metabolites that are potentially quite
toxic. One example of such a reaction that is fairly well documented is that referred to as methylthiolation. This process of the
introduction of a methylthio group into a xenobiotic molecule is
usually associated with compounds that are known to undergo glutathione conjugation and often the methylthio derivatives and mercapturic acid metabolites are identified as products of excretion (17,
18). Consequently, the methylthio metabolites of compounds such as
the chlorinated biphenyls, acetaminophen, naphthalene, and bromoben-
zene have been suggested as being derived from mercapturic acid
conjugates. Unlike the mercapturic acid metabolites, methylthio
compounds may have similar polarities as the parent xenobiotic and
may well have similar toxicological properties.
        There are several proposed pathways for methylthiolation via
the mercapturic acid route and the biochemistry of the reactions are
fairly well established (18). Two of the pathways that have been
proposed are shown in Figure 10. Direct attachment of a methylthio

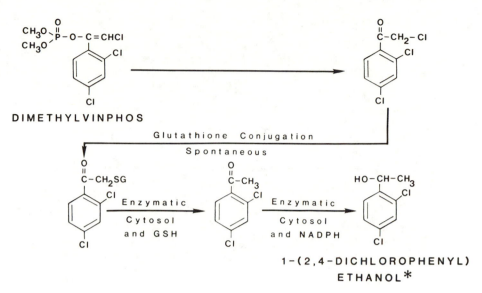

DIMETHYLVINPHOS

Glutathione Conjugation

Spontaneous

Enzymatic
Cytosol
and GSH

Enzymatic
Cytosol
and NADPH

1-(2,4-DICHLOROPHENYL)
ETHANOL*

*Of chemicals shown, only one detectable in excreta.

Figure 7. Proposed pathway of formation of 1-(2,4-dichloro-
phenyl)ethanol which is excreted in rats and dogs treated with
dimethylvinphos. In this case GSH is involved in metabolism but
does not become a part of the structure of the urinary metabolite.

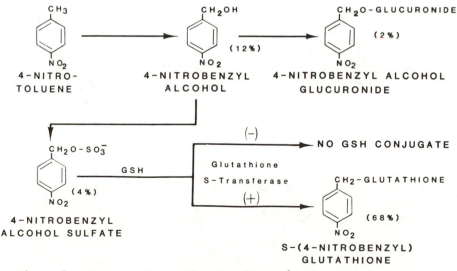

4-NITRO-
TOLUENE

4-NITROBENZYL
ALCOHOL

(12%)

4-NITROBENZYL ALCOHOL
GLUCURONIDE

(2%)

(−)

NO GSH CONJUGATE

Glutathione
S-Transferase

GSH

(+)

CH₂-GLUTATHIONE

(4%)

4-NITROBENZYL
ALCOHOL SULFATE

(68%)

S-(4-NITROBENZYL)
GLUTATHIONE

Figure 8. Example of a sulfate conjugate (4-nitrobenzyl alcohol
sulfate) serving as a substrate for glutathione conjugation.
Values are percent of indicated metabolite formed by rat hepatocytes (15).

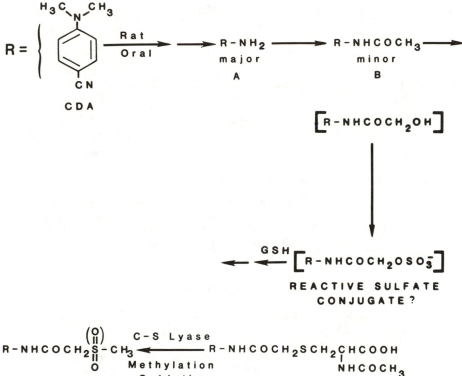

Figure 9. Identification of metabolites A through D in urine of rats treated with CDA led to the proposed hidden metabolites shown here in brackets. The reactive sulfate conjugate apparently undergoes glutathione conjugation, further metabolism via the mercapturic acid pathway and, finally, C-S cleavage, methylation and sulfur oxidation of C to yield D.

group has been proposed and early studies using radioactive methionine showed that the methylthio group was derived from methionine (17). However, most compounds are thought to derive the methylthio group by the second pathway summarized in Figure 10. In this case, the xenobiotic first reacts with glutathione which is subsequently converted to the mercapturic acid. Then, the cysteine conjugate is cleaved by enzymes (C-S lyases) present in tissues and in the intestinal microflora to yield the thiol. The latter is then methylated, presumably by a reaction catalyzed by S-methyltransferases using S-adenosylmethionine as the methyl donor. It is easy to understand why many other hidden metabolites may exist when one considers the many steps which are necessary in vivo just to end up with a simple methylthio derivative of the xenobiotic being voided in the urine.

An analogous situation to that just presented for methylthiolation via the mercapturic acid pathway has been proposed for the formation of N-(hydroxyacetyl)-aminoethanol from 1,1,2-trichloroethylene (Figure 11) in rats, mice and humans (19). One hypothesis is that the oxidative intermediate reacts directly with ethanolamine itself, while another proposes the reaction of the reactive intermediate with phosphatidylethanolamine. The latter hidden conjugate metabolite would then be metabolically degraded to the aminoethanol derivative. At this point the pathway is purely speculative but the investigators are pursuing their hypothesis. The importance of this proposed pathway is the recognition that there are many possibilities for hidden conjugates to exist and that innovative ideas are the first step in determining just what their nature might be.

Deconjugation and Further Metabolism

Hidden conjugates may occur as the result of in vivo deconjugation, usually in the gut, and excretion of the free exocon. Also, a conjugate may go undetected as an in vivo xenobiotic metabolite if the conjugate, a glucuronide for example, is deconjugated and the free exocon again conjugated, but with a different biochemical substrate such as sulfate.

It is also possible that deconjugation may yield a reactive intermediate that can cause tissue damage. For example, benzo(a)-pyrene (BP) is known to be metabolized to glucuronide conjugates which are excreted in the bile (Figure 12) and are mutagenic in the Ames assay after cleavage with ß-glucuronidase (BG) (20, 21). Liver S9 enzymes were not required for mutagenicity, nor did they render the biliary metabolites non-mutagenic.

As part of these investigations, it was demonstrated that a pure sample of one of the biliary metabolites, benzo(a)pyrene-3-0-glucuronide, was non-mutagenic unless ß-glucuronidase was added to the assay system. This implied that the aglycone was the mutagenic product. However, the 3-hydroxybenzo(a)pyrene required S9 activation before it was mutagenic. Consequently, the authors (21) concluded that BP-3-0-glucuronide was converted to a direct-acting mutagenic derivative upon cleavage with ß-glucuronidase. The nature of the mutagenic derivative was not proposed, nor was it stated whether or not the glucuronidase preparation would activate the aglycone in the absence of S9 as it did the glucuronide conjugate. In any event, the studies suggested that a hidden metabolite may be generated in the intestine when BP-3-0-glucuronide is hydrolyzed by the microflora.

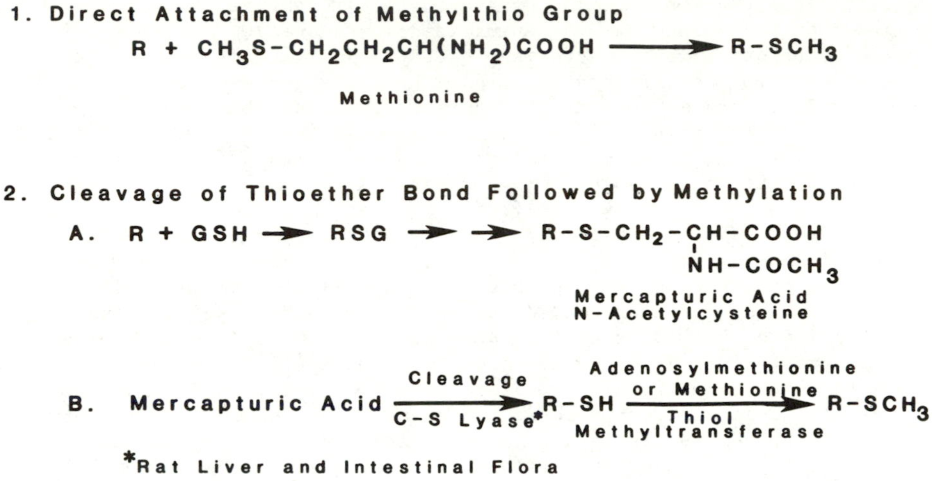

1. Direct Attachment of Methylthio Group
R + CH3S-CH2CH2CH(NH2)COOH → R-SCH3
Methionine
2. Cleavage of Thioether Bond Followed by Methylation
A. R + GSH → RSG → → R-S-CH2-CH-COOH
NH-COCH3
Mercapturic Acid
N-Acetylcysteine
B. Mercapturic Acid
Cleavage
C-S Lyase*
R-SH
Adenosylmethionine
or Methionine
Thiol
Methyltransferase
R-SCH3
*Rat Liver and Intestinal Flora

Figure 10. Proposed mechanisms of methylthiolation of xenobiotics (R).

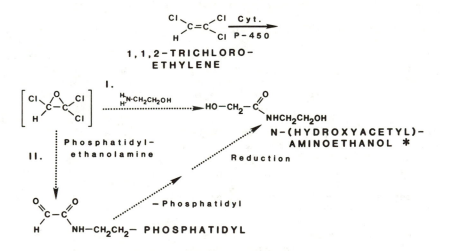

Cyt.
P-450
1,1,2-TRICHLORO-
ETHYLENE
I.
HO-CH2-C
NHCH2CH2OH
N-(HYDROXYACETYL)-
AMINOETHANOL *
Phosphatidyl-
ethanolamine
II.
Reduction
-Phosphatidyl
NH-CH2CH2- PHOSPHATIDYL
*Metabolite in Rats,Mice,and Humans

Figure 11. Phosphatidylethanolamine may be the biochemical donor responsible for the presence of N-(hydroxyacetyl)aminoethanol in the urine of mammals treated with tricholoethylene.

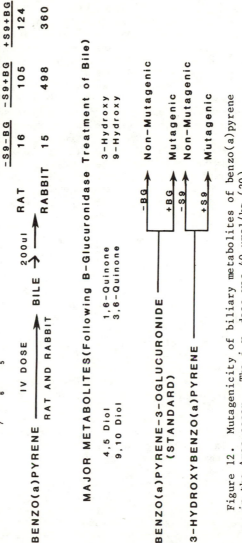

Figure 12. Mutagenicity of biliary metabolites of benzo(a)pyrene in the Ames assay. The i.v. dose was 40 μmol/kg (20).

There are many xenobiotics which are conjugated in mammalian systems but are not excreted as such because of deconjugation and further metabolism prior to elimination from the body. A typical example is presented in Figure 13 where 3-phenoxybenzoic acid (3PBA) metabolites in the bile of rats consisted predominately of glucuronide metabolites (22). However, in the urine these were minor products and the sulfate conjugate of hydroxy-3PBA was the major metabolite. The glucuronides are excreted into the gut via the bile and are then cleaved by glucuronidases of the intestinal microflora. The aglycones are subsequently reabsorbed and with each pass through the liver, the hydroxy-3PBA metabolites, at least in part, are conjugated as sulfates and voided in the urine.

With the naturally-occurring mutagen quercetin, the sulfate and glucuronide metabolites of the 3'-O-methyl ether (isorhamnetin, Figure 14) are excreted in the bile of rats, but are not present in any form, free or conjugated, either in the feces or urine (23). Quercetin is the aglycone of the glycoside, rutin, which is a natural component of many plant species. The glycoside is non-mutagenic in the Ames assay while the aglycone quercetin is highly mutagenic and is carcinogenic in rats. The 3'-O-methyl ether of quercetin, isorhamnetin, is weakly mutagenic and the authors suggest that metabolic 3'-O-methylation may be of importance in protecting the body against the carcinogenic action of quercetin. The latter is formed in the intestine when plants containing rutin are ingested. The actual toxicological importance of the 3'-O-methyl ether is yet unknown. Nonetheless, it is an excellent example of what would still be a hidden conjugate had the nature of the biliary metabolites not been evaluated. Moreover, the cleavage of the glycoside linkage of rutin to form quercetin is an excellent example of in vivo activation, a situation which has particular relevance to glucoside conjugates of insecticides and other pesticides formed when plants are treated with certain of these chemicals (24).

A very important role of conjugates once considered as hidden is that of transport of a carcinogen or procarcinogen to a site in the body where cancer is eventually induced. The process is presented in Figure 15 using 3,2'-dimethyl-4-aminobiphenyl (DMAB) as an example. This compound when fed to Syrian golden hamsters caused cancer of the bladder and the intestine, but in rats cancer was induced only in the intestine. Nussbaum and coworkers (25) reported that DMAB was metabolized in the liver to the N-hydroxy-N-glucuronide of DMAB. This compound was excreted in hamster urine and bile and in rat bile but not urine. The absence of bladder tumors in the rat was attributed to the preferential excretion of the N-hydroxy-N-glucuronide of DMAB in the bile rather than in the urine. In the rat, the tumors in the intestine resulted from the formation of the ultimate carcinogen (nitroso-DMAB) formed upon cleavage of the glucuronide linkage by the gut microflora. Because the N-hydroxy-N-glucuronide of DMAB was excreted in the bile and urine of the hamster, cleavage of the glucuronide and formation of nitroso-DMAB occurred both in the intestine and urinary bladder. Consequently, tumors were induced in the bladder and intestine of this species. Again, without careful examination of the biliary metabolites of DMAB and similar chemicals, the role of conjugate transport in the site-specific phenomenon just described would still be a mystery.

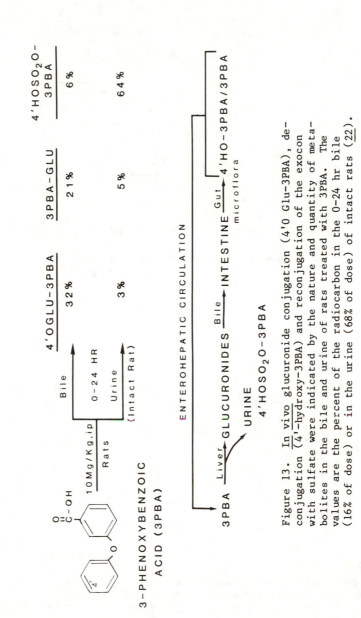

Figure 13. In vivo glucuronide conjugation (4'O Glu-3PBA), de-conjugation (4'-hydroxy-3PBA) and reconjugation of the exocon with sulfate were indicated by the nature and quantity of meta-bolites in the bile and urine of rats treated with 3PBA. The values are the percent of the radiocarbon in the 0-24 hr bile (16% of dose) or in the urine (68% of dose) of intact rats (22).

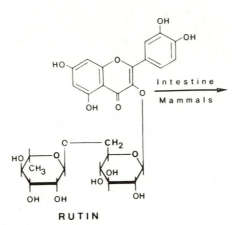

RUTIN

NATURAL PLANT PRODUCT

Non-mutagenic in Ames Assay

BILE COMPONENTS

GLUCURONIDES AND

SULFATES OF

QUERCETIN

and

3'-O-METHYL ETHER

ISORHAMNETIN

Weakly Mutagenic

QUERCETIN

Mutagenic in Ames Assay
Carcinogenic in Rats

Figure 14. Hidden conjugates of quercetin, a naturally-occurring mutagen, are the glucuronide and sulfate derivative of isorhamnetin. These products are not present in the excreta of mammals but are biliary metabolites.

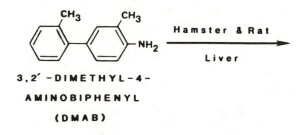

3,2´-DIMETHYL-4-
AMINOBIPHENYL
(DMAB)

Hamster & Rat

Liver

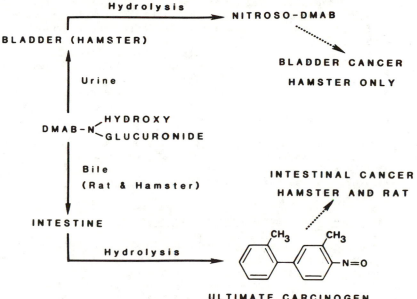

Hydrolysis

NITROSO-DMAB

BLADDER (HAMSTER)

BLADDER CANCER
HAMSTER ONLY

Urine

DMAB-N‹HYDROXY
       ‹GLUCURONIDE

Bile
(Rat & Hamster)

INTESTINAL CANCER
HAMSTER AND RAT

INTESTINE

Hydrolysis

ULTIMATE CARCINOGEN
NITROSO-DMAB

Figure 15. As shown here with DMAB, conjugates may serve as
procarcinogens whose sites of carcinogenicity differ among
species because of dissimilar routes of transport and excretion.

Species Differences

As pointed out in the previous paragraph, species handle xenobiotics differently and this may play a major role in the selective toxicity often observed among different animals. Hidden conjugates which are species specific are of particular importance when humans are one of the species being considered. Many toxicants are just too hazardous to test in humans and the metabolic pathways must be predicted using data obtained with various animal models. Ironically, the point being made is best explained using a chemical commonly used as a clinical tool in human medicine (Figure 16).

Dibromosulfophthalein (DBSP) is used in humans to assess organic anion transport in the liver because animal studies unequivocally showed that is was excreted unchanged in the bile and was not metabolized by the liver. While there are several clinical advantages of test chemicals having these characteristics, a very obvious one is that the compound should be useful in distinguishing factors affecting hepatic transport from those interfering with conjugation. Based on the findings of Meijer et al. (26), DBSP can no longer be considered as exhibiting in humans those characteristics just described for rats and other test animals. Their results show that in humans about 25% of the biliary "DBSP" is actually a metabolite. The identity has not been confirmed, but there is strong evidence that the product is a glutathione conjugate of DBSP. If animal data are to be used as the basis for predicting xenobiotic metabolism, or lack thereof, when unwarranted by toxicity considerations, and later it is discovered that humans differ significantly from the animal model, what must be the situation with the hundreds of toxicants tested in animals only? There is no perfect system for eliminating the possibility of hidden conjugates in man as a result of having to rely upon animal experimentation. However, the potential for such should be kept in mind and human data generated when possible (cases of poisoning, etc) so that a better data base will eventually be established.

Conjugate Alteration During Storage and Handling

Of all the potential causes of hidden conjugates, probably none is as common as chemical changes which occur during sampling, storage and analysis. Spontaneous cleavage often occurs with sulfate and glucuronide conjugates, enzymes are not always as substrate-specific as stated, and many components of biological extracts are potent inhibitors of most enzyme systems used in conjugate identification. These and many more problems exist which may lead to the incorrect identification of xenobiotic conjugates.

This was vividly demonstrated in a study by Dickinson et al. (27) using the ester glucuronide of valproic acid (VPA) as a model compound (Figure 17). With the conjugate contained in freshly collected rat bile having a pH of 8.2, the enzyme ß-glucuronidase cleaved over 98% of the conjugate to yeild free VPA. This was equal to the alkali treatment which was used to effect maximum hydrolysis of the conjugate without alteration of the aglycone. Storage of the bile under a variety of conditions and for varying periods of time changed the results dramatically. Results of one of these tests are presented in Figure 17. When the pH of the bile was adjusted to pH

DIBROMOSULFOPHTHALEIN (DBSP)

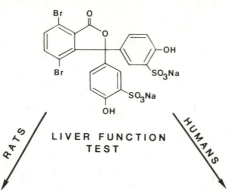

    LIVER FUNCTION
                  TEST

BILE   BILE
100% DBSP   75%DBSP
                              25% [DBSP-METABOLITE]
                                  GSH DERIVED?

Figure 16. Hidden metabolites may be the consequence of using
one species as a model for another. With DBSP, humans were
recently reported to differ from that assumed to be the case
based on earlier studies with rats.

$$(CH_2CH_2CH_2)_2CH\overset{\overset{\text{O}}{\|}}{C}-O-GLUCURONIDE$$

GLUCURONIDE OF VALPROIC ACID

(VPA-GLU)

VPA-GLU ─── IN FRESH ───┬─── ALKALI ───────► FREE VPA 98%
           RAT BILE   │
           pH 8.2   └── B-GLUCURONIDASE ─► FREE VPA 98%

VPA-GLU ─── STORAGE ────┬─── ALKALI ───────► FREE "VPA" 98%*
           OF BILE   │
           37°C,3HR,pH9 └── B-GLUCURONIDASE ─► FREE VPA 5%

*VPA + 6 Unknown Esters labile to alkaline hydrolysis

Figure 17. Chemical alteration during storage and handling of
samples may mask the presence of certain conjugates. For example, storage of bile containing VPA-Glu resulted in almost complete loss of glucuronidase-sensitive substrate after just
3 hrs.

9 and held for 3 hr at 37°C, only 5% of the VPA-glucuronide was
enzymatically cleaved to free VPA. Alkaline treatment, however,
still cleaved 98% of the conjugate. Further analysis of the "VPA"
generated by alkaline hydrolysis showed that the glucuronidase resistant conjugates were structural isomers formed by migration of
the xenobiotic acyl group around the pyranose ring of the exocon;
these products were not substrates for ß-glucuronidase. Consequently, such metabolites would normally be considered as conjugates
other than glucuronides.

Proper sample preparation and analysis are essential in all
studies of xenobiotic metabolism. With known standards of the free
metabolites and often with just a radioactive unknown, it is cus-
tomary to test their stability under conditions of analysis. The
same caution is not always possible with conjugates, but much can be
done to improve the validity of conjugate identification. Not only
will greater caution in sample handling improve the chances of
correctly identifying conjugates, it will also reduce the endless
hours of time spent by metabolism chemists in inventing bizarre
pathways to justify the formation of metabolites which never
existed in the first place.

Literature Cited

1. Timmerman, J. A., Jr.; Dorough, H. W.; Buttram, J. R.;
    Arthur, B. W. J. Econ. Entomol. 1961, 54, 441-4.

2. Gyrisco, G. G.; Norton, L. B.; Trimberger, G. W.; Holland,
    P. J.; McEnerney, P. J.; Muka, A. A. J. Agric. Food Chem.
    1959, 7, 707-11.

3. Dorough, H. W. J. Agric. Food Chem. 1967, 15, 261-6.

4. Nye, D. E.; Hurst, H. E.; Dorough, H. W. J. Agric. Food Chem.
    1976, 24, 371-7.

5. Boyland, E; Sims, P. Biochem. J. 1958, 68, 440-7.

6. Boyland, E; Ramsay, G. S.; Sims, P. Biochem. J. 1961, 78,
    376-84.

7. Jeffery, A. M.; Jerina, D. M. J. Am. Chem. Soc. 1975. 97,
    4427-8.

8. Chen, K. C.; Dorough, H. W. Drug Chem. Toxicol. 1979, 2,
    331-54.

9. Mitchell, J. R.; Jollow, D. J.; Potter, W. Z.; Davis, D. C.;
    Gillette, J. R.; Brodie, B. B. J. Pharmacol. Exptl. Ther.
    1973, 187, 185-94.

10. Rosen, G. M.; Singletary, W. V., Jr.; Rauckman, E. J.;
    Kellenberg, P. G. Biochem. Pharmacol. 1983, 32, 2053-9.

11. Miner, D. J.; Kissinger, P. T. Biochem. Pharmacol. 1979,
    28, 3285-90.

12. Hinson, J. A.; Monks, T. J.; Hong, M. H.; Highet, R. J.; Phol, L. R. Drug Metab. Dispos. 1982, 10, 47-50.

13. Branchflower, R. V.; Nunn, D. S.; Highet, R. J.; Smith, J. H.; Hook, J. B.; Phol, J. R. Toxicol. Appl. Pharmacol. 1984. 72, 159-68.

14. Crawford, M. J.; Hutson, D. H.; King, P. A. Xenobiotica 1976, 12, 745-62.

15. de Bethizy, J. D.; Rickert, D. E. Drug Metab. Dispos. 1984, 12, 45-50.

16. Hutson, D. H.; Lakeman, S. K.; Logan, C. J. Xenobiotica 1984, 14, 925-34.

17. Stillwell, W. G. Drug Metab. Distr., Cur. Rev. Biomed. 1983. 3, 99-103.

18. Bakke, J.; Gustafsson, J-A. Trends in Pharmacol. Sci. 1984, 5, 517-21.

19. Dekant, W.; Metzler, M.; Henschler, D. Biochem. Pharmacol. 1984, 34, 2021-7.

20. Chipman, J. K.; Millburn, P.; Brooks, T. M. Toxicol. Lett. 1983, 17, 233-40.

21. Chipman, J. K.; Millburn, P.; Brooks, T. M. Toxicol. Lett. 1983, 17, 361-2.

22. Huckle, K. R.; Chipman, D. H.; Hutson, D. H.; Millburn, P. Drug Metab. Disp. 1981, 9, 360-7.

23. Brown, S.; Griffiths, L. A. Experientia 1983, 39, 198-200.

24. Dorough, H. W. J. Toxicol. -Clin. Toxicol. 1983, 19, 637-59.

25. Nussbaum, M.; Fiala, E. S.; Kulkarni, B.; El-Bayoumy, K.; Weisburger, J. H. Environ. Hlth. Persp. 1983, 49, 223-31.

26. Meijer, D. K. F.; Weitering, J. G.; Bajema, B. L.; Vermeer, G. A. Eur. J. Clin. Pharmacol. 1983, 24, 707-9.

27. Dickinson, R. G.; Hooper, W. D.; Eadie, M. J. Drug Metab. Dispos. 1984, 12, 247-52.

RECEIVED August 19, 1985

BIOLOGICAL SIGNIFICANCE
OF XENOBIOTIC CONJUGATES

13

A Novel Lipophilic Cholesterol Ester Conjugate from Fenvalerate
Its Formation and Toxicological Significance

Junshi Miyamoto, Hideo Kaneko, and Yasuyoshi Okuno

Laboratory of Biochemistry and Toxicology, Takarazuka Research Center, Sumitomo Chemical Company Ltd., 4-2-1 Takatsukasa, Takarazuka, Hyogo 665, Japan

Fenvalerate[(RS)-α-cyano-3-phenoxybenzyl (RS)-2-
(4-chlorophenyl)isovalerate] consists of four optical
isomers due to the presence of two chiral carbons in
the acid and alcohol moieties. Of the four isomers,
one specific isomer ([2R,αS]) was preferentially
metabolized in mammals including rats and mice to a
cholesterol ester which was formed by condensation of
the acid moity with cholesterol. This conjugate does
not seem to be produced via three known endogenous
biosynthesis routes of cholesterol esters, but via
transesterification mediated by microsomal
esterase(s). Furthermore, this ester was
demonstrated to be a causative agent of granulomatous
changes observed in the liver, spleen and/or lymph
node of animals fed fenvalerate and the [2R,αS]-
isomer subacutely, but not fed the other isomers.

Fenvalerate is one of the synthetic pyrethroids now being used
worldwide for the control of various insect pests on cotton plants,
fruit and vegetable crops. Chemically fenvalerate is an ester of
[RS]-2-(4-chlorophenyl)isovaleric acid and [RS]-α-
cyano-3-phenoxybenzyl alcohol and therefore it is devoid of the
cyclopropane ring in the acid moiety which is common to
conventional pyrethroid insecticides.

 With respect to the mammalian toxicity of fenvalerate, enough
information is available, as shown in Table I, to warrant its safe
use. The only noteworthy long-term toxic effect is the occurrence
of microgranulomatous changes in experimental animals after
subacute and chronic administration of fenvalerate. Figure 1 shows
typical microgranulomatous changes induced in mouse liver by
chronic feeding of fenvalerate. Briefly, a granuloma is defined as
"a tissue response to injury caused by a poorly soluble substance"

Note: Full details of the studies have been submitted in 6 papers
to Toxicology and Applied Pharmacology (1-6).

0097-6156/86/0299-0268$06.00/0
© 1986 American Chemical Society

(7). The inducing agent stimulates the mononuclear phagocytic
system; the mononuclear cells change their form, proliferate and
migrate locally through the tissues. The resultant accumulations
of cells are called micro-granulomata or simply granulomata (8,9).
The granulomatous cells have a tendency to fuse, forming large
multinucleated cells (giant cells). In Figure 1 the arrows indicate
microgranulomata and giant cells in the liver.

Table I. Toxicological Profile of Fenvalerate

---

1. Acute oral LD_{50}: 370 mg/kg (male & female rats).
2. Peripheral nerve lesions: at lethal & sublethal doses.
3. Transient abnormal facial sensation: in human workers.
4. No skin sensitisation, slightly irritant to eye and skin.
5. No mutagenicity.
6. No teratogenicity.
7. Three-generation reproduction: no adverse effect in rat.
8. Subacute and chronic toxicities
   a) Microgranuloma formation in liver, lymph node,
      spleen and adrenals.
   b) Elevation of serum enzymes related to hepatotoxicity.
   c) Reduction of serum lipids.
   d) Slight anemia.
9. No carcinogenicity in rats and mice

Following subacute and chronic feeding of fenvalerate
granulomatous changes were induced in livers, spleen and lymph nodes
of rats and mice and in the adrenals of rats. After 6 month
feeding the liver of Beagle dogs also demonstrated similar changes
(Table II). Thus, although susceptibility differed among organs and
from one animal species to another, the mouse being most
susceptible, granulomata were commonly observed in a variety of
animal species. Therefore, an extensive investigation was initiated
to elucidate the molecular mechanism underlying the development of
these histological changes.

Since the commercial preparation of fenvalerate contains 4
chiral isomers in approximately equal ratio due to its two chiral
centers as shown in Figure 2, the first study was to discover a
possible relationship between granuloma formation and chirality of
fenvalerate isomers. (The purified racemic fenvalerate also
produced granulomatous changes.) Since mouse is most susceptible
to the changes, each isomer, abbreviated as Aα, Aβ, Bα or Bβ
hereafter, was given separately to groups of mice, and after
certain intervals they were sacrificed and examined histologically.

Figure 3 illustrates several hepatic microgranulomatus foci
including giant cell formation, indicated by arrows, induced by
administration of the Bα isomer. The histopathological features
are identical in every respect with those observed with the racemic
mixture. As Table III shows, of the four isomers only one specific
isomer, Bα, proved to be responsible for granuloma formation.
Thus, the Bα isomer fed at 125 ppm in diet induced granulomatous
changes in several mouse tissues after only 8 weeks, whereas

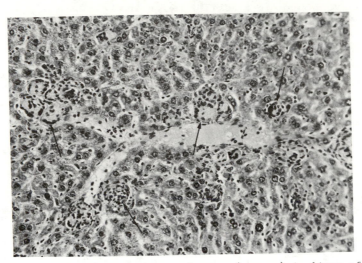

Figure 1. Microgranulomatous changes (arrows) in liver of a
female mouse fed 3000 ppm technical fenvalerate for 72 weeks
(H. and E. X170).

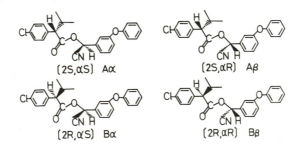

Figure 2. Four chiral isomers of fenvalerate.

Table II. Granuloma Formation in Animals Treated
with Fenvalerate

Dosages	Organs			
(ppm)	Lymph nodes	Liver	Spleen	Adrenals
20 month Mouse				
10, 30	−	−	−	−
100	+	+	−	−
300, 1000, 3000	+	+	+	−
2 year Rat				
50, 150	−	−	−	−
500	+	−	−	+
1500	+	+	+	+
6 month Dog				
250, 500, 1000	−	+	−	−

+; Positive   −; Negative

1000 ppm of the Bβ isomer for 13 weeks or 1000 ppm of the A isomer
or a 1:1 mixture of the Aα and Aβ isomers for one year produced no
granulomatous changes. The insecticidally active Aα isomer never
induced lesions in mice.

Table III.   Incidence of Fenvalerate-Induced Granulomatous
Changes in Mice

Compound	Dosages (ppm)a)	X	Duration (weeks)	Incidence (%)b)
Control	0	X	52	0
Racemic	500	X	13	100
Aα	500	X	52	0
A	1000	X	52	0
Bα	125	X	8	100
Bβ	1000	X	13	0

a) Concentration in diet
b) The number of mice with granulomatous changes in
   liver, lymph nodes and/or spleen is expressed as
   a percentage of the number of mice examined.

When the livers of mice fed the Bα isomer were subjected to
electron microscopic examination, many needle- or rod-like
crystalline inclusions were found scattered in the cytoplasm of the
multi-nucleated giant cells (Figure 4).
    In Figure 5, a large inclusion body from lower-left to
upper-right, together with many small needles is illustrated in the
ultrastructure of a giant cell in a lymph node of a mouse treated
with the Bα isomer for 4 weeks. Such crystalline inclusions were
never found in the tissues of mice treated with other isomers or in
the control animals.
    Thus it appeared clear that only the Bα isomer produced
granulomatous changes. The next step was to carry out a detailed

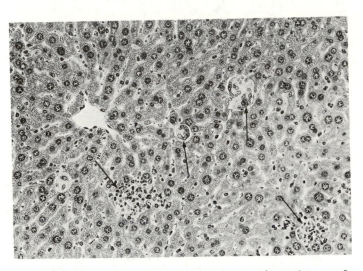

Figure 3. Microgranulomatous changes (arrows) in liver of a male
mouse fed 1000 ppm fenvalerate Bα isomer for 13 weeks (H. and E.
X170).

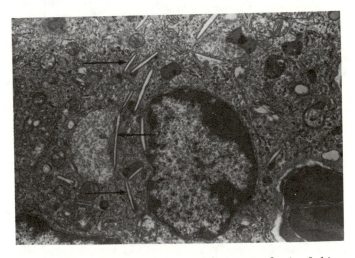

Figure 4. A giant cell in microgranulomatous foci of liver in a
male mouse fed 1000 ppm fenvalerate Bα isomer for 4 weeks
(X10000). Crystalline rods or needles are seen.

comparative metabolism study of these isomers in order to find out possible differences in metabolism which were associated with differences in their potential to induce granulomata. The metabolic pathways of fenvalerate in mammals including rats, mice and dogs have been established based on the identification of radioactive metabolites labelled at either acid or alcohol moiety or cyano position (Figure 6) (10-12). No significant differences were found in elimination of radiocarbon in rats and mice, or in metabolic pathways among the four chiral isomers. With respect to the metabolic pathways, there were no differences specific to the Bα isomer (Figure 7).

However, during analysis of residual tissue radiocarbon 6 days after single oral administrstion of 2.5 mg/kg acid-labeled fenvalerate isomers, it was found that the Bα isomer showed a quite different pattern (Table IV), although the residual amounts themselves were in the order of less than 1µ g equivalent/g wet tissue at the maximum. That is, in the analyzed tissues of both rats and mice, relatively higher residual radiocarbon from the Bα isomer was detected, as compared with the 3 other isomers, particularly in the adrenals, liver, mesenteric lymph node, and spleen.

Table IV. 14-C Tissue Residues at 6 Days after Single Oral Administration of 14-C-Chlorophenyl-Aα-, Aβ-, Bα- and Bβ-Isomers to Male ddY Mice and SD Rats at 2.5 mg/kg

Tissue	ng equivalent/g wet tissue							
	Rat				Mouse			
	Aα	Aβ	Bα	Bβ	Aα	Aβ	Bα	Bβ
Adrenal	12	14	371	23	38	66	597	53
Fat	511	326	304	756	431	496	117	314
Kidney	9	6	25	7	3	7	43	4
Liver	22	20	72	19	13	21	369	22
Mesenteric lymph node	45	68	315	94	59	47	205	33
Spleen	3	2	62	4	<3a)	<3	92	<3

The data showed mean values of four mice and two rats.
a) below the detection limits

To confirm such differences in the tissues contents, 500 ppm of the radioactive Aα, Bα and Bβ isomers were given to mice for 2 weeks, and several tissues were analyzed. As Figure 8 shows, the total radiocarbon from the Bα isomer is exceedingly high, as compared with that from the Aα or Bβ isomer. Moreover, the major portion of the radiocarbon from the Bα isomer (painted black in Figure 4) was found to be characteristic of the Bα isomer, no corresponding product being detected in residual radiocarbon from the 2 other isomers. The radioactive metabolite specific to the Bα isomer was less polar than the parent compound based on the chromatographic behavior. The remaining radiocarbon was composed of the metabolites common to the 3 isomers aside from chirality.

To determine the chemical structure of the residual radioactive

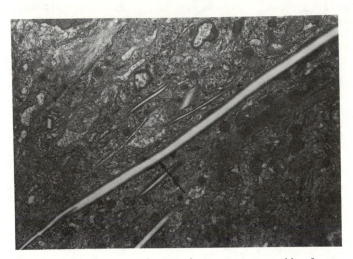

Figure 5.   Crystalline rods (arrows) in a giant cell of mesenteric lymph node of a male mouse fed 1000 ppm fenvalerate Bα isomer for 4 weeks (X10000).

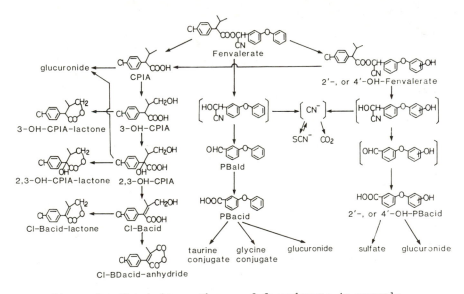

Figure 6.   Metabolic pathways of fenvalerate in mammals.

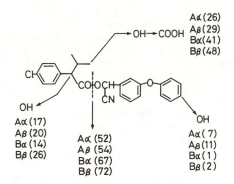

Figure 7. The sites of metabolic attack and percentage of metabolites in 1–2 day mouse excreta.

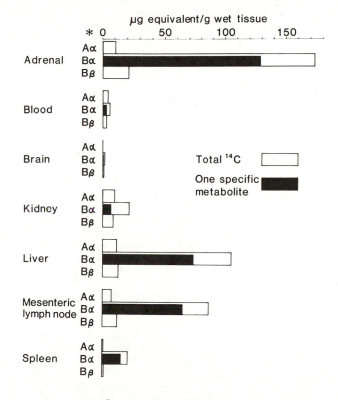

∗ : Compound administered

Figure 8. 14–C–Levels and one specific metabolite level in the tissues 2 weeks after exposure of mice to 500 ppm 14–C–chlorophenyl–Aα–, Bα–, and Bβ–isomers.

metabolite from the Bα isomer, the livers of 125 male mice given
the B isomer for 2 weeks and the acid-labeled racemic fenvalerate
for 1 subsequent week were extracted with a 2 to 1 mixture of
chloroform and methanol and the extract was partitioned between
n-hexane and acetonitrile. The radioactive metabolites in the
n-hexane layer were purified by means of a variety of different
chromatographic procedures, as shown in Figure 9. The structure of
the purified metabolite was identified as the cholesterol ester of
[2R]-2-(4-chlorophenyl)isovaleric acid, abbreviated here as
CPIA-cholesterol ester. The structure was confirmed by comparison
of mass spectra, P-NMR spectra, optical rotation and melting point
with those of the synthesized authentic reference compound.
Interestingly, although the Bβ isomer has also the [2R]
configuration in the acid moiety, the CPIA-cholesterol ester has
never been produced in vivo from the Bβ isomer (Figure 8).
Similarly, in other mouse tissues and also in rat tissues, the
presence of CPIA-cholesterol ester was demonstrated following
administration of the Bα isomer or fenvalerate. Thus,
CPIA-cholesterol ester proves now to be a novel type of conjugate
which is more lipophilic than the parent compound and CPIA.

Based on the above findings two characteristic points are
recognized that are specific to the Bα isomer. First, the Bα
isomer causes granulomatous changes in animals. Second, the Bα
isomer is metabolized to form CPIA-cholesterol ester, which is
lipophilic in nature. Thus it is possible that CPIA-cholesterol
ester might cause granuloma formation, or more specifically,
CPIA-cholesterol ester may be the causative agent for granulomatous
changes. The synthetic CPIA-cholesterol ester was therefore given
to male mice intravenously to determine if this lipophilic
conjugate induces granulomatous changes. Groups of male mice were
given intravenously 10 or 30 mg/kg CPIA-cholesterol ester, and at
intervals 5 mice from each group were sacrificed, and examined
histologically. The results are summarized in Table V. One week
after a single i.v. administration, microgranulomata, accumulations
of mononuclear phagocytes and multinucleate giant cells were
observed in liver by light microscopy; these were essentially
identical to those observed already in fenvalerate-treated mice.
Electron microscopic examination (Figure 10) revealed that the giant
cells contained cytoplasmic crystalline rods, as observed in
fenvalerate-treated mice.

To demonstrate further the association of such crystalline

Table V. Formation of Granulomatous Changes by
Single I.V. Injection of [2R]-CPIA-Cholesterol
Ester in Livers of Mice

Dosages	Weeks after injection		
(mg/kg)	1	4	8
0	−	−	−
10	+	+	+
30	++	++	++

−; Negative   +; Slight   ++; Moderate

Animal Treatment : Mouse, B(Bα+Bβ), 4000ppm, 2 weeks

Purification : Extraction from liver
  (chloroform/methanol (2/1))

Silica gel and florisil colomn chromatography
  (n–hexane/diethyl ether (9/1),
   n–hexane/diethyl ether (19/1))

Thin–layer chromatography (TLC)
  (petroleum ether/diethyl ether /
   acetic acid (90/10/1))

High performance liquid chromatography
  (HPLC)
  (C_{18}: acetonitrile/methanol/chloroform
   (1/1/1), acetonitrile/isopropylalcohol/
   chloroform (1/1/1))

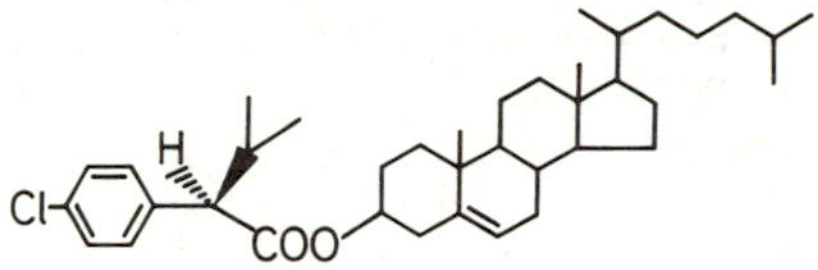

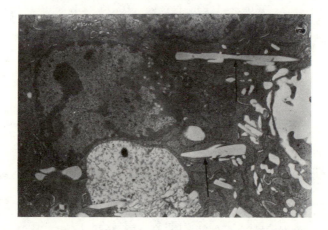

NMR spectra (TMS,$CDCl_3$) :
  δ(ppm), 0.65–2.48 (50H, m), 3.12 (1H, d),
    4.68 (1H, m), 5.39 (1H, m), 7.30 (4H, s)
Mass spectra : 580(M), 368(base peak)
Optical rotation : $[\alpha]_D^{23} = -38.5$
Melting point : 134–138°C

Figure 9. Purification and identification of [2R]–CPIA–cholesterol ester.

Figure 10. A giant cell in mouse liver treated with CPIA–cholesterol (X7000) 8 weeks after the single intravenous injection of 30 mg/kg of [2R]–CPIA cholesterol ester. Crystalline rods (arrows) are seen.

inclusions with CPIA-cholesterol ester, CPIA-cholesterol ester
labelled with tritium in the chlorophenyl ring was synthesized, and
given to mice. The livers were dissected out, and a thin section
of frozen liver was mounted onto a glass slide, and then coated
with SAKURA sensitizer emulsion, to carry out microautoradiography.
The sensitizer was developed after overnight exposure, while the
tissues were stained with hematoxylin. Figure 11 clearly shows the
tritium labeling in giant cells as indicated by 2 long arrows, and
in Kupffer cells (short arrow).

Similarly, to demonstrate the presence of CPIA-cholesterol
ester in the phagocytes, histochemical staining of cholesterol
esters was performed on the same liver used for the
microautoradiography. The staining was carried out by the
Schultz's method (13), the sliced liver preparation being treated
with ferric ammonium sulfate solution followed by freshly prepared
acetic-sulfuric acid solution. Positive staining in giant cells
(Figure 12) indicated the presence of cholesterol esters. The esters
were recognized as blue-colored materials (black grains indicated
by arrows in Figure 12) which contrasted with the brown color of
surrounding tissues.

Thus, the results of histochemistry and microscopic
autoradiography demonstrated that CPIA-cholesterol ester was
phagocytosed by the cells in microgranulomatous foci in mouse
liver. Thus it is concluded that the causative agent of the
microgranulomatous changes induced by racemic fenvalerate is the
CPIA-cholesterol ester which is produced from fenvalerate Bα isomer
in several tissues of rats and mice. This is certainly one of the
few examples of lipophilic conjugates of xenobiotic compounds
(14-18) demonstrated to have a toxicological significance in
animals.

Finally, the enzymatic nature of CPIA-cholesterol ester
formation will be briefly mentioned. None of the enzyme
preparations of three known biosynthetic pathways for cholesterol
esters, namely, acyl-CoA:cholesterol O-acyltransferase (ACAT),
lecithin:cholesterol O-acyltransferase (LCAT), nor cholesterol
esterase, was effective in producing CPIA-cholesterol ester from
the Bα isomer or CPIA. In contrast, the 9,000 g supernatant or
microsomal fractions from liver or kidney homogenate were found to
be capable of producing CPIA-cholesterol ester without the addition
of any cofactors. As substrate, only the Bα isomer was effective,
and none of the 3 other fenvalerate isomers nor free CPIA was
effective. The hepatic enzyme preparation also catalyzed
hydrolysis of fenvalerate, and in this case all the 4 isomers were
utilized as substrates. These facts imply that CPIA-cholesterol
ester is formed from the Bα isomer through a transesterification
reaction via intermediary acyl-enzyme complex.

The enzyme activity responsible for CPIA-cholesterol ester
formation is distributed in several tissues of animal species
including rats, mice, dogs and monkeys, as shown in Table VI. No
tissue enzymes were found to react with the Aα, Aβ or Bβ isomer.
It is premature to correlate the enzyme activity of tissues in any
animal species with the susceptibility to granuloma formation
since, for example, mouse kidney is not susceptible to granuloma
formation. Based on the preliminary trials, the enzyme in mouse
kidney preparation proved to become inactive if the microsomes were

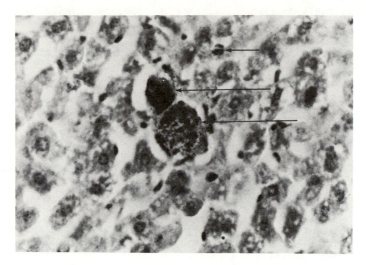

Figure 11. Microscopic autoradiography of liver of a mouse treated with tritium labelled CPIA-cholesterol (X540) 5 weeks after single intravenous injection of ca. 30 mg/kg of 3-H-[2R]-CPIA-cholesterol ester. 3-H label is localized in Kupffer cells (arrow and giant cells (long arrows).

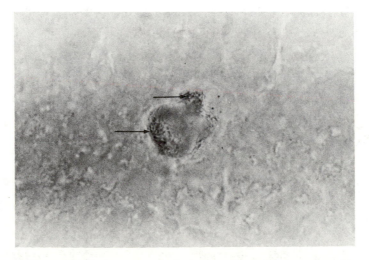

Figure 12. Histochemistry of cholesterol ester in mouse liver treated with 3-H-CPIA-cholesterol (X720) 5 weeks after single intravenous injection of ca. 30 mg/kg of 3-H-[2R]-CPIA-cholesterol ester. Positive coloring (arrows) is observed in giant cells.

Table VI. CPIA-Cholesterol Ester Formation from the Bα Isomer
in Various Tissue of Mice, Rats, Dogs and Monkeys

Tissues	% of CPIA-cholesterol ester relative to applied 14-C			
	Mouse	Rat	Dog	Rhesus monkey
Adrenal	-a)	0.6	0.0	-
Blood	0.0	0.0	-	-
Brain	9.4	0.7	1.6	0.8
Intestine	0.0	0.0	0.0	-
Kidney	10.9	0.3	0.0	0.1
Liver	1.5	0.6	0.5	0.0
Lymph node	0.2	-	0.0	-
Spleen	5.5	1.9	0.1	0.8

a) not determined
The data show mean values of at least two experiments.

solubilized by digitonin, and the solubilized enzyme possessed only
esterase activity. However, if the solubilized esterases were
incubated with artificial liposomes composed of egg lecithin and
cholesterol, then CPIA-cholesterol ester forming activity from the
Bα isomer was recovered (19). This means that cholesterol in the
liposomes will function as an acceptor of the acyl group from
intermediate CPIA-enzyme complex. The selective formation of
CPIA-cholesterol ester from fenvalerate Bα isomer by the
transesterification reaction will be of vital importance during the
whole process of granuloma formation, and the underlying enzymatic
mechanism is now under extensive investigation in the authors'
laboratory.

Literature Cited

1. Okuno, Y.; Ito, S.; Hiromori, T.; Murakami, M.; Miyamoto, J.
   Toxicol. Appl. Pharmacol. 1985, submitted.
2. Okuno, Y.; Seki, T.; Watanabe, T.; Kaneko, H.; Miyamoto J.
   Toxicol. Appl. Pharmacol. 1985, submitted.
3. Kaneko, H.; Takamatsu, Y.; Miyamoto, J. Toxicol. Appl.
   Pharmacol. 1985, submitted.
4. Kaneko, H.; Matsuo, M.; Miyamoto, J. Toxicol. Appl. Pharmacol.
   1985, submitted.
5. Kaneko, H.; Takamatsu, Y.; Miyamoto, J. Toxicol. Appl.
   Pharmacol. 1985, submitted.
6. Miyamoto, J.; Okuno, Y.; Kaneko, H.; Ito, S.; Yamada, T.
   Toxicol. Appl. Pharmacol. 1985, submitted.
7. Epstein, W.L. In "Advance in Modern Toxicology Vol. 4:
   Dermatotoxicology and Pharmacology" ; Marzulli, F.N. and
   Maiback, H.I. Eds.; Hemisphere Publ. Co.: Washington, 1977; pp.
   465-472.
8. Boyd, W. In "A Textbook of Pathology (8th ed.): Structure and
   Function in Desease" ; Lea & Febiger: Philadelphia, 1970; pp.
   76-113.
9. Adams, D.O. Am. J. Pathol. 1976, 84, 164-191.
10. Ohkawa, H.; Kaneko H.; Tsuji, H.; Miyamoto, J.
    J. Pesticide Sci. 1979, 4, 143-155.

11. Kaneko, H.; Ohkawa, H.; Miyamoto, J. J. Pesticide Sci. 1981, 6, 317-326.
12. Kaneko, H.; Izumi, T.; Matsuo, M.; Miyamoto, J. J. Pesticide Sci. 1984, 9, 269-274.
13. Schultz, A. Zentralbl. Allg. Pathol. 1924, 35, 314-317.
14. Schooley, D. A.; Quistad, G. B. In "Pesticide Chemistry: Human Welfare and the Environment"; Miyamoto, J.; Kearney, P. C., Eds.; Pergamon Press: Oxford, 1982; Vol. 3, pp. 301-306.
15. Quistad, G. B.; Staiger, L. E.; Schooley, D. A. J. Agric. Food Chem. 1976, 24, 644-648.
16. Fears, R.; Baggaley, K. H.; Walker, P.; Hindley, R. M. Xenobiotica 1982, 12, 427-433.
17. Gunnarsson, P. O.; Johansson, S.-Å.; Svensson, L. Xenobiotica 1984, 14, 569-574.
18. Hutson, D. H. In "Progress in Pesticide Biochemistry"; Hutson, D. H.; Roberts, T. R., Eds.; John Wiley & Sons, Chichester, 1982; Vol. 2, pp. 171-184.
19. Kaneko. H.; Takamatsu, Y.; Miyamoto, J., unpublished observation.

RECEIVED July 29, 1985

14

Bioactivation of Xenobiotics by Conjugation

Gerard J. Mulder, John H. N. Meerman, and Ans M. van den Goorbergh

Division of Toxicology, Center for Bio-Pharmaceutical Sciences, Sylvius Laboratories, University of Leiden, P.O. Box 9503, 2300 RA Leiden, The Netherlands

Toxification of xenobiotics by conjugation has been found in an increasing number of instances. The toxic effects are due either to the synthesis of a stable conjugate that binds reversibly to a receptor, or to the generation of chemically reactive intermediates that bind covalently to groups in macromolecules. Sulfation of hydroxamic acids or benzylic alcohols leads to the generation of nitrenium ions or carbonium ion, respectively; these may be involved in carcinogenesis. Glucuronidation of N-hydroxy-aromatic amines yields labile N-hydroxy-N-glucuronide conjugates that appear to be involved in bladder and bowel cancer. Glucuronidation of certain steroids in the D-ring results in a reversible cholestatic effect. The same effect also occurs with a phenolic drug, harmol, and certain bile salt conjugates. Glutathione conjugation of dihalogenated alkanes may lead to the formation of thiiranium ions, which are mutagenic. Therefore, conjugation can serve as a toxifying reaction for certain (classes of) compounds.

For a long time conjugation reactions have been regarded as purely detoxifying reactions. This is not surprising because many conjugates isolated from urine or bile have lost the biological effects of their precursors. They are, in general, highly water soluble, poorly lipid-soluble metabolites that are rapidly excreted.

However, in recent years quite a few compounds have been shown to be converted to conjugates that are more toxic than their precursors. In some cases this is due to a reversible interaction of the conjugates with a receptor, as in the case with cholestatic glucuronide conjugates of a number of steroids, and sulfate conjugates of some bile acids. Another possibility is that the conjugate is concentrated by excretory processes in urine or bile to such an extent that the resulting concentration exceeds the solubility of the conjugate, as occurs with harmol conjugates and acetylated conjugates of some sulfonamides.

0097–6156/86/0299–0282$06.00/0

Alternatively conjugation of certain substrates may result in extremely labile products that yield reactive intermediates, which subsequently may be involved in chemical carcinogenesis or tissue damage by these compounds. Finally, the further metabolism of conjugates, especially glutathione conjugates, may result in the formation of highly toxic derivatives, presumably because β-lyase activity in the kidney or the gut flora generates toxic thiol derivatives as is the case with several halogenated hydrocarbons. It is to be expected that in the years to come toxification by conjugation will be established for many more substrates than known at present, because the increasing awareness for this mechanism of toxification will stimulate research in this area. In this chapter the state of the art up till Spring 1985 will be reviewed for the three main conjugation reactions, sulfation, glucuronidation and glutathione conjugation.

Toxification by sulfation

Sulfation is one of the main conjugation reactions for phenolic, alcoholic and hydroxamic acid -OH groups; in addition, sulfamates of aromatic amine groups may be formed(1-2). the resulting conjugates are, in general, excreted in urine. Both filtration and active secretion of the sulfates in the kidney occurs (3). However, glomerular filtration may be limited by high protein binding of sulfate conjugates in blood: 4-methylumbelliferone sulfate is bound to 97%, as compared to 90% for 4-methylumbelliferone itself, and only 32% for the glucuronide conjugate (at 0.2 mM of the compound in rat plasma) (4).

   Although sulfate conjugates at the phenolic hydroxyl group are fairly stable at pH 7.4, those at the hydroxyl group in benzylic alcohols and especially in hydroxamic acids are very labile; the latter break down rapidly upon their formation to yield carbonium and nitrenium ions respectively (Figure 1). These can bind to nucleophilic groups in their environment, resulting in the formation of adducts to protein, RNA and DNA. Such adducts may lead to loss of structure and function of these macromolecules, which ultimately can result in cell death (necrosis) or tumor formation (5).

   The most thoroughly studied example up till now is sulfation of N-hydroxy-2-acetylaminofluorene (N-hydroxy-2AAF) (Figure 2). After the initial discovery that esters of N-hydroxy-2AAF were more reactive than N-hydroxy-2AAF itself (6-7), De Baun et al. (8) extensively characterized the potential role of sulfation in vivo in carcinogenesis by this compound. They showed that methionine was effective in trapping the reactive intermediate generated by sulfation; after alkaline hydrolysis the products identified were the 1- and 3- S-CH3 derivatives of 2AAF. It was also shown that the stability in water of the sulfate conjugate was very low (9), so that it completely broke down within a minute to a reactive intermediate. Subsequently, much more, sometimes rather circumstantial evidence for the involvement of sulfation in the toxicity of N-hydroxy-2AAF was published (see 10 and 11 for reviews).

   Recently, we have characterized some break-down products of the sulfate conjugate of N-hydroxy-2AAF, and the effect of glutathione and various other thiols on the process, as well as on the formation of adducts to RNA and DNA (12-13). N-Hydroxy-2AAF was not found upon break-down of the sulfate conjugate, which confirms that the break-

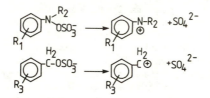

Figure 1. Break down of sulfate conjugates to nitrenium or carbonium ions.

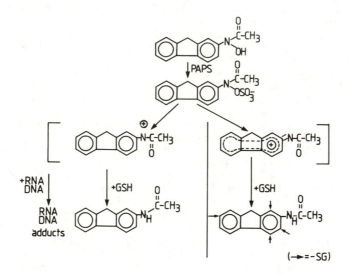

Figure 2. Break down of the sulfate conjugate of N-hydroxy-2-acetylaminofluorene. Two different reactive intermediates are proposed, that have different chemical properties. See text for further explanations. GSH = glutathione.

Table I.   Reaction of the Sulfate Conjugate of N-hydroxy-2AAF
with RNA in the Presence of Various Thiols

Thiols	RNA bound (%)	N-Hydroxy-2AAF (%)	2-AAF (%)
--	20 ± 1	1.3 ± 0.2	2 ± 0
Mercaptopropionic acid	9 ± 1	1.6 ± 0.2	21 ± 1
N-Acetyl cysteine	13 ± 1	1.3 ± 0.4	12 ± 1
Glutathione	11 ± 1	1.0 ± 0.1	14 ± 1
L-Cysteine	6 ± 0	1.1 ± 0.2	23 ± 1
Cysteamine	4 ± 0	1.7 ± 0.3	24 ± 1
Penicillamine	5 ± 1	0.9 ± 0.3	16 ± 0

*The reaction was performed in 100 mM sodium phosphate buffer, pH 7.4,
in the presence of 2.7 mg/ml RNA and 10 mM of the various thiols. A
concentration of 1 mM of the sulfate conjugate of ^{3}H-ring labeled N-
hydroxy-2AAF was added. RNA binding of radioactivity was determined
after extensive washing of the RNA; N-hydroxy-2AAF and 2AAF were
determined in diethyl ether extracts of the incubates by thin layer
chromatography. Percentages of radioactivity added that were recovered in the various fractions (means ± SEM) are given.*

down is an SN_1 reaction and not a hydrolysis. This is in contrast
with N-acetoxy-2AAF, where hydrolysis does occur, presumably because
the acetoxy ester is more stable than the sulfate ester, and therefore can be hydrolyzed by attacking water molecules (unpublished
data). In the presence of RNA much of the radioactivity becomes bound
to RNA (Table II); interestingly, many thiols decrease the RNA
binding, which corresponds with an increased formation of 2AAF (Table
I). Therefore, somehow the nitrenium ion that results initially from
break-down of the sulfate conjugate, is reduced to 2AAF. The RNA
binding is strongly influenced by the buffer used (Table II). In
phosphate buffer much less RNA binding is found than in Bis-tris
buffer. In addition, KCl decreased RNA binding very strongly, especially in Tris and Bis-tris buffer.

Table II.   Effects of 150 mM KCl on RNA binding of the Sulfate
Conjugate of N-hydroxy-2AAF in Various Buffers

Buffer	KCl	RNA bound (%)
No	−	48
No	+	22
Tris-HCl (50 mM)	−	30
Tris-HCl (50 mM)	+	17
Bis-tris-HCl (50 mM)	−	43
Bis-tris-HCl (50 mM)	+	18
Sodium phosphate (100 mM)	−	13
Sodium phosphate (100 mM)	+	10

*The ^{3}H-ring labeled sulfate conjugate of N-hydroxy-2AAF was added at
a final concentration of 1 mM to a RNA solution (2.7 mg/ml) in the
buffer indicated. See for further details, legend to Figure 1.*

Because adduct formation to guanosine monomers was equally affected
by 150 mM KCl effects on the level of tertiary RNA structure seem
not very likely. Interestingly, the formation of adducts to guanosine
monomers was equally effective in Tris as in phospate buffer, in
contrast to the situation with RNA.

Another important finding in these studies was that RNA does not
compete with glutathione for the same pool of reactive intermediate
(Table III); the amount of glutathione conjugates does not decrease
when RNA is added to the incubation medium. This leads to the hypothesis that two different reactive intermediates may be generated
from the sulfate conjugate, one which reacts only with glutathione
to form the glutathione conjugates, and the other which may react
with both RNA and glutathione. In the latter case the reaction with
glutathione leads to reduction of the nitrenium ion (to produce 2AAF
and other unidentified products), rather than to conjugate formation.
These results illustrate that the chemistry of such reactive intermediates is rather complicated.

Table III.   Mutual Effects of Glutathione and RNA on Adduct
            Formation by the Sulfate Conjugate of N-Hydroxy-2AAF

Additions in incubation medium	RNA bound (%)	GSH conjugates (%)
RNA	20	–
RNA + Glutathione	8	30
Glutathione	–	27

*The reaction was performed in 6 mM sodium phosphate buffer pH 7.4 in
the presence of 150 mM KCl; glutathione (7 mM) and/or RNA (2.7 mg/ml)
were present as indicated.*

The data in Table I show that glutathione is relatively little
effective in trapping of the reactive intermediate from the sulfate
conjugate of N-hydroxy-2AAF. This was confirmed in in vivo studies
(13) by depletion of glutathione in the rat in vivo (Table IV). Although the amount excreted as glutathione conjugates in bile decreased in the pretreated animals, there was little or no decrease
in total covalent binding of N-hydroxy-2AAF to liver macromolecules
(which is almost exclusively due to sulfation (Table IV).

Table IV.   The Effect of Glutathione Depletion in the Rat on the
            Excretion of Glutathione Conjugates of N-Hydroxy-2AAF
            and Covalent Binding of N-Hydroxy-2AAF to Macromolecules

Pre-treatment	1-GS-AAF (% of dose)	3-GS-AAF (% of dose)	Covalently bound (total) pmol/mg protein	DNA adducts (total) pmol/mg DNA
–	5.8 ± 0.3	3.6 ± 0.3	770 ± 90	322 ± 37
DEM	2.2 ± 0.2[+]	1.3 ± 0.2[+]	820 ± 70	258 ± 43[+]

*Rats were pretreated with 3.9 mmol/kg diethylmaleate (DEM) i.p. 45
min before an intravenous injection of 120 μmol/kg N-hydroxy-2AAF.
Bile was collected for 4 hours afterwards, and the glutathione conjugates were determined (1-GS-AAF and 3-GS-AAF). The covalently
bound material to total macromolecules and DNA was determined in the
liver taken at that time (see ref. 13 for details).*

The role of sulfation of N-hydroxy-2AAF in liver toxicity was
demonstrated straightforwardly by the use of the selective inhibitors
of sulfation, pentachlorophenol (PCP) and 2,6-dichloro-4-nitrophenol
(DCNP) (14-17). Pretreatment with these compounds completely pre-

vented periportal necrosis, the typical hepatotoxicity of N-hydroxy-2AAF (18), both as demonstrated by histochemistry and levels of the transaminases in blood that reflect liver damage. Sulfation of N-hydroxy-2AAF, therefore, occurs mainly in zone 1, the periportal zone. This was confirmed by autoradiography in liver slices incubated with ^{3}H-labeled N-hydroxy-2AAF with or without sulfate (19). Total covalent binding to macromolecules in the liver in vivo, in perfusion and in hepatocytes decreased by inhibition of sulfation (14,20).

It still remains a question at what level sulfation is involved in the carcinogenicity of N-hydroxy-2AAF. For some time it was be-lieved that sulfation was involved in initiation, because the sulfate conjugate binds to DNA through the nitrenium ion. However, later it was demonstrated that most of the in vivo adducts are de-acetylated adducts, that do not arise from sulfation, but from some other meta-bolic reaction, such as deacetylation to the hydroxylamine, or trans-acetylation to the acetoxy-2-aminofluorene (19,20). Recently, Meerman (21) showed that inhibition of sulfation by PCP during inhibition of preneoplastic foci in the Solt and Farber system by N-hydroxy-2-AAF resulted in an increase of the number of foci, suggesting that sul-fation does not play a role in initiation (Table V). Also epoxide hydrolyse activity, a putative preneoplastic marker in the rat liver, was increased in rats pretreated with PCP as compared to rats that received only N-hydroxy-2AAF (22). Interestingly, Meerman found that the generation of "oval cells", which occurs after N-hydroxy-2AAF administration, was prevented by PCP, implying that these oval cells are not involved in the development of the γ-glutamyltranspeptidase-positive preneoplastic foci (21).

Table V. The Effect of PCP on Formation of Preneoplastic Foci by N-Hydroxy-2AAF, and on Incorporation of ^{3}H-Thymidine in Various Liver Cell Types

Treatment	No of foci/cm^3	% of hepatocytes labeled	% of non-parenchymal cells outside portal triad labeled
–	60	2.7 ± 1.3	7.5 ± 1.4
PCP	50	3.9 ± 1.2	9.5 ± 0.7
N-OH-2AAF	170	3.0 ± 1.2	34.4 ± 2.5
NOH-AAF + PCP	269	1.8 ± 0.7	20.0 ± 4.1

Foci were identified on liver section by enzyme histochemical stain-ing for γ-glutamyltranspeptidase. The percentage labeled cells was determined by autoradiography. PCP (40 μmol/kg) was given i.p. 45 min before N-hydroxy-2AAF (90 μmol/kg i.v.). labeling was achieved by implanting osmotic minipumps that delivered ^{3}H-thymidine for 7 days after the pretreatment (or control treatment with solvents). The liver was used for autoradiography after 7 days. Foci were detected after a modified Solt & Farber procedure. Details can be found in ref. 21. The number of foci after N-hydroxy-2AAF alone was signif-icantly different from that of controls, while the number of the group pretreated with PCP before N-hydroxy-2AAF was significantly different also from the group that received only N-hydroxy-2AAF. (Data taken from 21).

Even more confusing is the finding that PCP did not prevent the induction of ornithine decarboxylase and tyrosine aminotransferase by N-hydroxy-2AAF (Figure 3) (23). Since these two responses have been suggested to reflect promoting activity, this finding may suggest that sulfation does not play a role in promotion by N-hydroxy-2AAF either. However, since the implications of the above tests are not yet fully characterized in terms of their meaning for chemical carcinogenesis, it is hard to draw a conclusion. Since there is much, albeit circumstantial evidence for a relationship between sulfation and carcinogenicity of N-hydroxy-2AAF (at least in the liver), sulfation must play a role although it is impossible at present to pinpoint this mechanistically. Clearly, however, sulfation of N-hydroxy-2AAF is responsible for the hepatotoxicity of this compound.

Hyperplastic noduli seem devoid of sulfotransferase activity for N-hydroxy-2AAF, as determined by autoradiography in liver slices (19). Total covalent binding to macromolecules in perfused livers from nodular rats was indeed lower than in normal rat livers, but PCP had approximately the same effect in both (Table VI). However, the disappearance rate of N-hydroxy-2AAF from the perfusion medium was the same. In rats fed N-hydroxy-2AAF a rapid decrease in sulfotransferase activity was seen (24); this was not specific, because other carcinogens showed the same effect.

Table VI. Effect of PCP on Total Covalent Binding to Macromolecules in Normal Rat Liver and in Nodular Rat Liver during Perfusion

PCP	Normal rats						Nodular rats	
	Protein pmol/mg	(%)	RNA pmol/mg	(%)	DNA pmol/mg	(%)	Total binding pmol/mg protein	(%)
–	785 ± 50		1155 ± 227		280 ± 47		456 ± 35	
PCP	206	(26)	298	(26)	177	(63)	122 ± 19	(26)

The data from normal rats were taken from ref. 14. Nodular rats were obtained from Dr. L.C. Erikson, Huddinge Hospital, Sweden; they were treated with 2AAF on an intermittent schedule as described by Epstein et al. (Cancer Res. 27 (1976) 1702). Total binding represents binding to protein (by far the major part), RNA and DNA.

Many other sulfate esters of hydroxamic acids or hydroxylamines have been shown to be reactive and to result in the formation of DNA adducts. This includes such diverse compounds as N-hydroxy-phenacetin (25-26), N-hydroxy-N-methylaminoazobenzene (27), N-hydroxy-N,N-diacetylbenzidine (28), N-hydroxy-4-acetylaminobiphenyl (29), N-hydroxy-trans-N-acetylaminostilbene (30), N-hydroxy-4-acetylamino-4'-fluorobiphenyl (17), N-hydroxyphenanthrene (31) and some more similar compounds which were only tested as substrates for sulfation to generate a reactive intermediate that bound to methionine or macromolecules (see 10 for review).

Sulfation of N-hydroxy-2-naphthylamine leads to a labile sulfamate that rapidly rearranges to the O-sulfate ester, and does not give rise to a reactive intermediate (33-34).

The group of Rickert (34) has thoroughly investigated the metabolism of dinitrotoluenes in the rat. This group found that sulfation

plays a role in the conversion of dinitrotoluene to reactive intermediates, because the sulfation inhibitors DCNP and PCP inhibited
both covalent binding to total macromolecules in the liver, and the
formation of DNA adducts (35). As yet, it is unclear whether the sulfation step involves the hydroxylamine group or the benzylic alcohol
group (as illustrated in Figure 4 for a metabolite of 2,6-dinitro-
toluene), which may result in the formation of a nitrenium ion or a
carbonium ion respectively. Elucidation of the structure of DNA
adducts will show which group was activated by sulfation.

Some hydroxylamine sulfates are stable, like that formed from
monoxidil (Figure 5), which is an aromatic N-sulfate (36).

In a number of cases sulfate conjugation of benzylic alcohol
groups leads to reactive conjugates, and may result in DNA binding.
Thus, the sulfate ester of the 7-hydroxylmethyl group in 7,12-di-
methylbenzanthracene (Figure 6) is mutagenic in the Ames test, and
binds to DNA (37-38). Similarly, a number of substituted benzylic
alcohols were excreted in vivo as mercapturates, presumably because
the sulfate esters formed reacted (spontaneously?) with glutathione
in vivo to form, ultimately, the mercapturates at the benzylic methyl
group. PCP treatment decreased drastically the amount of the mercapturates excreted (39).

Recently, Boberg et al. (40) demonstrated very elegantly that
sulfation is an obligatory step in the carcinogenesis by 1'-hydroxy-
safrole (Figure 7). They found that pretreatment with PCP decreased
covalent binding of ^{3}H-1'-hydroxy-safrole to protein, RNA and DNA
in mice, and also decreased carcinogenicity of 1'-hydroxy-safrole
very strongly (Table VII). They included in their studies a strain
of mice that is deficient in PAPS synthesis, and, by consequence, in
sulfation activity. They found that these brachymorphic mice had a
much lower tumor incidence upon 1'-hydroxy-safrole administration
than normal mice. Both sets of data are mutually confirmatory as to
the role of sulfation in carcinogensis by 1'-hydroxy-safrole.

Table VII. Effect of PCP Pretreatment on Carcinogenicity of
1'-Hydroxy-Safrole in Mice

Dose (µmol/g)	PCP	No of mice	Av. no. of hepatomas/mouse	No of mice without hepatomas at 10 mo.
0.05	–	34	0.5	22
0.05	+	36	0.1	32
0.1	–	36	2.2	5
0.1	+	32	0.2	28
0.2	–	33	4.4	1
0.2	+	31	0.1	28

*The data were taken from Boberg et al. (40). Preweanling mice were
given a single dose of 1'-hydroxy-saffrole on day 12. As indicated,
they received an i.p. injection of PCP (0.04 µmol/g). After 9 to 10
months they were killed and examined for tumors*

A very special case of toxicity by a sulfate conjugate is the
effect of ethanolamino-O-sulfate, which is an irreversible "suicide"
inhibitor of 4-aminobutyrate aminotransferase. This sulfate conjugate is a substrate analogue for the enzyme, which, because of the
electronegativity of the sulfate group, is converted by β-elimination

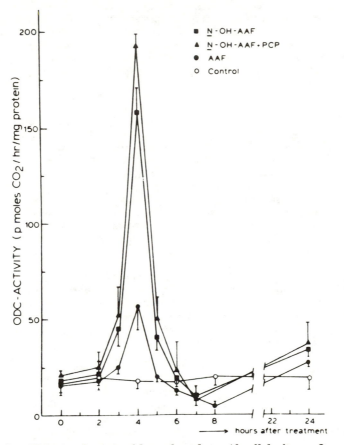

Figure 3. Effect of pentachlorophenol on the N-hydroxy-2-acetyl-aminofluorene-induced increase in ornithine decarboxylase (ODC) activity in rat liver. N-hydroxy-2AAF or 2AAF was administered i.p. (94 and 63 μmol/kg i.p. respectively). One hour prior to N-hydroxy-2AAF, pentachlorophenol (40 μmol/kg) was given. ODC was measured in liver cytosol. (Reproduced with permission from Ref. 23. Copyright 1984, Marcel Dekker.)

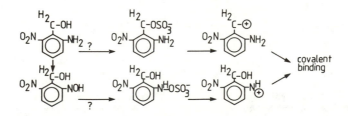

Figure 4. Proposed mechanisms for the generation of a reactive intermediate by sulfation from 2-amino-6-nitro-hydroxytoluene, a metabolite of 2,6-dinitrotoluene.

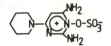

Figure 5.   Monoxidil-N-sulfate.

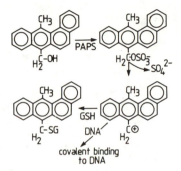

Figure 6.   Generation of a reactive carbonium ion by sulfation
from 7-hydroxymethyl-12-methylbenzanthracene.

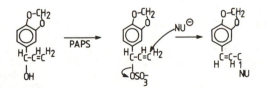

Figure 7.   Generation of a reactive intermediate by sulfation
from 1'-hydroxy-safrole.

instead of transamination; the ethanol moiety becomes covalently
bound to the active site of the enzyme (41).

Some bile salts are quite toxic in the liver, leading to cholestasis and morphologic changes in bile canalicular membranes associated with this toxicity. Some of these toxic bile salts, such as
lithocholate and 3β-hydroxy-5α-cholanic acid are not detoxified by
sulfation, but the sulfate conjugates retain their toxic effects
almost completely. Although strictly speaking this is no toxification
by sulfation, yet sulfation does not alleviate toxicity of these
compounds (42-45).

Glucuronidation

For glucuronidation toxification by the generation of reactive intermediates seems to play a lesser role. The main reason is that glucuronic acid is a much poorer leaving group than sulfate (46). Thus,
the chance that glucuronidation results in a reactive intermediate
is much lower; for instance, around pH 7 the N-O-glucuronide of N-
hydroxy-2AAF is stable. Yet glucuronides can give rise to reactive
intermediates, but cases identified so far involve mainly further
metabolism or rearrangement of glucuronides.

An example is the N-glucuronide of N-hydroxy-2-naphtylamide (33).
This compound is synthesized in the liver from 2-naphtylamine; first
N-hydroxylation occurs, immediately followed by N-glucuronidation
because the affinity of UDP-glucuronosyltransferase for N-hydroxy-2-
naphtylamine is very high. The glucuronide is subsequently excreted
in bile and urine (Figure 8). When it is excreted in bile it may be
hydrolyzed by the gut microflora, yielding the reactive hydroxylamine in the gut; this may bind there, or may be further reduced to
the amine. In urine it is exposed to the local pH in the bladder.
In several species such as man and the dog, the urinary pH is
slightly acidic, down to pH 5; under those conditions the glucuronide
is no more stable. It rapidly breaks down to the hydroxylamine, which
subsequently may bind to DNA in the bladder epithelial cells, and
bladder tumor may be the result. A similar reaction may occur with
some other aromatic amines.

An alternative activation mechanism for glucuronides that may
occur with hydroxamic acids is the de-acetylation of N-acetyl-N-O-
glucuronides, as has been demonstrated for 2AAF (47). Deacetylation
of the N-O-glucuronide of N-hydroxy-2AAF yields the N-O-glucuronide
of 2-aminofluorene, which is very reactive and binds spontaneously
to DNA. Whether such a mechanism operates in vivo is still unclear.

For as yet unexplained reasons the hydrolysis of the glucuronide
of 3-hydroxybenzo[a]pyrene yields a reactive intermediate that spontaneously binds to DNA (18). Less reactive glucuronides can be found
by conjugation of carboxylic acids with glucuronic acid to yield
acyl-glucuronides. These are reactive as was demonstrated for the
acyl glucuronide of clofibrate, which reacts spontaneously with
glutathione and other SH-group containing reagents (49). As yet it
is not clear whether these acyl glucuronides play a role in the
toxicity of compound like clofibrate, as reviewed recently by
Faed (50).

Several glucuronide conjugates of steroid hormones have a cholestatic activity. This concerns glucuronide conjugates at the D-ring

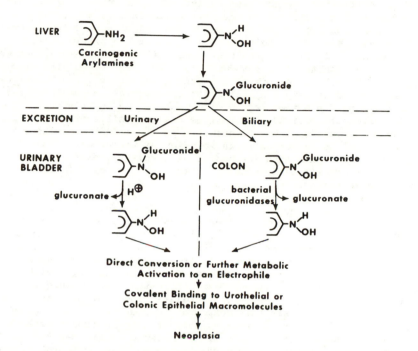

Figure 8. Hypothesis for arylamine-induced urinary bladder and colon carcinogenesis. (Reproduced with permission from Ref. 33. Copyright 1981, Elsevier Scientific Publishers.)

of the steroid: estradiol-17β-glucuronide, ethynylestradiol-17β-
glucuronide, estriol-17β-glucuronide, testosterone-17β-glucuronide
and estriol-16α-glucuronide all showed a reversible cholestatic
effect in the rat (51-53). Sulfation at the 3-position (were available) eliminated the cholestatic activity. Moreover, sulfate groups
in the D-ring positions do not show cholestatic activity. The glucuronides inhibit both bile acid secretion and bile-acid independent
flow (Figure 9). The cholestasis was not due to precipitation of the
glucuronides in bile. Further studies with estradiol-17β-glucuronide
in the rat and the monkey showed that the clearance of indocyanine
green (ICG), which occurs almost exclusively by biliary excretion,
was retarded by this conjugate; the glucuronide conjugate of estra-
diol at the 3-position had no effect at all on ICG clearance. The
effects of the steroid glucuronides are reversible. The cholestasis
induced by estradiol-17β-glucuronide in the isolated perfused rat
liver could be prevented by an infusion of taurocholate (54), due to
the fact that this micelle-forming bile salt increased the rate of
excretion of this glucuronide in bile. Infusion of dehydrocholate,
which does not form micelles, could not reduce toxicity of the
glucuronide, and did not result in an enhanced elimination. The
mechanism of the cholestatic action of the steroid D-ring glucuronides has not yet been resolved. Since the biliary permeability to
^{14}C-sucrose was increased by estradiol-17β-glucuronide, Adinolfi et
al. (54) have suggested that these glucuronides may cause alterations
of canalicular membrane permeability. No effects on ATPase activities
in liver plasma membrane fraction were observed, which might have
been another side of action of the cholestatic glucuronides. The
glucuronide of a bile salt, lithocholate glucuronide, also has a
cholestatic action. In this compound the glucuronide acid group is
at the 3-position in the A-ring, so that this contrasts with the
choleretic action of steroids glucuronidated at the 3-position. The
lithocholate glucuronide is highly water soluble, but precipitates
in the presence of calcium ions. It is not yet clear whether this
precipitation is the cause for the cholestatic effect (55).

The glucuronide conjugate of another compound, harmol, has a
similar cholestatic action, but in this case this is an irreversible
effect. This glucuronide is excreted to very high concentrations in
rat bile, up to 25 mM. At that concentration it is no more soluble,
and crystals of the conjugate are formed in bile, just before cholestasis occurs. Although an effect at another site cannot yet be
excluded, this precipitation of crystals alone could explain the
irreversible cholestasis, due to plugging of and damage to bile
canaliculi (56).

Glutathione conjugation

Glutathione contains a nucleophilic -SH group which in many cases
detoxifies electrophilic reactive intermediates through formation
of stable glutathione conjugates, as in the case of paracetamol.
These glutathione conjugates are further metabolized by either the
intestinal microflora (when they are excreted in bile), or by the
kidney, to mercaptures (Figure 10).

However, in recent years two different pathways for toxification
by glutathione conjugation have been identified. In the first path-

way, toxification occurs because a compound contains two functional groups that may react with the same S-atom in glutathione, forming reactive thiiranium ion. In the other case a toxic metabolite arises through β-lyase activity of the cysteine conjugates; presumably, an -SH derivative is involved in this toxicity (Figure 10).

The first report on activation of a compound by glutathione conjugation to a toxic metabolite was by Rannug et al. (57), who observed that glutathione increased mutagenicity of 1,2-dichloroethane in the Ames test. This was soon confirmed by Van Bladeren et al. (58) for a related compound, 1,2-dichlorocyclohexane. The authors demonstrated that only the *cis*-1,2-dichloro isomer was converted by glutathione to a mutagenic compound, while the *trans* isomer was not activated. This difference could be explained by the postulation of an intermediate thiiranium ion which could only be formed from the *cis*-isomer of the dichloride (Figure 11), because only in that case the glutathione group would be in the *trans*-position to the remaining chlorine atom, a requirement for the formation of the thiiranium ion. This work was further extended to 1,2-dibromoethane (59-60), where a similar increase of mutagenicity in the Ames test was observerd. It was shown that in vivo glutathione conjugation, rather than oxidation of dibromoethane to the aldehyde, was responsible for the formation of the reactive species. The same applies to 1,2-dichloroethane (61). Recently, proof for this came when it was demonstrated that covalent binding of 1,2-dibromoethane to DNA in the presence of ^{35}S-glutathione led to the incorporation of ^{35}S-radioactivity into the adduct to DNA (62-63). This proves that the thiiranium ion is the reactive intermediate in DNA binding and, presumably, in mutagenicity in the Ames test. For another analogue, 1,2-dibromo-3-chloropropane, the situation was more complex, and glutathione actually decreased DNA binding in the presence of cytosol (63). Labile glutathione conjugates could also be formed from vinylidene chloride (64).

Similar work with a series of dihalogen-subsituted alkanes with varying chain length and distance between the halogen atoms revealed that the distance between the halogen atoms is critical, because of neighbouring group participation in nucleophilic displacement reactions of the different halogen atoms. For instance, if there is one CH_2 group between the halogen-carrying carbon atoms, the compound is much less mutagenic than when there is either none, or two or three. Furthermore, the order of leaving group (I > Br > Cl), parallels the order of mutagenicity (65).

Another pathway of toxification by glutathione conjugation is the generation of certain substrates for C-S-lyase activity that become highly toxic upon their conversion (66). Already in 1965 it was noted that cleavage of 1,2-dichlorovinylcysteine resulted in the generation of a highly reactive thiovinyl intermediate which bound covalently to macromolecules and was very toxic for microorganisms (67). This reaction was postulated to play a role in kidney toxicity of 1,2-dichlorovinylcysteine (68). Recently, Elfarra & Anders (69) showed that β-lyase activity was required for activation of such a conjugate by substituting the β-hydrogen atom for a methyl group (Figure 12); this conjugate cannot be cleaved by β-lyase, and was not nephrotoxic in the rat (69).

Several halogenated alkenes are potent nephrotoxins; an example

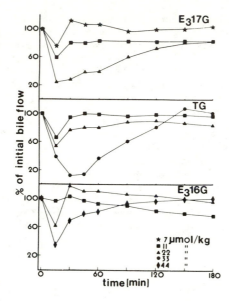

Figure 9. Effect of various steroid D-ring glucuronides on bile
flow in the rat. E₃ 17G = estriol-17β(β-D-glucuronide);
TC = testosterone glucuronide; E₃ 16G = esteriol-16α(β-D-glucuronide). (Reproduced with permission from Ref. 51. Copyright 1981,
American Society for Pharmacology and Experimental Therapeutics.)

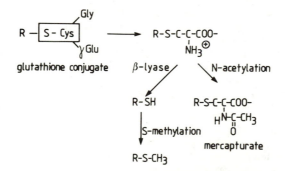

Figure 10. Further metabolism of glutathione conjugates.

Figure 11. Formation of a thiiranium ion from the glutathione
conjugate of cis-1,2-dichlorocyclohexane. (Reproduced with permission from Ref. 58. Copyright 1979, Pergamon Press.)

is hexachlorobutadiene. The major biliary metabolite of this compound is the glutathione conjugate, which is formed by replacement of a chlorine atom. Rat kidney slices convert this glutathione conjugate and also the corresponding N-acetylcysteine conjugate by β-lyase activity (70). The cysteine conjugate of hexachlorobutadiene causes inhibition of active transport of p-aminohippurate and tetraethyl-ammonium ions in isolated kidney tubule preparation from rabbit kidney (71), but little DNA damage was observed. In mice the cysteine conjugate was highly nephrotoxic. Both the glutathione conjugate and the N-acetylcysteine conjugate were potent mutagens in the Ames test (72).

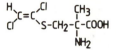

Figure 12.   Structure of S-(1,2-dichlorovinyl)-DL-α-methyl-cysteine.

Several conjugations occur with chlorotrifluorethylene, another nephrotoxin used in plastics manufacturing, and tetrafluoroethylene (73-74). In both cases the glutathione was added over the double bond, and all halogen atoms were retained in the conjugate. The synthetic cysteine conjugate of tetrafluorethylene had a kidney toxicity (when administered orally to rats) identical to that of tetrafluoroethylene itself. Odum & Green (73) suggest that the nephrotoxicity of tetrafluoroethylene (and similar compounds) derives from the hepatic glutathione conjugation and subsequent biliary excretion of the conjugate. In the gut the glutathione conjugate is degraded to the cysteine conjugate which is subsequently reabsorbed and metabolized by β-lyase in the kidney to a cytotoxic species.

A final example of toxification by glutathione conjugation of unknown mechanism was recently reported by Monks et al. (74), who found that an as yet unidentified glutathione conjugate of 2-bromo-hydroquinone was very nephrotoxic in the rat. They propose that this conjugate is synthesized in the liver and subsequently transported to the kidney.

Conclusion

The above results show that quite a few classes of compounds can become toxic upon conjugation by the various pathways discussed. Since the mechanism of toxicity may be both a reversible inter-action with a receptor, and irreversible binding due to the formation of a reactive intermediate, it is difficult to predict toxicity, even within a series of structural analogues as demonstrated with the dihalogenated alkanes. Unless, of course the exact mechanism of action at the molecular level is known and there is a firmer base for structure activity relationship studies.

Literature Cited

1. Mulder, G.J. "Sulfation of Drugs and Related Compounds"; CRC Press: Boca Raton, FL., 1981.
2. Mulder, G.J. In "Progress of Drug Metabolism"; Bridges, J.W. and Chasseaud, L.F., Ed.; Taylor & Francis: London, 1984; Vol. 8, p. 35.
3. Møller, J.V.; Sheikh, M.I. Pharmacol. Revs. 1983, 34, 315.
4. Mulder, G.J.; Brouwer, S.; Weitering, J.G.; Scholtens, E.; Pang, K.S. Biochem. Pharmacol. 1985, 34, 1325.
5. Chasseaud, L.F. Adv. Cancer Res. 1979, 29, 176.
6. DeBaun, J.R.; Rowley, J.Y.; Miller, E.C.; Miller, J.A. Proc. Soc. Exp. Biol. Med. 1968, 129, 268.
7. King, C.M.; Phillips, B. Science 1968, 159, 1351.
8. DeBaun, J.R.; Miller, E.C.; Miller, J.A. Cancer Res. 1970, 30, 577.
9. Maher, V.M.; Miller, E.C.; Miller, J.A.; Szybalski, W. Mol. Pharmacol. 1968, 4, 411.
10. Mulder, G.J. In "Sulfation of Drugs and Related Compounds"; Mulder, G.J., Eds.; CRC Press: Boca Raton, FL., 1981; p. 213.
11. Mulder, G.J.; Meerman, J.H.N. In "Extrahepatic Drug Metabolism and Chemical Carcinogenesis"; Rydström, J.; Montelius, J.; Bengtsson, M., Eds.; Elsevier: Amsterdam, 1983; p. 143.
12. Van den Goorbergh, J.A.M.; Meerman, J.H.N.; De Wit, H.; Mulder, G.J. Submitted.
13. Meerman, J.H.N.; Tijdens, R.B. Cancer Res. 1985, 45, 1132.
14. Meerman, J.H.N.; Mulder, G.J. Cancer Res. 1980, 40, 3772.
15. Meerman, J.H.N.; Mulder, G.J. Life Sci. 1981, 29, 2361.
16. Meerman, J.H.N.; Beland, F.A.; Mulder, G.J. Carcinogenesis 1981, 2, 413.
17. Meerman, J.H.N.; Mulder, G.J. In "Sulfate Metabolism and Sulfate Conjugation"; Mulder, G.J.; Caldwell, J.; Van Kempen, G.M.J., Eds.; Taylor & Francis: London, 1982, p. 145.
18. Thorgeirsson, S.S.; Mitchell, J.R.; Sesame, H.A.; Potter, W.Z. Chem. Biol. Interact. 1976, 15, 139.
19. Shirai, T.; King, C.M. Carcinogenesis 1982, 3, 1385.
20. Loretz, L.J.; Pariza, M.W. Carcinogenesis 1984, 5, 895.
21. Meerman, J.H.N. Carcinogenesis 1985, in press.
22. Meerman, J.H.N.; Beland, F.A.; Fullerton, N.F.; Dooley, K.L. Proc. Am. Soc. Cancer Res. 1984, abst. 513.
23. Bisschop, A.; Bakker, O.; Meerman, J.H.N.; Van Wijk, R.; Van der Heijden, C.A.; Stavenuiter, J.F.C. Cancer Invest. 1984, 2, 267.
24. Ringer, D.P.; Kampschmidt, K.; King, R.L.; Jackson, S.; Kizer, D.D. Biochem. Pharmacol. 1983, 32, 315.
25. Mulder, G.J.; Hinson, J.A.; Gillette, J.R. Biochem. Pharmacol. 1978, 27, 1641.
26. Vaught, J.B.; McGarvey, P.B.; Lee, M.S.; Garner, C.D.; Wang, C.Y.; Linsmaier-Bednar, E.M.; King, C.M. Cancer Res. 1981, 41, 3424.
27. Kadlubar, F.F.; Miller, J.A.; Miller, E.C. Cancer Res. 1976, 36, 2350.
28. Morton, K.C.; Beland, F.A.; Evans, F.E.; Fullerton, N.F.; Kadlubar, F.F. Cancer Res. 1980, 40, 751.

29. King, C.M.; Traub, N.R.; Cardona, R.A.; Howard, R.B. Cancer Res. 1976, 36, 2374.
30. Glatt, H.R.; Oesch, F.; Neumann, H.G. Mutation Res. 1980, 73, 237.
31. Scribner, J.D.; Naimy, N.K. Cancer Res. 1973, 33, 1159.
32. Boyland, E.; Nery, R. J. Chem. Soc. 1962, 5217.
33. Kadlubar, F.F.; Unruh, L.E.; Flammang, T.J.; Sparks, D.; Mitchum, R.K.; Mulder, G.J. Chem. Biol. Interact. 1981, 33, 129.
34. Rickert, D.E.; Butterworth, B.E.; Propp, J.A. CRC Crit. Revs. Toxicol. 1984, 13, 217.
35. Kedderis, G.J.; Dyroff, M.C.; Rickert, D.E. Carcinogenesis 1984, 5, 1199.
36. Johnson, G.A.; Barsuhn, K.J.; McCall, J.M. Biochem. Pharmacol. 1982, 31, 2949.
37. Watabe, T.; Ishizuka, T.; Isobe, M.; Ozawa, N. Science, 1982, 215, 403.
38. Watabe, T.; Izhizuka, T.; Hakamata, Y.; Aizawa, T.; Isobe, M. Biochem. Pharmacol. 1983, 32, 2120.
39. Rietveld, E.C.; Plate, R.; Seutter-Berlage, F. Arch. Toxicol. 1983, 52, 199.
40. Boberg, E.W.; Miller, E.C.; Miller, J.A.; Poland, A.; Liem, A. Cancer Res. 1983, 43, 5163.
41. Fowler, L.J.; John, R.A. Biochem. J. 1981, 197, 149.
42. Yousef, I.M.; Tuchweber, B.; Vonk, R.J.; Massé, D.; Andet, M.; Roy, C.C. Gastroenterology 1981, 80, 233.
43. Vonk, R.J.; Tuchweber, B.; Massé, D.; Perea, A.; Audet, M.; Roy, C.C.; Yousef, I.M. Gastroenterology 1981, 80, 242.
44. Dorvil, N.P.; Yousef, I.M.; Tuchweber, B.; Roy, C.C. Am. J. Clin. Nutr. 1983, 37, 221.
45. Mathis, U.; Karlanganis, G.; Preisig, R. Gastroenterology 1983, 85, 674.
46. Irving, C.C. Xenobiotica 1971, 1, 387.
47. Cardona, R.A.; King, C.M. Biochem. Pharmacol. 1976, 25, 1051.
48. Kinoshita, N.; Gelboin, H.V. Science 1978, 199, 307.
49. Stogniew, M.; Fenselau, C. Drug Metab. Disp. 1982, 10, 609.
50. Faed, E.M. Drug Metab. Revs. 1984, 15, 1213.
51. Meijers, M.; Slikker, W.; Vore, M.J. J. Pharmacol. Exp. Ther. 1981, 218, 63.
52. Vore, M.; Hadd, H.; Slikker, W. Life Sci. 1983, 32, 2989.
53. Slikker, W.; Vore, M.; Bailey, J.R.; Meijers, M.; Montgomery, C. J. Pharmacol. Exp. Ther. 1983, 225, 138.
54. Adinolfi, L.E.; Utili, R.; Gaeta, G.B.; Abernathy, C.O.; Zimmerman, H.J. Hepatology 1984, 4, 30.
55. Oelberg, D.G.; Chari, M.V.; Little, J.M.; Adcock, E.W.; Lester, R. J. Clin. Invest. 1984, 73, 1507.
56. Krijgsheld, K.R.; Koster, H.; Scholtens, E.; Mulder, G.J. J. Pharmacol. Exp. Ther. 1982, 221, 731.
57. Rannug, U.; Sundvall, A.; Ramel, C. Chem. Biol. Interact. 1978, 20, 1.
58. Van Bladeren, P.J.; Van der Gen, A.; Breimer, D.D.; Mohn, G.R. Biochem. Pharmacol. 1979, 28, 2521.
59. Van Bladeren, P.J.; Breimer, D.D.; Rotteveel-Smijs, G.M.T.; De Jong, R.A.W.; Buijs, W.; Van der Gen, A.; Mohn, G.R. Biochem. Pharmacol. 1980, 28, 2521.

60. Van Bladeren, P.J.; Breimer, D.D.; Rotteveel-Smijs, G.M.T.;
    De Knijff, P.; Mohn, G.R.; Buijs, W.; Van Meeteren-Wälchli, B.;
    Van der Gen, A. Carcinogenesis 1981, 2, 499.
61. Storer, R.D.; Conolly, R.B. Toxicol. Appl. Pharmac. 1985, 77,
    36.
62. Ozawa, N.; Guengerich, F.P. Proc. Natl. Acad. Sci. USA 1983,
    80, 5266.
63. Inskeep, P.B.; Guengerich, F.P. Carcinogenesis 1984, 5, 805.
64. Liebler, D.C.; Meredith, M.J.; Guengerich, F.P. Cancer Res.
    1985, 45, 186.
65. Buijs, W.; Van der Gen, A.; Mohn, G.R.; Breimer, D.D. Mutation
    Res. 1984, 141, 11.
66. Bakke, J.; Gustafsson, J.A. Trends in Pharmacol. Sci. 1984,
    5, 517.
67. Anderson, P.M.; Schultz, M.O. Arch. Biochem. Biophys. 1985,
    109, 593.
68. Derr, R.F.; Schultz, M.O. Biochem. Pharmacol. 1963, 12, 465.
69. Elfarra, A.A.; Anders, M.W. Biochem. Pharmacol. 1984, 33,
    3729.
70. Jaffe, D.R.; Hassall, C.D.; Brendel, K.; Gandolfi, A.J.
    J. Toxicol. Evironm. Health 1983, 11, 857.
71. Green, T.; Nash, J.A.; Odum, J.; Howard, E.F. In "Extrahepatic
    Drug Metabolism and Chemical Carxinogenesis"; Rydström, J;
    Montelius, J.; Bengtsson, M., Eds.; Elsevier: Amsterdam, 1983;
    p. 623.
72. Dohn, D.R.; Anders, M.W. Biochem. Biophys, Res. Commun. 1982.
    109, 1329.
73. Odum, T.; Green, T. Toxicol. Appl. Pharmacol. 1984, 76, 306.
74. Monks, T.J.; Lau, S.S.; Gillette, J.R. Abstracts First Inter-
    natl. Symp. on Foreign Compound Metabolism, ISSX, 1983, p. 48.

RECEIVED November 15, 1985

Catabolism of Glutathione Conjugates

J. E. Bakke

Metabolism and Radiation Research Laboratory, Agricultural Research Service, U.S. Department of Agriculture, Fargo, ND 58105

The intracellular conjugation of xenobiotics with glutathione (GSH) results in their detoxication and adjusts their polarity to facilitate excretion from the cell (for recent reviews of GSH conjugation, see 1 and 2). Here the disposition of these GSH conjugates will be discussed.

The best known route for disposition of GSH conjugates in vivo is biotransformation to mercapturic- or premercapturic acids in the respective pathways (MAP and preMAP) as defined in figure 1, however, several other pathways are known to function with MAP and

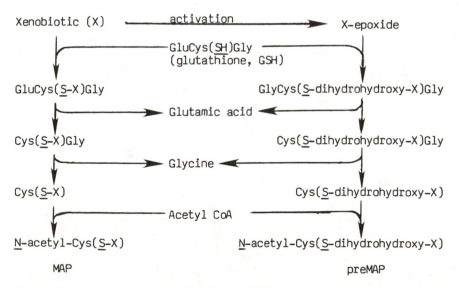

Figure 1. Mercapturic acid pathway (MAP) and premercapturic acid pathway (preMAP).

preMAP intermediates. Often these alternative pathways can be quantitatively more important than mercapturic acid and premercapturic
acid synthesis. A list of the pathways for disposition of GSH conjugates that will be discussed is given in Table I.

Table I. Pathways for Biotransformation of Glutathione Conjugates

1. Mercapturic acid and premercapturic acid synthesis
2. Thiiranium ion formation leading to toxicity [proposed mechanism
    for dibromoethane toxicity (18)]
3. Diconjugate formation
4. Oxidative transamination
5. Sulfoxidation
6. Sulfoxide reduction (microflora)
7. Defuntionalization reactions
    a) reductive dehalogenation
    b) reductive desulfuration
8. Aromatization of preMAP metabolites
9. Thiol formation

     Most of the pathways listed in Table I have been demonstrated
from in vivo studies without the availability of MAP and preMAP
intermediates necessary for precursor/product studies. Substrate
specificities which mediate entry into these various pathways have
been determined in only a few cases. More than one of these
biotransformations can occur in parallel or sequentially during
metabolism, and species variations occur. The organ locations,
topologies and specificities of the enzymes involved in the MAP and
preMAP must also be considered when discussing GSH conjugate disposition. These enzyme factors govern the interorgan translocation of
intermediates of the MAP, preMAP, and subsequent processes, and
therefore control the availability of substrates for biotransformation.

Interorgan Translocation of MAP and PreMAP Metabolites

The interorgan translocation of MAP and preMAP metabolites is
assumed to mimic that deduced for GSH and oxidized GSH (GSSG) metabolism (for reviews see 2 and 3). The in vivo time-course of the
translocation of ^{14}C from S-carbamido-[^{14}C]-methyl glutathione (4)
support a parallelism between GSH and GSH conjugate translocation
and biotransformation. GSH synthesized in cells is translocated
into the blood; GSSG is translocated primarily into the bile.
Circulating GSH is both filtered and secreted (5) into the lumen of
the renal tubule where γ-glutamyltranspeptidase (GTPase) and dipeptidases located on the lumenal side of the renal brush boarder
sequentially hydrolyze the glutamyl and glycyl bonds. The free
amino acids are absorbed preventing their loss through excretion.
     Many tissues are capable, at least in vitro, of conjugation of
xenobiotics with GSH (1); the liver is assumed to be the most
active, especially with ingested xenobiotics, however GSH transferase activity has been detected in the intestinal mucosa (6). The
interorgan translocation of GSH conjugates formed in the intestinal

mucosa upon absorption is not known. Do they become substrates for
liver enzymes and biliary excretion, or do they translocate to the
kidney for excretion?

As with GSH, the GSH conjugates are translocated from the cell
into the circulatory system, with hepatocytes, translocation into
bile also occurs. Circulating GSH conjugates are translocated to
the lumen of the renal tubule and processed as described for GSH.
The terminal xenobiotic-containing hydrolysis product, the cysteine
conjugate, can either be excreted with the urine, or, more commonly,
absorbed, acetylated (either in liver or kidney) and the resulting
mercapturic or premercapturic acid excreted with urine or bile. The
mercapturic acid can also be oxidized to the sulfoxide prior to
excretion.

MAP and preMAP metabolites excreted with the bile undergo the
same hydrolytic biotransformations as those translocated into the
blood, however, the translocation of the intermediates is obviously
different. In contrast to the tissue systems where synthesis of
mercapturic and premercapturic acids are considered as the end products of the MAP and preMAP, respectively, in the intestine the
corresponding cysteine conjugates are the end products. γ-GTPase
associated with epithelial cells lining the biliary tree (7), present in bile (8) and in the intestinal lumen remove the glutamic
moiety to yield cysteinylglycine conjugates which are hydrolyzed to
cysteine conjugates by peptidases present in gastrointestinal secre-
tions as well as the microflora. Mercapturic acid sulfoxides are
reduced by the flora to mercapturates, and mercapturates can be
deacetylated to cysteine conjugates (9). Mercapturates and cysteine
conjugates (including their preMAP analogues) formed in the intestinal lumen can then be either translocated as such to tissue systems
for further metabolism and/or excretion, i.e. enterohepatic circulation, or become substrates for microfloral metabolism (10,11).

Properties of MAP and preMAP conjugates which influence the
extent of the enterohepatic circulation of intact MAP and preMAP
metabolites are not known. Differences in enterohepatic circulation
can be deduced from levels of excretion of MAP and preMAP metabolites by control rats and rats with cannulated bile ducts. Data
from 2-chloro-N-isopropylacetanilide (10) and naphthalene (11) metabolism studies are given in Table II. It is apparent that the
acetanilide biliary MAP metabolites were not absorbed from the
intestine for excretion as mercapturic acid, but were absorbed as
intestinal catabolites of MAP metabolites (23%) and excreted with
the urine.

The preMAP metabolites from naphthalene were extensively reabsorbed (~ 24%) to be excreted with the urine as the premercapturic
acid. The lack of enterohepatic circulation of intact MAP metabolites of the acetanilide is attributed to microfloral catabolism and
not to an inability of these MAP metabolites to be translocated to
the tissues (10), because MAP metabolites of the acetanilide were
extensively absorbed (~ 60%) from the gastrointestinal tracts of
germfree and antibiotic treated rats and were excreted unchanged
(except for mercapturic acid formation and sulfoxidation) in the
urine or bile. Similar studies with naphthalene preMAP metabolites
in control rats showed that 70 to 80% of the doses (oral) were
absorbed and excreted in the urine mainly as the premercapturic acid
(~ 60%) with about 17% as microfloral catabolites (11). When

Table II. MAP and PreMAP Metabolite Excretion from Control and Bile-
Duct-Cannulated Rats Dosed Orally with 2-Chloro-N-isopropylaceta-
nilide and Naphthalene

| | % of dose excreted as MAP and preMAP metabolites | | | |
| | 2-chloro-N-isopropyl acetanilide | | Naphthalene | |
	control[a]	cannulated[b]	control[a]	cannulated[b]
URINE	17	18	38	14
BILE	-	67	-	27
total	17	85	38	41
enterohepatic circulation as MAP (urine, a - b)	~ 0		~ 24	
amount catabolized (total, b - a)	~ 67%		~ 3	

naphthalene preMAP metabolites were injected directly into the
cecum, the urinary excretion of microfloral catabolites increased
(36 to 43% of the doses) probably reflecting increased microfloral
catabolism before translocation of preMAP metabolites to the tissue
could occur. The results from these two xenobiotics indicate that
either the capabilities for catabolism of preMAP (naphthalene) metabolites in rats resides lower in the gastrointestinal tract than
those for the MAP (acetanilide) metabolites or, the capability for
absorption of preMAP metabolites resides higher in the tract. In
either case, a compartmentalization of activities is apparent.
    Endogenous synthesis of GSH conjugates and excretion with the
bile is not the only source for gastrointestinal MAP metabolites;
they can also be present in the diet. S-Alkylcysteine conjugates
are constituents of many plants (mustards, legumes, onions, cabbage,
see reference 12), also, economic plants metabolize xenobiotics in
the MAP (13). MAP metabolites have also been inadvertently added to
diets by processing. Dichlorovinylcysteine was present in soybean
meal that had been defatted using trichloroethylene (14). This
synthesis was discovered because of the toxicity of this cysteine
conjugate to calves fed the defatted meal (15). In addition, biological activity has been attributed to many cysteine conjugates, or
to the plants containing them (16,17).

Biotransformations Occurring with Glutathione Conjugates (Table I)

Thiiranium (also termed "episulfonium") Ion Formation. Thiiranium
ion formation (Equation 1) has been proposed to explain the toxicity
of dibromoethane (18).

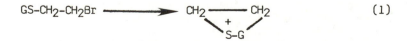

$$GS-CH_2-CH_2Br \longrightarrow \underset{\underset{S-G}{+}}{CH_2 \text{---} CH_2} \qquad\qquad (1)$$

The reactivities of such postulated intermediates were well documented in vitro, and are discussed elsewhere in this series.

Thiiranium ions have also been proposed as intermediates in the decomposition of preMAP metabolites (see "Aromatization of preMAP Metabolites" below).

Diconjugate Formation. Examples of biosynthesis of diconjugates containing glutathione are listed in Table III. No significance has been associated with the formation of diconjugates. The excretion

Table III. Diconjugates of Xenobiotics Containing Elements from GSH

Xenobiotic	Diconjugate	Source	Ref
$(ClCH_2CH_2)_2S$	$(GS^a-CH_2CH_2)_2S$	rat urine	19
$BrCH_2CH_2Br$	$GS-CH_2CH_2-GS$	rat liver	20
$CHCl_3$; $BrCCl_3$; CCl_4	$(GS)_2CO$	rat bile	21
Pentachloronitrobenzene	$(GS)_2(C_6Cl_4)^b$	plant tissues	22
$Br(CH_2)_4CH_3$	$^-O_3SO(CH_2)_5$ $SCH_2\overset{\displaystyle HN-C(O)CH_3}{\underset{\displaystyle }{CHCOOH}}$	rat urine	23
$(C_6H_5)^b-N_2-(C_6H_4)^b-N(CH_3)_2$	$^-O_3SO-(C_6H_4)^b-N_2-(C_6H_4)^b-NH-GS$		24
Acetaminophen	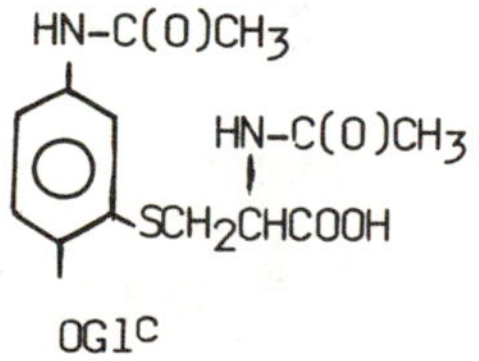	human urine	25
Naphthalene	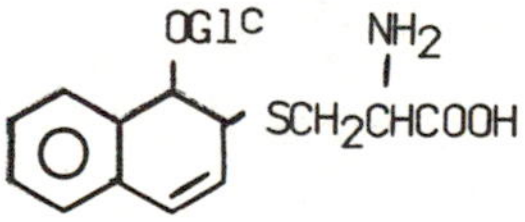	rat bile	26
$CHBr(CH_2Br)_2$	$CHBr[CH_2-(S)-MAP]_2^d$		27

a GS = glutathione
b benzene rings
c Gl = glucuronide
d MAP = mercapturic acid pathway metabolites

of the bisGSH conjugate from bis(β-chloroethyl)sulfide is the only reported excretion of a GSH conjugate with the urine (19). With our present knowledge of kidney metabolism of GSH conjugates to thiols, this excretion may have resulted from the C-S lyase mediated kidney damage that will be considered later under "Thiol Formation".

For the mixed diconjugates, the order in which the conjugations take place is not known. The order can be predicted for acetaminophen and naphthalene because of the proposed pathways for formation of the GSH conjugates. For acetaminophen, the proposed intermediate for GSH conjugation is a quinoneimine (28) which could not be formed with a glucuronide in the para position. For naphthalene, displacement of a hydroxyl in the dihydrodiol glucuronide with GSH is difficult to rationalize.

A sequential diconjugation involving oxidative transamination (see "Oxidative Transmination" below) followed by glycine conjugation has been observed in $\underline{S}$-methylcysteine metabolism (Equation 2) (29).

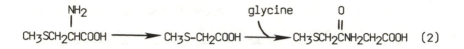

$$CH_3SCH_2CHCOOH \longrightarrow CH_3S-CH_2COOH \xrightarrow{glycine} CH_3SCH_2\overset{O}{\overset{\|}{C}}NH_2CH_2COOH \quad (2)$$

Another type of sequential diconjugation that has been observed involves catabolism of a cysteine conjugate to hippuric acid (Equation 3) (30,31), and a proposed sulfate ester to a MAP metabo-

$$(C_6H_5)CH_2N=C=S \longrightarrow (C_6H_5)CH_2NH\overset{S}{\overset{\|}{C}}-SCH_2\overset{NH_2}{\overset{|}{C}}H \longrightarrow (C_6H_5)\overset{O}{\overset{\|}{C}}NHCH_2COOH \quad (3)$$
$$COOH$$

lite (Equation 4) (32). Intermediary metabolism such as this is not detected $\underline{in}$ $\underline{vivo}$ without the availability of the proper intermediates for precursor/product studies.

$$NC(C_6H_4)\overset{CH_3}{\underset{CH_3}{\overset{|}{N_2}}} \longrightarrow [NC(C_6H_4)\overset{O}{\overset{\|}{N}}CCH_2OSO_3^-] \longrightarrow NC(C_6H_4)\overset{O}{\overset{\|}{N}}CCH_2S(MAP) \quad (4)$$

Oxidative Transamination. MAP and preMAP metabolites undergo transamination to α-keto acids which can then be oxidatively decarboxylated to $\underline{S}$-substituted-2-mercaptoacetic acids and/or reduced to $\underline{S}$-substituted-3-thiolactic acids (Equation 5). MAP and preMAP

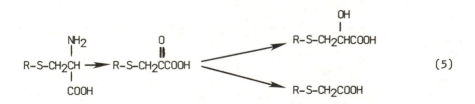

$$R-S-CH_2\overset{NH_2}{\overset{|}{C}}H \longrightarrow R-S-CH_2\overset{O}{\overset{\|}{C}}COOH \begin{cases} R-S-CH_2\overset{OH}{\overset{|}{C}}HCOOH \\ \\ R-S-CH_2COOH \end{cases} \quad (5)$$
$$COOH$$

cysteine conjugates are assumed to be the precursors for this pathway. The pyruvate intermediate has been isolated in at least two
instances [styrene, 33, and S-pentyl-L-cysteine, 34], but the most
usual products excreted from this pathway are the thiolactic and
thioacetic acids. Examples where oxidative transamination of MAP
and preMAP metabolites have occurred are given in Table IV.

No biological significance has been attached to the existence
of this pathway other than as a step to further increase the
polarity, and therefore excretability, of cysteine conjugates.

Table IV.   Xenobiotics and Cysteine Conjugates That Are Excreted as
Transaminated Metabolites

Compound	Species	Ref
S-Methyl-L-Cysteine	rat	(35)
	man	(12)
S-Pentyl-L-cysteine	guinea pig, mouse, rabbit	(34)
1,1-Dichloroethylene	rats	(35)
S-Propylcysteine	rats	(36)
S-Carboxymethyl-L-cysteine	man, rat, dog,	
	monkey, rabbit	(37,38,39)
Benzylisothiocyanate	guinea pig, rabbit	(31)
2,4',5-Trichlorobiphenyl	rat	(40)
Naphthalene	mouse	(41)
Phenanthrene	rat, guinea pig	(42)
Styrene oxide	guinea pig	(33)
2-Chloro-N-isopropylacetanilide	mouse	(unpublished)

Reactions subsequent to transamination have been reported, one
being the diconjugate formation shown in equation 2. Spontaneous
cyclization of a pyruvate intermediate was proposed as the mechanism
for formation of the N,N-disubstituted hemiaminal (equation 6) found
as a metabolite of benzylisothiocyanate (31). Another common
biotransformation for transamination products is sulfoxidation.

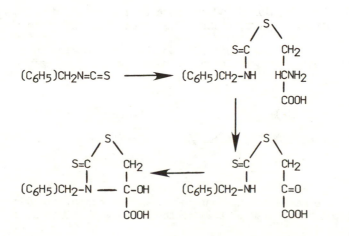

(6)

<u>Sulfoxidation</u>. Sulfoxidation is common in sulfur biochemistry
(43,44), and has been observed with many of the transamination products listed in Table III, with mercapturic acids (45,46,47) and
cysteine conjugates (12). Sulfoxidation is also common with the
methylthio-containing metabolites produced from cysteine conjugates
(see "Thiol Formation" below). Sulfoxidation increases the polarity
of compounds and therefore affects their excretability.
    Sulfoxidation also may be involved in other aspects of the
metabolism of methylthio-containing compounds. Methylsulfoxyl- and
methylsulfonyl-groups are better leaving groups for displacement by
GSH, and may play a role in the methylthio-turnover observed in the
metabolism of bis(methylthio)tetrachlorobenzene (48). Methylthio-
turnover will be considered later.

<u>Sulfoxide Reduction</u>. Reduction of the sulfoxide of the mercapturic
acid of 2-chloro-<u>N</u>-isopropylacetanilide by intestinal microflora has
been shown <u>in vitro</u> (9), but has not been demonstrated to occur in
tissues. The excretion of this mercapturate sulfoxide, which is
excreted into the intestine with the bile, in feces from germfree
rats dosed with the acetanilide is <u>in vivo</u> evidence for the reductive function of the intestinal microflora (49) in MAP catabolism.
This reduction, in addition to deacetylation, is another source for
cysteine conjugates which can be translocated to the tissues for
metabolism or excretion, or remain in the intestine and further
catabolized by the microflora.

<u>Defunctionalization</u>. The mechanism for GSH-mediated reductive
dehalogenation of 2,4-dichlorophenacylchloride (Equation 7) was
described by Hutson et al. (50) and has been confirmed (51). The

$$(C_6H_3Cl_2)\overset{\overset{O}{\|}}{C}CH_2Cl \xrightarrow{GSH} (C_6H_3Cl_2)\overset{\overset{O}{\|}}{C}CH_2\text{-SG} \xrightarrow[GSSG]{GSH} (C_6H_3Cl_2)\overset{\overset{O}{\|}}{C}CH_3 \qquad (7)$$

reaction is enzyme catalyzed and the mercapturic acid did not serve
as an intermediate, however, other MAP intermediates prior to the
mercapturate were not tested.
    2-Chloro-<u>N</u>-isopropylacetanilide, which also contains an α-halo-
carbonyl structure undergoes reductive dehalogenation in chickens
(Equation 8, unpublished) and probably in rats (52) and soil (53).

$$(C_6H_5)\overset{\overset{HC(CH_3)_2}{|}}{N}C(O)CH_2Cl \xrightarrow[GSSG]{2\ GSH} (C_6H_5)\overset{\overset{HC(CH_3)_2}{|}}{N}C(O)CH_3 \qquad (8)$$

In chickens, the mechanism did not involve amide hydrolysis (which
occurred in the metabolism of this acetanilide) with subsequent <u>N</u>-

acetylation because the product retained the carbonyl-carbon-label
present in the dosed compound. The cysteine conjugate served as a
precursor for this dehalogenated product in chickens. This
precursor-product relationship indicates a broader specificity for
MAP metabolites as intermediates in reductive dehalogenation than
just the GSH conjugate.

  Recent work with N-acetyl-S-(pentachlorophenyl)-cysteine also
implicates a reductive defuntionalization process in the metabolism
of hexachlorobenzene and pentachloronitrobenzene (54) as shown in
Equation 9. In addition to pentachlorobenzene, other reduction

$$\begin{array}{c} C_6Cl_6 \\ \searrow \\ \\ \\ C_6Cl_5(NO_2) \end{array} \quad (C_6Cl_5)\text{-S(MAP)} \longrightarrow C_6Cl_5H \qquad (9)$$

products, tetrachlorothiophenol, tetrachlorothioanisole and tetrachlorobenzene were also detected as metabolites from pentachlorophenyl-mercapturic acid. These authors also showed that
pentachlorothiophenol and pentachlorothioanisole were also *in vivo*
precursors for several of these reduction products, and that
tetrachlorothiophenols were *in vivo* precursors for tetrachloroben-
zenes. A "reductive desulphuration" of the corresponding thiophenols was proposed as the route of formation (Equation 10), but no
details or quantitative data were presented.

$$(C_6Cl_4)SCH_3 \longrightarrow (C_6Cl_4)SH \longrightarrow C_6Cl_4H \qquad (10)$$

  An alternative or expanded pathway based on the above GSH-
dependent reductive dehalogenation model is proposed in figure 2
which can account for formation of all the metabolites reported for
pentachlorophenyl-mercapturic acid (54). Key intermediates for formation of three of the four reduction products (rxns E, N, Q, and
R), S-[(methylthio)tetrachlorophenyl]-MAP metabolites, have been
isolated in high yields from bile from rats dosed with either
pentachlorothioanisole (rxn F) (55) or bis(methylthio)tetrachloro-
benzene (rxn I) (48). These latter two compounds are both formed by
the C-S lyase mediated pathway described under "Thiol Formation".
The pathways shown in figure 2 can also account for the formation of
reduction products from thiophenols and thioanisoles. A further
attraction of the pathway shown in figure 2 is the sequential reductions leading to the formation of tetrachlorobenzene, and the cycles
shown can be expanded to include formation of further reductive
dehalogenation products that have been reported (56). The reduction
products described in figure 2 could undergo oxidation and therefore
be precursors for the phenols and anisoles that have also been
reported as metabolites (56). The proposed MAP intermediate for
tetrachlorobenzene formation (rxn R) has probably not been observed
due to its low concentration; quantitative data were not available

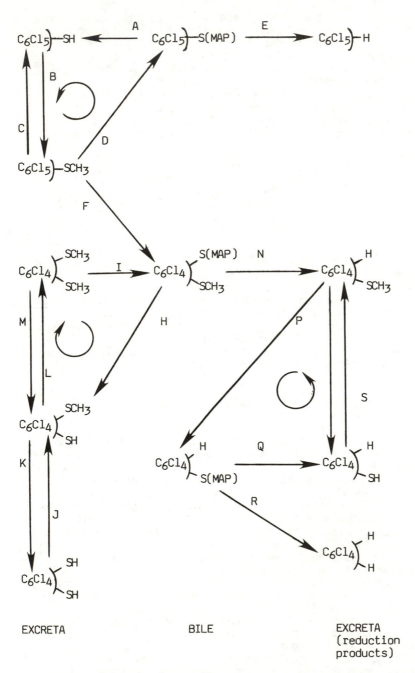

Figure 2. Proposed pathway for reductive defunctionalization (see text) and methylthio-group turnover observed in hexachlorobenzene metabolism.

for the formation of the reduction products but they are probably
very minor metabolites.

Thiol Formation. Thiol formation from cysteine conjugates has been
known since 1951 (57); toxicity of the thiol formed during
dichlorovinylcysteine metabolism was reported in 1957 (15); and the
enzyme involved was termed a "thionase" in 1965 (58). The enzyme(s)
is now called a cysteine conjugate-β-lyase (C-S lyase) and has been
purified from rat liver (59,60), partially purified from microorganisms (61,62) shown to function in the intestinal microflora (9,49),
and in kidney (63). The general reaction catalyzed by C-S lyase is
shown in Equation 11.

$$X-S-CH_2\overset{\overset{\displaystyle NH_2}{|}}{C}HCOOH \xrightarrow{\text{C-S lyase}} X-SH + CH_3\overset{\overset{\displaystyle O}{\|}}{C}COOH + NH_4 \qquad (11)$$

Of the MAP metabolites tested in vitro, the enzymes studied to
date have a specificity for L-cysteine conjugates. A preMAP metabolite (the cysteine conjugate of naphthalene) has been tested only in
microorganisms (61) and was a substrate. Tissue C-S lyase exhibits
an additional specificity for cysteine conjugates in which the carbon of the xenobiotic moiety attached to the cysteine sulfur be
unsaturated (vinyl or aromatic), therefore preMAP metabolites, which
lack this aromaticity, should not be substrates for tissue C-S
lyase. In vivo evidence that this is the case was obtained in the
metabolism of naphthalene (11). C-S lyases associated with the
intestinal microflora, which may involve many different enzymes in a
number of different microorganisms, do not exhibit this specificity
for aromatic compounds.

Thiols are generally considered to be biologically active, if
not toxic, and are seldom excreted in quantity without biotransformation. The detoxifying biotransformations for thiols are usually
methylation followed by oxidation to sulfoxides and sulfones and S-
glucuronides are also formed (58,64,65). The biotransformation of
vinyl halogens to compounds exhibiting nephrotoxicity exemplify an
adverse biological activity of thiols (Equation 12). The nephrotoxicity is effected when the circulating GSH conjugate is translocated to the kidney tubule, sequentially hydrolyzed by the brush

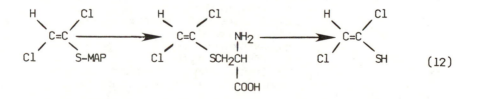

boarder enzymes (γ-GTPase and dipeptidases) to the cysteine conjugate which is cleaved to the reactive thiol by renal C-S lyase.
The reactive thiol becomes bound into the tissues. Dichlorovinylcysteine (DCVC) has been used as a model for this binding process

not only in kidney (66,67) but also in bovine liver homogenates (67)
and E. coli (68). No known beneficial biological function for C-S
lyases has been proposed.

A similar pathway may be operating in the metabolism of
bis(β-chloroethyl)sulfide (19). In this study, damage to renal
tubules was indicated because of the rapid excretion of the doses as
bis-glutathione conjugates (45% of the dose in 12 hrs). GSH conjugates are rarely excreted, this being the only reported excretion
of GSH conjugates in the urine. If renal C-S lyase is functioning,
it would indicate an exception to the specificity of tissue enzymes
for aromaticity in the cysteine conjugate.

The C-S lyases in the intestinal microflora and their lack of
specificity for cysteine conjugates complicates the disposition of
MAP- and preMAP-metabolites that enter the gastrointestinal tract
with the bile, the diet, and through secretion. Metabolic forces
associated with bile, bile duct, intestinal enzymes and intestinal
flora function to produce cysteine conjugates from these MAP and
preMAP metabolites. If these cysteine conjugates are not translocated to the tissues, they become substrates for C-S lyase-
containing microorganisms.

Four examples will be used to show the complexities introduced
into GSH conjugate disposition because of the presence of C-S lyase
in the intestinal microflora. The first example (shown in figure 3)
is 2-chloro-N-isopropylacetanilide. This acetanilide is quantitatively conjugated with GSH in the rat (49) and the resulting MAP

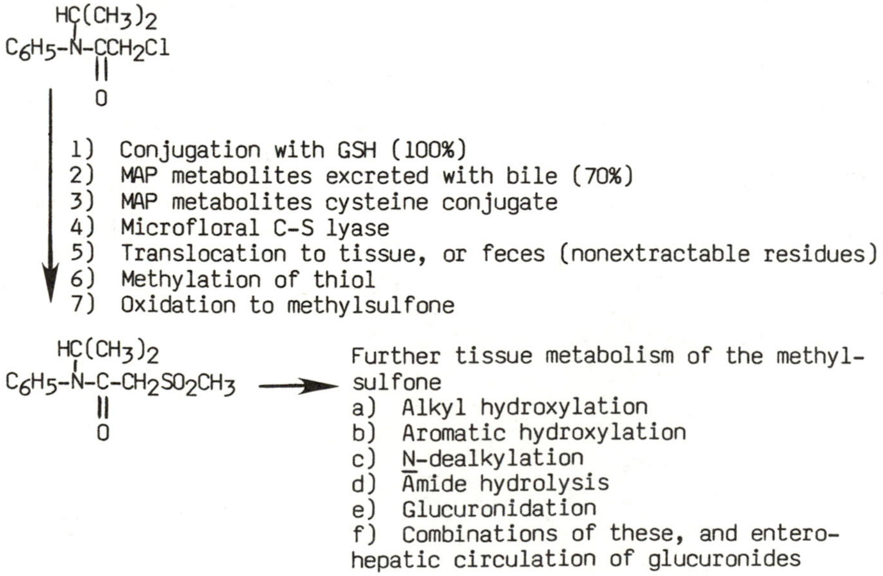

$$
\begin{array}{l}
\text{HC(CH}_3)_2 \\
\text{C}_6\text{H}_5\text{-N-CCH}_2\text{Cl} \\
\quad\quad \| \\
\quad\quad \text{O}
\end{array}
$$

1) Conjugation with GSH (100%)
2) MAP metabolites excreted with bile (70%)
3) MAP metabolites cysteine conjugate
4) Microfloral C-S lyase
5) Translocation to tissue, or feces (nonextractable residues)
6) Methylation of thiol
7) Oxidation to methylsulfone

$$
\begin{array}{l}
\text{HC(CH}_3)_2 \\
\text{C}_6\text{H}_5\text{-N-C-CH}_2\text{SO}_2\text{CH}_3 \\
\quad\quad \| \\
\quad\quad \text{O}
\end{array}
$$
$\longrightarrow$ Further tissue metabolism of the methyl-
sulfone
a) Alkyl hydroxylation
b) Aromatic hydroxylation
c) N-dealkylation
d) Amide hydrolysis
e) Glucuronidation
f) Combinations of these, and enterohepatic circulation of glucuronides

Figure 3. Biotransformations involved in 2-chloro-N-isopropyl-
acetanilide metabolism.

metabolites are extensively excreted with the bile (*69*); the mercapturate (*52*), cysteine conjugate (*70*), and sulfoxide of the mercapturate (*45*) are excreted in the urine. The cysteine conjugate
formed from the biliary MAP metabolite is translocated to the flora
and cleaved to the thiol which is, in part, translocated to the
tissues where it is methylated and oxidized. The methylation is
presumed to occur in the tissues because S–methyltransferase has not
been detected in the intestinal contents (71). The formation of the
methylsulfonylacetanilide results in conversion of polar, excretable, MAP metabolites into a xenobiotic exhibiting a polarity similar to that of the parent chloroacetanilide. The new lipophilic
compound undergoes further tissue functionalization and conjugation
transformations to render it excretable. This metabolism includes a
second cycle of enterohepatic circulation of the glucuronides formed
(10).

   A similar 2-methylsulfonyl acetanilide was formed in the metabolism of p–dimethylaminobenzonitrile (*32*). In this case very
complex metabolism went on prior to GSH conjugation (Figure 4).

R-N(CH$_3$)$_2$
   ↓ 1) Demethylation

R-NH$_2$
   ↓ 2) Acetylation

R-NHC(O)CH$_3$
   ↓ 3) Hydroxylation

R-NHC(O)CH$_2$OH
   ↓ 4) Conjugation

[R-NHC(O)CH$_2$OSO$_3$]
   ↓ 5) GSH conjugation

R-NHC(O)CH$_2$S-G
   ↓ 6) [Biliary secretion]
   ↓ 7) [Microfloral] C-S lyase

[R-NHC(O)CH$_2$SH]
   ↓ 8) [Translocation to tissue]
   ↓ 9) Methylation

[R-NHC(O)CH$_2$SCH$_3$]
   ↓ 10) Oxidation

R-NHC(O)CH$_2$SO$_2$CH$_3$

Figure 4. Proposed pathway for metabolism of p–dimethylaminobenzonitrile to a methylsulfonyl containing acetanilide. Bracketed [ ]
intermediates and processes are proposed.   R= p-NC(C$_6$H$_4$)

Although biliary secretion of MAP metabolites was not determined, the known specificities of the C-S lyases and the studies outlined in Fig. 3 indicate that it was necessary for the formation of the end product.

In the metabolism of 2,4',5-trichlorobiphenyl (TriCB, Figure 5), biliary MAP metabolites appeared in the feces and lungs (lung residues in Table V) as TriCB methylsulfones (72,73); and the mercapturate was shown to be a precursor for the triCB methylsulfones (73). Studies in germfree rats and rats with cannulated bile ducts showed that the lung residues are probably formed, for the greatest part, by the microfloral C-S lyase mediated scenario presented in figure 5. The fecal TriCB methylsulfones are assumed to be formed in the same pathway as the lung residues except that instead of translocation of triCB-SCH$_3$ and oxidized products to the lungs they are secreted back into the intestinal lumen for excretion. Methylation of the microfloral C-S lyase cleavage product is assumed to occur after translocation to the tissues because S-methyltransferase is not present in the intestinal lumen (71).

Routes other than biliary secretion of triCB MAP metabolites must also be considered in formation of triCB-methylsulfones from triCB mercapturic acid because rats with cannulated bile ducts, dosed either with triCB (72) or triCB mercapturate (40), also had

Table V. Rat Lung Residues of 4-Methylsulfonyl-2,4',5-trichloro-
biphenol Resulting from Metabolism of 2,4',5-Trichlorobiphenyl
or Its Mercapturate

animal	compound dosed (route)	ppm ± S.D.	Ref.
control	triCB (oral)	12.7 ± 1.8	72
cannulated bile duct	triCB (oral)	0.6 ± 0.3	73
germfree	triCB (oral)	1.3 ± 0.3	unpublished
control	triCB mercapturate (intracecal)	19.3 ± 6.3	unpublished
cannulated bile duct	triCB mercapturate (intravenous)	4.0 ± 0.3	unpublished

lung residues (72), as did germfree rats dosed with triCB (unpublished). Three alternative pathways to the formation of triCB-methylsulfones can be functioning. Considering the amounts of material involved in rat lung residues (0.13 to 14 µg of triCB-SO$_2$CH$_3$), these mechanisms need involve only minor amounts of triCB.

The germfree experiment showed that minor amounts of triCB-SO$_2$CH$_3$ are formed by a nonmicrofloral mediated pathway. This could be mediated by tissue C-S lyase activity as shown in Figure 5(C), or by the pathway shown in Equation 13. This epoxide mediated pathway

$$\text{-iCB} \longrightarrow \text{triCB-epoxide} \xrightarrow{\quad \text{R-SCH}_3 \quad} \text{triCB-S(CH}_3\text{)R} \longrightarrow \text{triCB-SCH}_3 \qquad (13)$$

has been demonstrated in vitro for 2,5,2',5'-tetrachlorobiphenyl
(74). Also, arene oxide formation is indicated in the metabolism of
triCB from the excretion of dihydrodiols and vic-diols as metabolites (40), however, preMAP metabolites were not detected (72).
PreMAP metabolites of triCB could, however, undergo dehydration
prior to excretion even though the opposite aromatization process
(elimination of HS-R) appears to function in the metabolism of
naphthalene preMAP metabolites to yield naphthol (11).

Translocation of triCB-S(MAP) metabolites into the intestine by
a nonbiliary route could also be functioning [Fig. 5(B)]. Rats with
cannulated bile ducts dosed intravenously with [14C]-triCB mercapturic acid also had lung residues of triCB-methylsulfones (Table V)
which showed the mercapturate was a precursor for triCB-methylsulfones without biliary excretion. These residues could have been
formed by the tissue C-S lyase mediated pathway [Fig. 5(C)], but, in
this experiment, 14C from the mercapturate (2% of the dose) was also
translocated by some nonbiliary route into the intestine for excretion with the feces (73). Therefore, the microflora mediated pathway is also a possible source of the lung residues after secretion
of precursors into the intestine by a nonbiliary route.

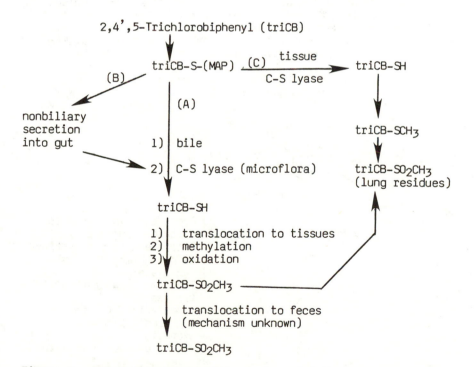

Figure 5. Metabolism and possible routes of interorgan translocation of 2,4[1],5-trichlorobiphenyl.

Another variation in GSH conjugate biotransformation has been
observed in metabolism of preMAP metabolites of naphthalene (11).

As outlined in figure 6, biliary preMAP metabolites are precursors for naphthol and methylthio-containing metabolites excreted in the urine primarily as glucuronides. These methylthio-containing metabolites of naphthalene were first reported by Stillwell et al. (75). The formation of both of these aglycones (naphthol and 1,2-dihydro-1-hydroxy-2-methylthionaphthalene) was mediated by the intestinal microflora because they were not excreted by germfree rats or rats with cannulated bile ducts that were dosed with naphthalene (11). PreMAP metabolites were shown to be in vivo precursors for both naphthol glucuronide and the glucuronide of 1,2-dihydro-1-hydroxy-2-methylthionaphthalene by dosing rats orally and intracecally with the preMAP metabolites (11). PreMAP metabolites of phenanthrene are also precursors for dihydrohydroxy-methylthio-containing metabolites of phenanthrene by the same pathway (76).

The mechanism and precursors for microfloral naphthol formation are not known. An obvious possibility is the elimination of the amino acid or peptide moiety as the thiol. The preMAP metabolites need not be the immediate precursors because C-S lyase products could also be aromatized by the elimination of the appropriate thiols (i.e., H_2S, $HSCH_3$).

The mechanism for formation of the methylthio-containing aglycone of naphthalene is the same as that for the acetanilide (figure 3). Tissue C-S lyase activity does not function with preMAP [either naphthalene (11) or phenanthrene (74)], presumably because preMAP structures lack the aromaticity required.

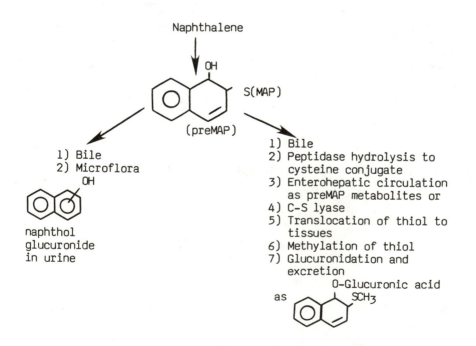

Figure 6. Catabolism of biliary preMAP metabolites of naphthalene.

Aromatization of preMAP Metabolites. Another spontaneous aromatization process has been observed for preMAP metabolites. PreMAP
metabolites of naphthalene (77) and phenanthrene (78) have been
shown to spontaneously aromatize to the parent hydrocarbon. The
decomposition of 9,10-dihydro-9-hydroxy-10-(S-cysteinyl)phenanthrene
has been proposed to occur through a thiiranium ion as shown in
Equation 14. The isolation of the 9,10-dihydrodiol, and a phenol

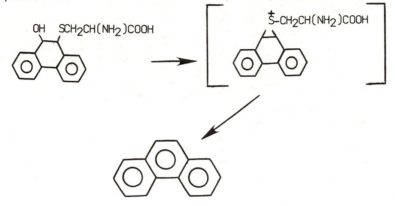

(14)

other than the 9-isomer as metabolites from rats dosed with the
above cysteine conjugate of phenanthrene (76) (Equation 15) indicates that this aromatization, or at least the reformation of the 9,10
double bond, occurs in vivo.
     This latter aromatization process may be the source of the
benzo(a)pyrene formed when biliary metabolites of benzo(a)pyrene
were incubated with human and rat feces (79). This bile most certainly contained preMAP metabolites of benzo(a)pyrene (80). This

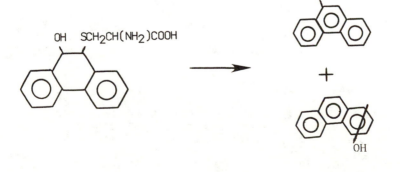

(15)

decomposition (enzymatic or spontaneous) of biliary preMAP metabolites may be a mechanism for translocation of polyaromatic hydrocarbons to the intestine where mucosal cytochrome P_{450} systems could
activate them.

The introduction of methylthio-groups into xenobiotics during metabolism has been observed with approximately 40 compounds. Some of these are listed in reference (81). Many of these compounds are conjugated with GSH, therefore, many of the methylthio metabolites are probably formed in C-S lyase mediated pathways.

Methylthio-Group Turnover. The sequence of events which result in the introduction of methylthio-groups into xenobiotics also results in metabolic turnover of methylthio-groups. Methylthio-group turnover is outlined in the three cyclic processes shown in figure 2. This process was shown to occur in the metabolism of bis(methylthio)tetrachlorobenzene (48), and is also functioning in the metabolism of pentachlorothioanisole (unpublished).

Methylthio turnover could be responsible for the high residues of pentachlorothioanisole and bis(methylthio)tetrachlorobenzene found in mussels gathered around the estuary of the river Rhine (82). These residues could result from the metabolism of pentachloronitrobenzene or hexachlorobenzene in the MAP in some species upstream, and the resultant methylthio-containing metabolites translocated downstream undergoing GSH-mediated methylthio-turnover in various species, the end result being an apparent bioaccumulation.

Acknowledgment

No warranties are herein implied by the U.S. Department of Agriculture.

Literature Cited

1. Chasseaud, L.F. In "Glutathione: Metabolism and Function"; Arias, I.M.; Jakoby, W.B., Eds.; Raven Press: New York, 1975; p. 79.
2. Reed, D.J.; Meredith, M.J. In "Drugs and Nutrients"; Roe, D.A.; Campbell, T.C., Eds.; Marcel Dekker, Inc.: New York and Basel, 1984; pp. 179-224.
3. Meister, A. Science 1983, 220, 472-477.
4. Inoue, M.; Okajima, K.; Morino, Y. Hepatology 1982, 2, 311-316.
5. Anderson, M.E.; Bridges, R.J.; Meister, A. Biochem. Biophys. Res. Commun. 1980, 96, 848-853.
6. Pinkus, L.M.; Ketley, J.N.; Jakoby, W.B. Biochem. Pharmacol. 1977, 26, 2359-2363.
7. Tanaka, M. Acta Pathol. Jap. 1974, 24, 651.
8. Rosalki, S.B. Adv. Clin. Chem. 1975, 17, 53.
9. Larsen, G.L.; Bakke, J.E. Xenobiotica 1983, 13, 115-126.
10. Bakke, J.E.; Rafter, J.; Larsen, G.L.; Gustafsson, J.-A.; Gustafsson, B.E. Drug Metab. Dispos. 1981, 9, 525-528.
11. Bakke, J.E.; Struble, C.; Gustafsson, J.-A.; Gustafsson, B.E. Proc. Natl. Acad. Sci. USA. 1985, 82, 668-671.
12. Mitchell, S.C.; Smith, R.L.; Waring, R.H.; Aldington, G.F. Xenobiotica 1984, 14, 767-779.
13. Lamoureux, G.L.; Rusness, D.G. In "Sulfur in Pesticide Metabolism and Function"; Rosen, J.D.; Magee, P.S.; Casida, J.E., Eds.; ACS Symposium Series No. 158, American Chemical Society: Washington, D.C., 1981; p. 133.

14. McKinney, L.L.; Weakley, F.B.; Eldridge, A.C.; Campbell, R.E.; Cowan, J.C.; Picken, J.C., Jr.; Biester, H.E. J. Am. Chem. Soc. 1957, 79, 3932.
15. Schultze, M.O.; Klubes, P.; Perman, V.; Mizuno, N.S.; Bates, F.W.; Sautter, J.H. Blood 1959, 14, 1015-1025.
16. Virtanen, A.I. Phytochemistry 1965, 4, 207-228.
17. Smith, R.H. Reports of the Rowett Inst. 1974, 30, 112-131.
18. van Bladeren, P.J.; Breimer, D.D.; Rotteveel-Smijs, G.M.T.; DeJong, R.A.W.; Buijs, W.; van der Gen, A.; Mohn, G.R. Biochem. Pharmacol. 1980, 29, 2975-2982.
19. Davison, C.; Rozman, R.S.; Smith, P.K. Biochem. Pharmacol. 1961, 7, 65-74.
20. Nachtomi, E. Biochem. Pharmacol. 1970, 19, 2853-2860.
21. Pohl, L.R.; Branchflower, R.V.; Highet, R.J.; Martin, J.L.; Nunn, D.S.; Monks, T.J.; George, J.W.; Hinson, J.A. Drug Metab. Dispos. 1981, 9, 334-339.
22. Lamoureux, G.L.; Rusness, D.G. J. Agr. Food Chem. 1980, 28, 1057-1070.
23. James, S.P.; Needham, D. Xenobiotica 1973, 3, 207-218.
24. Coles, B.; Srai, S.K.S.; Ketterer, B.; Waynforth, B.; Kadlubar, F.F. Chem.-Biol. Interact. 1983, 43, 123-129.
25. Mrochek, J.E.; Katz, S.; Christie, W.H.; Dinsmore, S.R. Clin. Chem. 1974, 20, 1086-1096.
26. Bakke, J.E. Biomed. Mass Spectrom. 1982, 9, 74-77.
27. Jones, A.R.; Fakhouri, G.; Gadiel, P. Experientia 1979, 35, 1432.
28. Miner, D.J.; Kissinger, P.T. Biochem. Pharmacol. 1979, 28, 3285.
29. Barnsley, E.A. Biochim. Biophys. Acta 1964, 90, 24-36.
30. Brüsewitz, G.; Cameron, B.D.; Chasseaud, L.F.; Grler, K.; Hawkins, D.R.; Koch, H.; Mennicke, W.H. Biochem. J. 1977, 162, 99-107.
31. Görler, K.; Krumbiegel, G.; Mennicke, W.H.; Siehl, H.-U. Xenobiotica 1982, 12, 535-542.
32. Hutson, D.H.; Lakeman, S.K.; Logan, C.J. Xenobiotica 1984, 14, 925-934.
33. Nakatsu, K.; Hugenroth, S.; Sheng, L-S.; Horning, E.C.; Horning, M.G. Drug Metab. Disp. 1983, 11, 463-470.
34. James, S.P.; Needham, D. Xenobiotica 1973, 3, 207-218.
35. Reichert, D.; Werner, H.W.; Metzler, M.; Henschler, D. Arch. Toxicol. 1979, 42, 159-169.
36. Jones, A.R.; Walsh, D.A. Xenobiotica 1980, 10, 827-834.
37. Turnbull, L.B.; Teng, L.; Kinzie, J.M.; Pitts, J.E.; Pinchbeck, F.M.; Bruce, R.B. Xenobiotica 1978, 8, 621-628.
38. Waring, R.H.; Mitchell, S.C. Drug Metab. Disp. 1982, 10, 61-62.
39. Waring, R.H. Xenobiotica 1978, 8, 265-270.
40. Bakke, J.E.; Bergman, A.L.; Feil, V.J. Xenobiotica 1983, 13, 555-564.
41. Horning, M.G.; Stillwell, W.G.; Griffin, G.W.; Tsang, W.-S. Drug Metab. Dispos. 1980, 8, 404-414.
42. Lertratanagkoon, K.; Horning, M.G.; Middleditch, B.S.; Tsang, W.-S.; Griffin, G.W. Drug Metab. Dispos. 1982, 10, 614-623.
43. Ziegler, D.M. In "Metabolic Basis of Detoxication"; Jakoby, W.B.; Bend, J.R.; Caldwell, J., Eds.; Academic Press: New Yor 1982; pp. 171-182.

44. Mannervik, B. In "Metabolic Basis of Detoxication"; Jakoby,
    W.B.; Bend, J.R.; Caldwell, J., Eds.; Academic Press: New York,
    1982; p. 185.
45. Feil, V.J.; Bakke, J.E.; Larsen, G.L.; Gustafsson, B.E.
    Biomed. Mass Spectrom. 1981, 8, 1-4.
46. Nery, R. Biochem. J. 1971, 122, 317-326.
47. Moldéus, P. Biochem. Pharmacol. 1978, 27, 2859-2863.
48. Bakke, J.E. Chemosphere 1983, 12, 793-798.
49. Bakke, J.E.; Gustafsson, J.-Å.; Gustafsson, B.E. Science 1980,
    210, 433-435.
50. Hutson, D.H.; Holmes, D.S.; Crawford, M.J. Chemosphere 1976,
    2, 79-84.
51. Brundin, A.; Ratnayake, J.H.; Sunram, J.M.; Anders, M.W.
    Biochem. Pharmacol. 1982, 31, 3885-3890.
52. Bakke, J.E.; Price, C.E. J. Environ. Sci. Health 1979, 14,
    424-441.
53. Lee, J-K.; Minard, R.D.; Bollag, J-M. J. Korean Agr. Chem.
    Soc. 1982, 20, 44.
54. Renner, G.; Nguyen, P.-T. Xenobiotica 1984, 14, 693-704.
55. Bakke, J.E.; Aschbacher, P.W.; Feil, V.J.; Gustafsson, B.E.
    Xenobiotica 1981, 11, 173-178.
56. Renner, G. Xenobiotica 1981, 11, 435-446.
57. Parke, D.V.; Williams, R.T. Biochem. J. 1951, 48, XXVII.
58. Colucci, D.F.; Buyske, D.A. Biochem. Pharmacol. 1965, 14,
    457-466.
59. Tateishi, M.; Suzuki, S.; Shimizu, H. J. Biol. Chem. 1978,
    253, 8854-8859.
60. Stevens, J.L.; Jakoby, W.B. Mol. Pharmacol. 1983, 23,
    761-765.
61. Larsen, G.L.; Larson, J.D.; Gustafsson, J.-Å. Xenobiotica
    1983, 13, 689-700.
62. Tomisawa, H.; Suzuki, S.; Ichihara, S.; Fukazawa, H.; Tateishi,
    M. J. Biol. Chem. 1984, 259, 2588-2593.
63. Bonhaus, D.W.; Gandolfi, A.J. Life Sciences 1981, 29,
    2399-2405.
64. Bakke, J.E.; Rafter, J.J.; Lindeskog, P.; Feil, V.J.;
    Gustafsson, J.Å.; Gustafsson, B.E. Biochem. Pharmacol. 1981,
    30, 1839-1844.
65. Aschbacher, P.W.; Feil, V.J. J. Agr. Food Chem. 1983, 31,
    1150-1158.
66. Hassall, C.D.; Gandolfi, A.J.; Brendel, K. Drug Chem. Toxicol.
    1983, 6, 507-520.
67. Anderson, P.M.; Schultze, M.O. Arch. Biochem. Biophys. 1965,
    111, 593-602.
68. Saari, J.C.; Schultze, M.O. Arch. Biochem. Biophys. 1965, 109,
    595-602.
69. Larsen, G.L.; Bakke, J.E. Xenobiotica 1981, 11, 473-480.
70. Bakke, J.E.; Price, C.E. J. Environ. Sci. Health 1979, 14,
    291-304.
71. Weisiger, R.A.; Pinkus, L.M.; Jakoby, W.B. Biochem. Pharmacol.
    1980, 29, 2885-2887.
72. Bakke, J.E.; Bergman, Å.L.; Larsen, G.L. Science 1982, 217,
    645-647.
73. Bakke, J.E.; Bergman, Å.L., Brandt, I.; Darnerud, P.; Struble,
    C. Xenobiotica 1983, 13, 597-605.

74. Preston, B.D.; Miller, J.A.; Miller, E.C. Amer. Assoc. Cancer Res. Abstracts 1982, 23, 61.

75. Stillwell, W.G.; Bouwsma, O.J.; Thenot, J-P.; Horning, M.G.; Griffin, G.W.; Ishikawa, K.; Takaku, M. Res. Comm. Chem. Pathol. Pharmacol. 1978, 20, 509-530.

76. Struble, C.B.; Larsen, G.L.; Feil, V.J.; Bakke, J.E. In "Polynuclear Aromatic Hydrocarbons: Chemistry, Characterization and Carcinogenesis"; Battelle Press: Columbus, OH, 1985; in press.

77. Bakke, J.; Struble, C.; Gustafsson, J-A.; Gustafsson, B. In "Extrahepatic Drug Metabolism and Chemical Carcinogenesis"; Rydström, J.; Montelius, J.; Bengtsson, M., Eds.; Elsevier Science Publishers B.V., 1983; pp. 257-266.

78. Feil, V.J.; Huwe, J.K.; Bakke, J.E. In "Polynuclear Aromatic Hydrocarbons: Chemistry, Characterization and Carcinogenesis"; Battelle Press: Columbus, OH, 1985; in press.

79. Renwick, A.G.; Drasar, B.W. Nature 1976, 263, 234-235.

80. Chipman, J.K.; Hirom, P.C.; Frost, G.S.; Millburn, P. Biochem. Pharmacol. 1981, 30, 937-944.

81. Bakke, J.; Gustafsson, J.-A. Trends in Pharmacol. Sci. 1984, 5, 517-521.

82. Quirijns, J.K.; Van Der Paauw, C.G.; Ten Noever De Brauw, M.D.; De Vos, R.H. Science of the Total Environment 1979, 13, 225-233.

RECEIVED November 15, 1985

16

The Disposition of Plant Xenobiotic Conjugates in Animals

Valerie T. Edwards and David H. Hutson

Shell Research Ltd., Sittingbourne, Kent ME9 8AG, England

The fate of plant xenobiotic conjugates i.e. glycosides, malonates, N-acyl-amino acids, alkyl/aryl
glutathiones and derivatives, lipophilic conjugates
and polymer conjugates (bound residues), in animals
is reviewed. Some classes are reasonably well-
studied but no information is available for others.
The conjugates are hydrolyzed in mammals and
metabolized as their aglycones (exocons), are eliminated unchanged, or are handled in both of these
ways. The toxicological significance of plant conjugates depends largely on the nature of the aglycone
and its rate and site of release from the conjugate.
Bound residues tend to have limited bioavailability
but this depends on the digestibility of the material.
Studies of the metabolic fate of conjugates and bound
residues are valuable in the assessment of their
toxicological significance.

The conjugation of xenobiotic compounds in plants has been described
in relation to that in some other life forms in the first section of
this Symposium. It is apparent that the processes are similar among
among the various taxa but that some notable differences are found
both in the chemistry and the disposition of the conjugates. This
chapter deals with a relationship between plants and animals, mostly
mammals, and therefore a comparison of conjugation in plants and
animals is shown in Table I. The observed differences are important
because it is from these differences that the requirement to study
the fate of xenobiotic plant conjugates in animals is derived.
    Most of our recent knowledge on the fate of xenobiotics in
plants has been gained from studies aimed at (i) discovering and
optimizing the mode of action of herbicides and (ii) confirming
the environmental acceptability of pesticides. The study of plant
metabolites of xenobiotics in mammals (and some other animals),
whilst originally a subject of academic interest with potential
ramifications for the discovery of therapeutic agents, is now

0097–6156/86/0299–0322$06.00/0

Table I. Comparative Conjugation of Xenobiotics in Mammals and Plants

Mammal	Plant	(Variations, Comments)
Glucuronides	Glucosides	Di-, Tri- & Oligoglucosides Xylosyl-glucosides Malonyl-glucosides
Sulphates	-	Very rare
Methylation	Methylation	Not common
Acetylation	Acetylation	Malonylation
Amino acids	Amino acids	Possibly wider range
Glutathione	Glutathione	Sequence of catabolism different
N-Acetylcysteine	N-Malonylcysteine	
Lipid conjugates	(One example)	Too little information

mainly pursued by pesticide scientists. When a pesticide metabolite is discovered in plants and shown to be absent from mammals,
the question of the toxicity of the metabolite to the latter is
raised. Of the qualitative differences between plants and mammals in
xenobiotic metabolism, most arise at the level of conjugation (Table
I) and, therefore, this question is posed more frequently about conjugates than about primary (phase I) metabolites. Plant xenobiotic
conjugates are terminal metabolites which are stored in plant tissues (there being no mechanisms of elimination for non-volatile
metabolites). Thus, when plant material is used as a source of diet
for humans, either directly or via farm animals, the imputation of
the conjugate as a contaminant - a pesticide residue - generates
pressure for toxicological studies. However, the conjugated metabolite of a pesticide is clearly not the parent pesticide. Furthermore, the conjugate is diluted in the plant matrix and is further
diluted by processing of the tissue prior to consumption. These
factors, plus the supposition that the toxicology of the parent
pesticide in mammals covers all eventualities, may be used to argue
against the need for further studies. Thus two extreme views on how
to assess the significance of xenobiotic plant conjugates for
mammals have been presented over the last ten years. One view is
that two-year feeding studies should be carried out on all "plant-
unique" metabolites (or on mixtures representative of the profile
found in the plant). The opposing view is that nothing should be
done further than measuring the concentration of parent compound and
major aglycone (after hydrolysis). The first view leads to insup-
portable costs; the second view propagates ignorance. A compromise
between these two views is emerging and metabolism studies are
providing the means to reach it. In many cases, the significance of
conjugated xenobiotic metabolites can be assessed simply via a two-
stage process: (i) their isolation and identification and (ii) a
study of their fate in mammals (monogastric and/or ruminant depending on the primary use of the plant material). In the event that the
conjugate(s) is difficult or impossible to identify, or if it is a
macromolecular conjugate (a bound residue), biosynthetic material
derived from the radioactively-labelled pesticide can be used for

metabolism study. The information required is two-fold: (i) is the
phase I metabolite (the 'aglycone' or the 'exocon') released during
metabolism and is it common to the metabolic pathway of the parent
pesticide in mammals? (ii) does the plant conjugate afford residues
in the edible products of treated farm animals? If it should prove
that the phase I plant metabolite is not present in the mammalian
metabolic pathway of the parent and should there be measurable
residues in food crops, then consideration must be given to further
toxicological studies. If the plant conjugate is not hydrolyzed by
the mammal, its chemical reactivity, distribution in, and rate of
elimination from the mammal will provide useful information for risk
assessment.

The anticipated fate of xenobiotic plant conjugates in mammals
is that they will be hydrolyzed at the conjugating bond and processed as phase I metabolites (or possibly as the parent if an alcohol,
phenol, carboxylic acid, etc.). Relatively few such studies have
been reported to date but the purpose of this chapter is to review
the available results and to judge whether or not the expectation is
justified and whether or not generalizations may be useful. Macromolecular conjugates (bound residues) are also considered but, in
the absence of enough definitive data, only approaches to their
study are suggested.

The fate of plant conjugates in vertebrates was last reviewed
by Harvey (1) in the Proceedings of the 5th International Congress
of Pesticide Chemistry. It was clear then (early 1982) that only a
few studies were reported. We highlight some of these and report on
information published since that time. The toxicological significance of pesticide conjugates has been considered recently by
Dorough (2).

Plant Conjugates

Before describing the disposition of plant conjugates in animals,
these metabolites must be seen in the context of the vast amounts of
natural plant conjugates which we, and other animals, ingest in our
normal diet. The categorization between 'natural' and 'xenobiotic'
may require a temporary re-alignment while we consider these. Many
plant metabolites and conjugates are xenobiotic (or anutrient) with
respect to animals. Pharmacology was built upon, and early toxicology was a response to, products derived from the plant kingdom.
Still however, in the lay mind, a natural product is perceived as
safe, e.g. pyrethrum - the natural insecticide. This popular perception even extends to synthetic derivatives of natural products e.g.
the pyrethroids. Relatively safe these may be, but the generalizations are obviously not. There is nothing benign about cycasin
(1, methylazoxymethyl β-D-glucopyranoside) which is the toxic constituent of cycad meal obtained from the nuts of Cycas species. This
glucoside is hepatotoxic and carcinogenic to rats (3). The aglycone,
methylazoxymethanol (2) is itself toxic, whereas when the glucoside
(1) is dosed intraperitoneally to rats it is not toxic and is eliminated virtually unchanged. These findings illustrate two points: the
importance of enterobacterial hydrolysis of the glucoside and the
importance of the structure of the aglycone to the toxicity of a
glycoside.

Miserotoxin (3, 3-nitro-1-propyl β-D-glucopyranoside), a product of <u>Astragalus</u> species (Leguminosae), is poisonous to livestock such as cattle and sheep. In these, and probably other, ruminants the glycoside is rapidly hydrolyzed to 3-nitropropanol which is the precursor of 3-nitropropanoic acid, thought to be the ultimate toxicant (via the inhibition of mitochondrial respiration) (4). Interestingly, miserotoxin is relatively innocuous to rats, presumably because of a lower rate of microbial hydrolysis in the gastrointestinal tract. This suggests another general feature: that ruminants and monogastric animals differ in their capacity in this respect and one cannot be used as a model for the other.

The digitalis glycosides, e.g. digoxin (4) exert their therapeutic action (an increase in the force of myocardial contraction) via their aglycones, digoxigenin (5) in the case of digoxin. These drugs are still very important in the treatment of heart failure and some other less acute heart conditions but they have rather low margins of safety and cause a variety of side-effects via a variety of mechanisms. Intoxication can be fatal (5).

Thus plant glycosides appear to be readily hydrolyzed by mammals. Judging by the general acceptability of plant products, their glycosides do not seem to cause many problems. However, given a toxic aglycone, glycosidation may afford little protection. This is important to note in the context of xenobiotic plant conjugates but it must not be given the status of a generalization. There will be many exceptions as, for example, in the case of miserotoxin in rats cited above.

The sites of location of the hydrolysis of conjugates may vary with animal species, aglycone and conjugate type and, where known, this will be indicated below. In principle, if hydrolysis does not occur via the action of bacteria in the lumen of the intestine, it will occur in:

(i)   the intestinal wall during absorption
(ii) the liver following absorption
(iii) the blood during transport
or   (iv) the kidney during elimination.

If resistant to hydrolysis at all of these sites, the conjugate is likely to be eliminated reasonably efficiently via the kidneys (though if neutral, e.g. a glucoside, it may be partially retained by passive resorption from the glomerular filtrate). For reasons of structural analogy with a specific mammalian product or the existence of a particularly active hydrolase, unexpected hydrolysis in a specific organ may occur. This is perhaps the major threat (and opportunity) that residual conjugates present - the liberation of a bioactive aglycone at a specific site which is ill-equipped to deal with the product. Attempts to target drugs by administration of their conjugates has met with only limited success but the principle remains valid. In the absence of such specific situations however, the toxicity of a xenobiotic conjugate will usually depend upon the dose, the toxicity of the exocon and its rate of release from the conjugate.

Sugar Conjugates

Glycosides are the most widely observed class of xenobiotic conjugates in plants and within this class glucose is by far the most common sugar involved (6). Glucosides can be usefully classed by the nature of the sugar-aglycone linkage. Most widespread are the ether-linked O-glucosides which often arise via conjugation at a phenolic hydroxyl group. Ester linked O-glucosides and N-glucosides are often observed, whereas S-glucosides are less common. The isolation of intact sugar conjugates from plants is a lengthy process and often their structures are deduced from enzyme hydrolysis studies. However, it is becoming clear that following the initial conjugation with glucose further conjugation reactions can take place. For instance disaccharide and, more recently, trisaccharide conjugates have been discovered (7). Perhaps, more usual is the 6-O-malonylation of glucosides (8-10); this step prevents any additional sugar groups being added to the conjugate. Malonylation, which can also be a primary conjugation mechanism, appears to be unique to plants and will be dealt with later in this chapter.

All of the studies concerning the fate of plant sugar conjugates in animals have concentrated on O-glucosides or plant material containing glucosides together with unidentified water-soluble metabolites. With the exception of the glucoside of 3-phenoxybenzoic acid (11), these studies have concerned ether-linked glucosides. The ester-linked glucoside (6) was readily hydrolyzed by rats and the metabolites excreted in the urine were similar both qualitatively and quantitatively to those found following treatment with 3-phenoxybenzoic acid itself. The major urinary metabolite in both cases was the sulphate conjugate of ring hydroxylated 3-phenoxybenzoic acid.

In the case of the ether-linked glucosides, studies have been carried out on rodents (as models for man) and ruminants (as models for farm animals generally). The work on ruminants has been reviewed by Harvey (1) and it is clear that cows, goats and sheep will effectively hydrolyze glucoside conjugates in the gastro-intestinal (GI) tract, presumably as a result of the action of rumen microorganisms. The fate of the released aglycone depends on its structure as it can pass unchanged through the GI tract and appear in the feces or be absorbed and further metabolized generally following the same pathways as in the rat.

Although more studies have been carried out with rodents than with ruminants, it is less easy to generalise about (and hence predict) the behaviour of ether-linked O-glucosides in the monogastric animal. Much of this work has been previously reviewed (1) but a brief summary here will put the more recently published work into context.

For instance, when 1-naphthyl glucoside (7) was administered orally to rats it was quickly absorbed and 17% of the dose was eliminated unchanged (12). Similarly, the major glucoside metabolite (8) of oxamyl is relatively resistant to degradation in the rat both in vivo following oral administration and in vitro in rat liver microsomal preparations. In the former, at least 30% of the dose was recovered as the glucoside in the urine (13). A similar result was btained with the glucoside of N-(3-chloro-4-hydroxymethylphenyl)

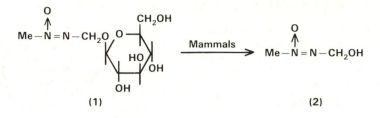

(1)   (2)

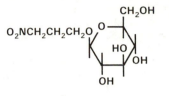

(3)

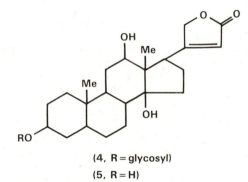

(4, R = glycosyl)
(5, R = H)

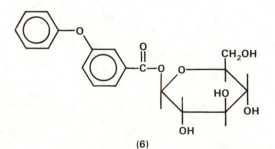

(6)

urea, a plant metabolite of the herbicide chlortoluron, when orally
administered to rats. The glucoside was the major component (63%) in
the rat urine and, in a parallel experiment with quail, the glucoside accounted for 32% of the radioactivity in the excreta (14). By
contrast, however, when the glucoside of 3-hydroxy-carbofuran (3-OH-
C-Glu; see also Figure 1) was fed to rats as the major radiolabelled component of a bean plant residue, none of it survived. A
follow-up study with [^{14}C-ring]-3-OH-C-Glu in rats has been recently
reported (15). 3-Hydroxy-carbofuran is a primary metabolite of
carbofuran which is common to plants and animals and has similar
toxicity to carbofuran. From the initial study it was clear that
radioactivity from 3-OH-C-Glu was absorbed from the gastro-intestinal tract as evidenced by the urinary excretion of sulfate and
glucuronide conjugates. The question which remained was whether the
toxic aglycone was released in the body and if so what were its
toxicokinetics compared with those following administration of the
aglycone. In other words, is the glucosylation likely to render
3-OH-C more or less toxic to rats? Marshall and Dorough administered
the radiochemical as a single oral dose (0.1 mg kg^{-1}) and found that
the rate of excretion of radioactivity in the urine and in the bile
was faster from the rats treated with 3-OH-C than from those treated
with the glucoside. The difference was most marked in the urine 1.5
hours after dosing, although still measurable at 24 hours. With both
substrates, glucuronide and sulfate metabolites of 3-OH-C and its
hydrolytic degradation products were present in the urine but no
glucoside was observed. These biotransformations are summarized in
Figure 1. The fact that absorption was slower for the glucoside was
confirmed by examining portions of the GI tract. After 1.5 hours,
only 5% of the radioactivity from the 3-OH-C dose remained in the
stomach, whereas the corresponding figure for the glucoside was 25%.
The radioactivity in the small intestine from the 3-OH-C dose was
composed mainly of glucuronides and indeed none of the administered
compound was recovered from anywhere in the GI tract. The glucoside,
in contrast, moved intact through the small intestine and even as
far as the cecum. Over 50% of the dose was recovered in the cecum 3
hours after treatment, the majority being unchanged glucoside. Only
1% was detected in the colon, however, suggesting that microbial
action in the cecum was responsible for the cleavage of the glucoside and hence absorption of the aglycone. In summary therefore,
although the ultimate fate(s) of 3-OH-C and its glucoside are the
same, conjugation increases the residence time of the toxic aglycone
in the animal. The authors suggest that this 'slow-release' may be
an advantage by avoiding the saturation of detoxification mechanisms.
    A study of the metabolism in mouse of 1-naphthyl glucoside (7)
in comparison with 1-naphthyl glucuronide and the aglycone has been
published (16). The dose rates used were rather high for the main
comparative study (100 mg kg^{-1} for the conjugates) but in a secondary study the effect of dose rate on the metabolism and excretion of
1-naphthyl glucoside was investigated. Interesting differences in
the route of excretion and the urinary metabolite profiles were
observed. At 10 mg kg^{-1}, 85% of the dose was recovered in the urine
24 hours after treatment (7% in the feces) and at 100 mg kg^{-1} the
corresponding figures were 62% (urine) and 12% (feces). The results
for 50 mg kg^{-1} were intermediate. Thus, the lower the dose, the

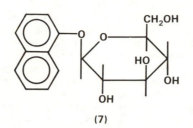

(7)

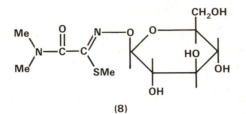

(8)

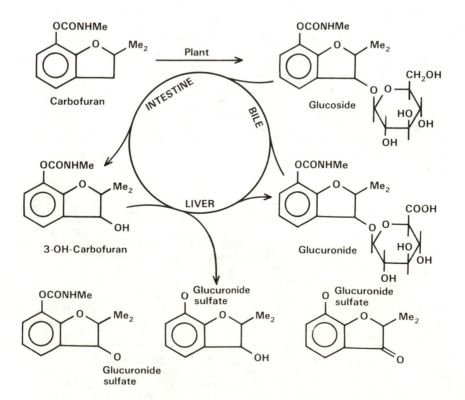

Figure 1.   Fate of 3-hydroxycarbofuran glucoside in the rat (15).

higher was the percentage excretion in the urine. The reduced urinary excretion at high dose rates can be partly accounted for by
limited absorption and excretion in the feces. The major metabolite
in the urine of the mice dosed at 10 mg kg^{-1} was unchanged glucoside, whereas at the higher dose rates it was 1-naphthyl sulfate.
The glucuronide conjugate of 1-naphthol was present at low levels at
all dose rates (8-12% of total in urine). Thus, at lower doses, a
higher percentage of the dose was absorbed and bypassed all sites of
hydrolysis. It is likely that rate-limiting absorption from the GI
tract at higher doses and the longer residence time leads to more
hydrolysis before absorption and consequent sulfation when the
naphthol reaches the intestinal cells or the liver.

Similar results were obtained in a study recently carried out
in our laboratory (17). When phenyl glucoside was dosed orally to
rats (3 mg kg^{-1}), excretion in the urine was very rapid (88% of dose
in 24 hours). As in the previous study, the major component in the
urine was intact glucoside (77%), the remainder being phenyl sulfate
(23%) with only trace amounts of glucuronide. This profile was
markedly different from the urinary metabolite profiles obtained
from rats treated with phenol or phenyl malonyl-glucoside (see next
section). One might conclude that phenyl glucoside is more rapidly
absorbed than 1-naphthyl glucoside and therefore more survives to be
excreted. Carbofuran glucoside, on the other hand, is not absorbed
intact but is only absorbed following hydrolysis and therefore the
metabolic profile resembles that following dosing of the aglycone.

In summary, the metabolic fate of glucoside conjugates in
animals is very dependent on the structure of the aglycone. It
appears that the glucosides which are most rapidly absorbed are most
likely to be excreted intact whereas, where intestinal hydrolysis
takes place prior to (or at the same time as) absorption, the elimination products are the same as those of the aglycone. It should be
noted that phenol itself (and possibly other phenols) is efficiently
conjugated in the intestinal wall (18) and thus phenol liberated
from, for example, phenyl glucoside within the intestine or in the
intestinal cells, need not necessarily appear as free phenol in the
systemic circulation.

Malonates

As mentioned in the Introduction, malonylation of xenobiotics is
apparently unique to plants. The most commonly observed conjugates
are 6-0-malonyl esters of glucosides (8-10). However, N-malonylation
of anilines and more commonly N-malonylation of xenobiotic amino
acids has been observed (19). Often N-malonylation of cysteine will
take place as a step in the catabolism of glutathione conjugates.

To our knowledge there has been no published work on the fate
in animals of these malonylated xenobiotic conjugates. Therefore, we
have recently carried out a study of the metabolism of phenyl 6-0-
malonyl-glucoside, phenyl glucoside and phenol in rats (17). The
radiolabelled conjugates were synthesised from [^{14}]phenol and dosed
orally to rats in equimolar quantities, equivalent to 1.2 mg kg^{-1} of
phenol. As expected, all 3 compounds were rapidly excreted in the
urine with at least 80% of the radioactivity being recovered in the
24 hours following administration. Phenol was eliminated mainly as

phenyl sulfate (68%) and partly as phenyl glucuronide (12%). Phenyl glucoside was mainly eliminated unchanged (68%) with some phenyl sulfate (20%) as described in the previous section. The fate of phenyl malonyl-glucoside, in contrast, was similar to that of phenol with only small amounts of unchanged phenyl malonyl-glucoside (5%) and phenyl glucoside (4%) being recovered in the urine (see Table II and Figure 2).

Table II. Fate of Phenyl 6-O-Malonyl-Glucoside compared with that of Phenyl Glucoside and Phenol in the Rat

	Compound dosed [a]		
Yields	Phenyl Malonyl-glucoside	Phenyl Glucoside	Phenol
Percent dose in			
0-24 hr urine	79	88	80
Phenyl sulfate	63	20	68
Phenyl glucuronide	7	-	12
Phenyl glucoside	4	68	-
Phenyl malonyl-glucoside	5	-	-

[a] Dose equivalent to 1.2 mg kg^{-1} phenol for all compounds.

It is difficult to explain why the malonylation of the sugar should render the glucoside link with phenol more susceptible to hydrolysis unless it causes a difference in the disposition of the two conjugates. In line with the reasoning in the previous section, the malonyl-glucoside may be less readily absorbed from the GI tract than the glucoside and therefore more likely to be exposed to hydrolytic enzymes which will not only cleave the ester bond but also presumably, subsequently and rapidly cleave the link to the aglycone. The majority of the radioactivity is thus found as phenyl sulfate (63%) and phenyl glucuronide (7%), the normal products of the metabolism of phenol. It is interesting that even a small amount of phenyl malonyl-glucoside survives to be excreted in the urine. This shows that the conjugate can be absorbed intact and as it passes through the liver and kidney it remains partially resistant to hydrolytic processes. The small amount of phenyl glucoside found with it in the urine could have been generated in the body but more likely was formed in the intestine and excreted without further metabolism.

This experiment demonstrates that although much information can be gained from examining urinary metabolites 24 hours after dosing, much more information can be generated by the methods of Marshall and Dorough (15) where several time points within the first 24 hours were examined. In that experiment, differences in absorption rate and topography of absorption between the conjugate and the aglycone were observed, whereas in the work described above these differences could only be inferred.

As for malonates linked to the xenobiotic via an amide link (N-malonylamines) there appears to be no published information on their

fate in animals. This is an obvious gap in our knowledge and one
which is not easy to cover by analogy with other structures.

Amino acid conjugates

Xenobiotics with a carboxylic acid function can be conjugated with
an amino acid in plants. The most studied class of pesticides which
are conjugated with amino acids is that comprised of the phenoxy-
acetic acid herbicides and, in particular, 2,4-dichlorophenoxyacetic
acid (2,4-D) has been extensively investigated. 2,4-D forms conjugates in plants mainly with aspartic and glutamic acids but alanine,
leucine, phenylalanine, tryptophan and valine conjugates have been
identified in plant tissue culture experiments (20). When Harvey
(1) reviewed this subject in 1982 there was no published information
on the fate of amino acid conjugates in animals. Recent work by Buly
and Mumma (21) is very useful in starting to fill the gap. The
aspartate (asp) and valine (val) conjugates of [^{14}C]2,4-D (9, 10)
were each dosed orally to rats at 10 and 100 mg kg^{-1}. Urine and
feces samples were collected every 6 hours up to 72 hours after
treatment. With both compounds, over 90% of the radioactivity
was excreted within the first 24 hours predominantly in the urine as
2,4-D. Urinary excretion was greatest in the first 6 hours, whereas
faecal excretion reached its maximum level between 6 and 12 hours
after dosing. Overlying this picture of rapid cleavage and excretion
there are interesting differences between the two conjugates and the
two dose levels.

With 2,4-D-asp, as expected, at the higher dose rate more of
the dose was excreted via the feces than at the lower dose. This can
be explained by the limited absorptive capacity of the intestinal
tract. However, this effect was not observed with 2,4-D-val where
at the lower dose, the urine contained a somewhat smaller proportion
of the dose (69%) compared with the higher dose (76%). The authors
suggested that 2,4-D-val may be absorbed more readily than 2,4-D-
asp.

Another interesting difference between the two conjugates was
in their biotransformation. With 2,4-D asp, only unchanged conjugate
and free 2,4-D were found in the urine and feces. With 2,4-D-val, in
addition to unchanged conjugate and 2,4-D, a minor metabolite (2%)
was found in the urine and at least six minor metabolites were
excreted in the feces (total <5%). These were not identified but
were all more polar than 2,4-D and only one appeared to be less
polar than 2,4-D-val.

Although not a plant xenobiotic conjugate, the metabolism of
the glycine conjugate of salicylic acid in isolated perfused rat
kidney has been studied (22). The authors were surprised that 20-30%
of the dose was hydrolyzed to salicylic acid in that organ, whereas
in parallel experiments the liver did not posses this activity.

Clearly, more studies are required in this area before one can
usefully generalise, although it does appear from the limited data
available that amino acid conjugates of small polar molecules may
not necessarily be terminal metabolites; in animals they may be
hydrolyzed, and the exocon excreted without further conjugation.

Glutathione Conjugates and Derivatives

Glutathione (GSH) conjugation is common to each of the life forms covered in the early chapters. The contributions by Lamoureux (this Volume) and Bakke (preceding Chapter) provide the background on which to base predictions of the fate of plant GSH conjugates and their derivatives in mammals. It is now known that the pathway from GSH conjugate to mercapturic acid in the mammal is an oversimplification. GSH conjugates and their degradation products are subject to extensive re-cycling and biotransformation and thus many GSH-derived metabolites are common to both plants and animals. Two important differences, however, must be considered. These derive from plant-mammal differences in the primary catabolic pathway of GSH conjugates and in the N-acylating moiety.

Plants differ from mammals in the sequence in which the glycine and glutamic acid groups are removed from the GSH conjugate. This has been found to be the case with corn, sorghum and sugar cane; it is not known if the sequence applies to all plants (23). Thus S-substituted γ-glutamylcysteine conjugates are major (intermediary) metabolites in plants which are not found in mammals. In practice, these peptides are not generally found as important terminal residues, therefore we do not need to consider them in detail. It is likely that mammalian peptidases will catalyse their conversion to S-substituted cysteines; however, this does not appear to have been studied for a xenobiotic conjugate.

The common terminal metabolite of a glutathione conjugate in a mammal is a mercapturic acid — an N-acetyl-S-substituted-L-cysteine. The equivalent in the plant — the phytomercapturic acid — is an N-malonyl-S-substituted-L-cysteine (Figure 3). The susceptibility of the N-malonyl group to hydrolysis in the mammal has not, to our knowledge, been studied. If it is stable then the phytomercapturic acid is likely to be eliminated unchanged (or perhaps S-oxidized). If readily hydrolyzed, it will be converted to a mercapturic acid and eliminated as such but it would also be subject to the complex reactions outlined in the preceding chapter. In particular, the product of hydrolysis of an N-malonyl-cysteine conjugate is the substrate for enterobacterial, hepatic and renal C-S lyases (Figure 4). In certain, albeit rare, circumstances the xenobiotic thiol derived from the action of C-S lyase is toxic. The now well-documented example of the renal toxicity induced by hexachlorobutadiene (24) is a case in point. Such toxicity should be predictable from mammalian metabolism/toxicity studies but it would seem to be advisable to have some information on the fate of N-malonyl cysteine conjugates in general in mammals. In a case where a potentially toxic cysteine conjugate is found (N-malonylated) in plants it would be necessary to study its fate in mammals.

Our perception that the toxicology (to rat) of a plant metabolite has been studied if the metabolite is produced from the parent pesticide by the rat is, of course, a compromise. The thinking is flawed because it ignores the fact that the disposition of a metabolite depends upon its site and rate of biotransformation. Thus, when the mammalian glutathione-derived metabolites of naphthalene (the GSH conjugate, the premercapturic acid and the mercapturic acid) were dosed to rats the ratio of metabolic products and extent of

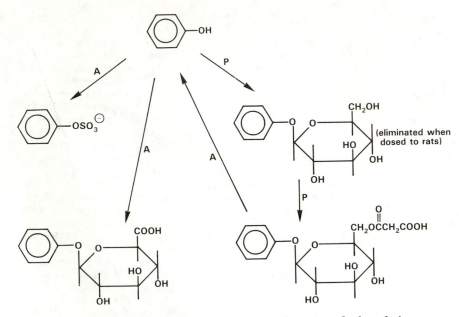

Figure 2. Relationships between conjugates of phenol in plants (P) and animals (A).

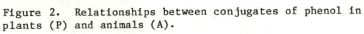

(9, R = CH₂COOH)
(10, R = CHMe₂)

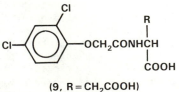

(A)

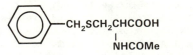

(P)

Figure 3. A Mercapturic acid (A) and a phytomercapturic acid (P).

urinary elimination were different for each and different from that derived from the parent naphthalene (2).

Lipophilic Conjugates

The mammalian lipophilic conjugates (xenobiotic lipids) of the type described by Chang et al. (this Volume) and by Miyamoto (this Volume), and reviewed in references (25) and (26), probably have their equivalents in plants. There is only limited evidence at present to confirm this, for example: the 2-naphthylmethyl esters of C16, C18, C20 and C22 fatty acids (11) found in (on) apples following spraying of the trees with 2-naphthylmethyl cyclopropanecarboxylate (27), the chain elongated products of cyclopropanecarboxylic acid derived from oranges treated with cycloprate (28) and the angelic acid conjugate of 3-hydroxy-carbofuran found in carbofuran-treated carrots (12) (29). There is no information on the fate of these conjugates in mammals. However, as they are all relatively simple esters, it is likely that they will all be hydrolysed to their component alcohols and acids. The xenobiotic portion will then be metabolised and eliminated in the usual way. This perception should, however, be tested in individual cases. The synthetic derivative of (+)catechin, 3-palmitoyl-(+)catechin (13), a potential therapeutic agent utilizing the action of catechin on acute viral hepatitis, is somewhat persistent in the liver when dosed orally to rats. About 0.6% of the radioactivity derived from a single oral dose of 30 mg kg^{-1} was still present at 28 days (30). The xenobiotic triacylglycerols containing 3-phenoxybenzoic acid, when present in the fat or skin of rats, are eliminated with half-lives of 7-10 days (31). However, when such metabolites are dosed orally to mammals they would be hydrolyzed, if not within the intestine by pancreatic lipase or by enterobacterial enzymes then by enzymes within the intestinal cells (as are natural triglycerides).

Polymeric Conjugates

These conjugates, known commonly as bound residues, were first reviewed at the Vail Conference organised by the Pesticide Chemistry Division of the ACS in 1975 (32). These ill-defined plant "metabolites" are justifiably classified with conjugates in the context of the toxicology of plant metabolites but they will not be described at length in this paper. The current definition (33, 34) is still operational and pragmatic rather than molecular, being based on the conditions under which they survive extraction. The subject has recently been reviewed by Pilmoor and Roberts (35) and has formed the subject of a report by the IUPAC Pesticide Chemistry Commission (34). Their complexity, and the dominant effect of structure on xenobiochemistry, combine to render generalisations about bound residues of limited value. It is doubtful, for example, whether the study of the fate of a series of model xenobiotic - polymer adducts (e.g. with starch, cellulose or lignin) would be of general use. An assessment of their significance to mammals is probably best approached via their bioavailability. The bound residue should be biosynthesised with a suitable (xenobiotic) ^{14}C-label under as near normal conditions as possible, prepared for presentation to animals

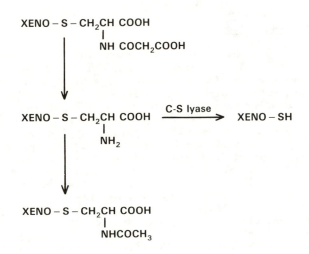

$$XENO-S-CH_2CH\ COOH$$
$$|$$
$$NH\ COCH_2COOH$$

$$\downarrow$$

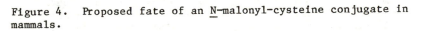

$$XENO-S-CH_2CH\ COOH \xrightarrow{\text{C-S lyase}} XENO-SH$$
$$|$$
$$NH_2$$

$$\downarrow$$

$$XENO-S-CH_2CH\ COOH$$
$$|$$
$$NHCOCH_3$$

Figure 4.   Proposed fate of an N-malonyl-cysteine conjugate in mammals.

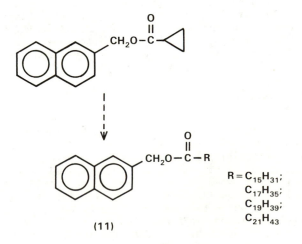

$$R = C_{15}H_{31};$$
$$C_{17}H_{35};$$
$$C_{19}H_{39};$$
$$C_{21}H_{43}$$

(11)

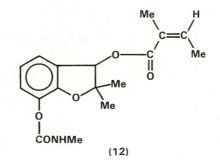

(12)

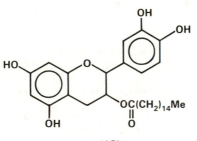

(13)

using relatively gentle techniques and then administered to animals
in feed. The test species used must be monogastric or a ruminant, or
both types used depending on the practical situation being modelled.
The bioavailability of the radiolabel will almost certainly be
highly dependent on the digestibility of the natural polymer matrix.
This is more likely to be complete in the ruminant as opposed to the
monogastric animal. If completely digestible, as for example a
protein, the xenobiotic should be available to the animal as a small
molecule, possibly unconjugated or as a peptide. The rate of elimination of ^{14}C and the structure of the eliminated metabolites should
allow some assessment of the chemistry and the toxicology of the
molecule absorbed.

It is worth emphasizing that a plant bound residue that has
relied upon high chemical reactivity for its formation (e.g. an
electrophile reacting with a nucleophilic centre) will be devoid of
that reactivity. Xenobiotics incorporated via energy-dependent
biochemical mechanisms, however, cannot be viewed in this way.

Relatively few studies have been carried out in vivo. However,
the results available, for example on carbaryl and carbofuran (36),
propham (37), propanil (38), and atrazine (39), support the general
conclusion of the IUPAC Pesticide Chemistry Commission (34) that
covalently-bound non-extractable residues in plants are not generally bioavailable to animals. This conclusion, however, is based
upon limited data and it would be helpful to have the results of
some more experiments in which sympathetically-treated radiolabelled
bound residues have been fed to animals.

Conclusions

The following conclusions may be drawn:
(i)   the tendency overall is for plant conjugates to be hydrolyzed
        by animals.
(ii)   ruminants are more effective than mono-gastric animals at
        conjugate hydrolysis.
(iii)   conjugation in some cases offers protection from the effects
        of a toxic exocon by delaying absorption (and, therefore,
        hydrolysis).
(iv)   rapidly absorbed conjugates (ether glucosides in particular)
        may be partly elminated unchanged.
(v)   there are some gaps in our knowledge of the fate of plant
        conjugates (e.g. N-malonyl derivatives) in animals.
(vi)   metabolism studies offer valuable cost-cutting opportunities
        for the assessment of the toxicology of plant conjugates.

The dilution of residues when formed in plants, and the processing of plant feedstuff prior to consumption must afford a large
measure of protection to the mammal. The xenobiotic conjugates, as
well as being diluted, are presented to animals in admixture with a
vast array of plant products many of which are xenobiotic with
respect to the recipient and the former must be seen in this
context. It is an irony that the enormous biological and social
damage curently caused by the legal and illegal use of the products
of plants, yeasts and fungi far outweighs that caused by the use of
xenobiotic pesticides on plants.

Literature Cited

1. Harvey, J. Jr. In "Pesticide Chemistry: Human Welfare and the Environment"; Miyamoto, J.; Kearney, P. C., Eds.; Vol. 3: "Mode of Action, Metabolism and Toxicology"; Matsunaka, S.; Hutson, D. H.; Murphy, S. D., Eds.; Pergamon Press: Oxford, 1983; pp. 369-374.
2. Dorough, H. W. J. Toxicol.-Clin. Toxicol. 1982-83, 19, 637-659.
3. Yang, M. G.; Mickelsen, O. In "Toxic Constitutents of Plant Foodstuffs"; Liener, I. E., Ed.; Academic Press: New York, 1969; pp. 159-167.
4. Muir, A. D.; Majak, W.; Pass, M. A.; Yost, G. S. Toxicol. Letters 1984, 20, 137-141.
5. Hoffman, B. F.; Bigger, J. T. In "The Pharmacological Basis of Therapeutics"; Gilman, A. A.; Goodman, L. S.; Gilman, A., Eds.; Macmillan: New York, 1980; pp. 729-760.
6. Edwards, V. T.; McMinn, A. L.; Wright, A. N. In "Progress in Pesticide Biochemistry"; Hutson, D. H.; Roberts, T. R., Eds.; Wiley, Chichester, 1982; Vol. 2, pp. 71-125.
7. Mikami, N.; Wakabayashi, N.; Yamada, H.; Miyamoto J. Pestic. Sci. 1984, 15, 531-542.
8. Dutton, A. J.; Roberts, T. R.; Wright, A. N. Chemosphere 1976, 3, 195-200.
9. Frear, D. S.; Swanson, H. R.; Mansager, E. R. Pestic. Biochem. Physiol. 1983, 20, 299-310.
10. Frear, D. S.; Mansager, E. R.; Swanson, H. R. Pestic Biochem. Physiol. 1982, 19, 270-281.
11. Crayford, J. V.; Hutson, D. H. Xenobiotica 1980, 10, 355-364.
12. Dorough, H. W.; McManus, J. P.; Kumar, S. S.; Cardona, R. A. J. Agric. Food Chem. 1974, 22, 642-5.
13. Harvey, J. Jr.; Han, J. C-Y. J. Agric. Food Chem. 1978, 26, 902-910.
14. Chandurkar, P. S.; Menzer, R. E. Paper No. 42, presented at the 178th National Mtg. ACS, Washington, D.C. (1979).
15. Marshall, T. C.; Dorough, H. W. J. Agric. Food Chem. 1984, 32, 882-886.
16. Chern, W. H.; Dauterman, W. C. Toxicol. and Appl. Pharmacol. 1983, 67, 303-309.
17. Edwards, V. T.; Jones, B. C.; Hutson, D. H., unpublished data.
18. Cassidy, M. K.; Houston, J. B. Drug Metab. Disposit. 1984, 12, 619-624.
19. Berlin, J.; Witte, L.; Hammer, J.; Kukoschke, K. G.; Zimmer, A.; Pape, D. Planta 1982, 155, 244-250.
20. Feung, C. S.; Hamilton, R. H.; Mumma, R. O. J. Agric. Food Chem. 1973, 21, 637-640.
21. Buly, R. L.; Mumma, R. O. J. Agric. Food Chem. 1984, 32, 571-577.
22. Bekersky, I.; Colburn, W. A.; Fishman, L.; Kaplan, S. A. Drug Metab. Disposit. 1980, 8, 319-324.
23. Lamoureux, G. L.; Bakke, J. E. In "Foreign Compound Metabolism"; Caldwell, J.; Paulson, G. D., Eds.; Taylor and Francis: London, 1984; pp. 185-199.

24. Lock, E. A.; Green, T. In "Proceedings of the Ninth IUPHAR
     International Congress of Pharmacology"; Paton, W.; Mitchell,
     J.; Turner, P., Eds.; Macmillan: London, 1984; pp. 197-202.
25. Hutson, D. H. In "Progress in Pesticide Biochemistry"; Hutson,
     D. H.; Roberts, T. R., Eds.; Wiley, Chichester, 1982; Vol. 2,
     pp. 171-184.
26. Schooley, D. A.; Quistad, G. B. In "Pesticide Chemistry: Human
     Welfare and the Environment"; Miyamoto, J.; Kearney, P. C.,
     Eds.; Vol. 3: "Mode of Action, Metabolism and Toxicology";
     Matsunaka, S.; Hutson, D. H.; Murphy, S. D., Eds.; Pergamon
     Press: Oxford, 1983; pp. 301-306.
27. Pryde, A.; Hanni, R. P. J. Agric. Food Chem. 1983, 31, 564-567.
28. Quistad, G. B.; Staiger, L. E.; Schooley, D. A. J. Agric. Food
     Chem. 1978, 26, 66-70.
29. Sonobe, H.; Kamps, L. R.; Mazzola, E. P.; Roach, J. A. G. J.
     Agric. Food Chem. 1981, 29, 1125-1129.
30. Hackett, A. M.; Griffiths, L. A. Xenobiotica 1982, 12, 447-456.
31. Hutson, D. H.; Dodds, P. F.; Logan, C. J. Biochem. Soc. Trans
     1985, in press.
32. Kaufman, D. D.; Still, G. G.; Paulson, G. D.; Bandal, S. K.
     "Bound and Conjugated Pesticide Residues"; ACS Symposium
     Series, No. 29, 1976.
33. Huber, R.; Otto, S. In "Pesticide Chemistry: Human Welfare and
     the Environment"; Miyamoto, J.; Kearney, D. C., Eds.; Vol. 3,
     "Mode of Action, Metabolism and Toxicology"; Matsunaka, S.;
     Hutson, D. H.; Murphy, S. D., Eds.; Pergamon Press: Oxford,
     1983; pp. 357-362.
34. Roberts, T. R. Pure and Appl. Chem. 1984, 56, 945-956.
35. Pillmoor, J. B.; Roberts, T. R. In "Progress in Pesticide
     Chemistry and Toxicology"; Hutson, D. H.; Roberts, T. R., Eds.;
     Wiley: Chichester, 1985; Vol. 4, pp. 85-101.
36. Marshall, T. C.; Dorough, H. W. J. Agric. Food Chem. 1977, 25,
     1003-1009.
37. Paulson, G. D.; Jacobsen, A. M.; Still, G. G. Pestic Biochem.
     Physiol. 1975, 5, 523-535.
38. Sutherland, M. L. in Reference 32; pp. 153-155.
39. Bakke, J. E.; Shimabukuro, R. H.; Davison, K. L.; Lamoureux,
     G. L. Chemosphere 1972, 1, 21-24.

RECEIVED November 15, 1985

INDEXES

Author Index

Subject Index

Production by Meg Marshall
Indexing by Susan F. Robinson
Jacket design by Pamela Lewis

Elements typeset by Hot Type Ltd., Washington, DC
Printed and bound by Maple Press Co., York, PA